ECE/ENG.AUT/35

ECONOMIC COMMISSION FOR EUROPE
Geneva

ENGINEERING INDUSTRIES
DYNAMICS OF THE EIGHTIES

The Long-Term Assessment of the Role and Place of Engineering Industries in National and World Economies 1979-1986

UNITED NATIONS
New York, 1988

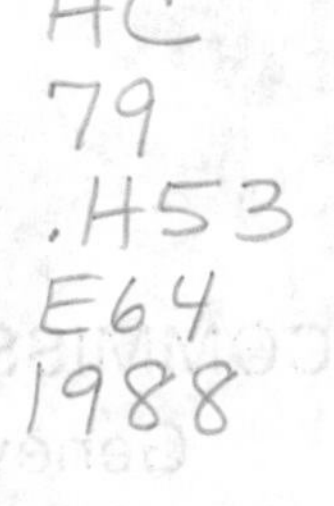

NOTE

The designations employed and the presentation of the material in this publication do not imply the expression of any opinion whatsoever on the part of the Secretariat of the United Nations concerning the legal status of any country, territory, city or area, or of its authorities, or concerning the delimitation of its frontiers or boundaries.

In the present study, the designation "developed" and "developing" economies (countries) is intended for statistical convenience and does not necessarily express a judgement about the stage reached by a particular country or area in the development process.

ECE/ENG.AUT/35

UNITED NATIONS PUBLICATION

Sales No. E.88.II.E.34

ISBN 92-1-116438-9

04800P

CONTENTS

LIST OF TABLES

Chapter I

Chapter II

Chapter III

Chapter IV

Chapter V

Chapter VI

LIST OF FIGURES

Chapter II

Chapter III

Chapter IV

Chapter V

MAIN ABBREVIATIONS USED IN THE STUDY

AC/DC	Alternating current/direct current
ADP	Automated data processing
AGV	Automated guided vehicle
AH	Activity heading
BRITE	Basic Research in Industrial Technologies for Europe (EEC)
CACM	Central American Common Market
CAD	Computer-aided design
CAE	Computer-aided engineering
CAM	Computer-aided manufacturing
CCITT	International Telegraph and Telephone Consultative Committee (ITU)
CEUCA	Customs and Economic Union of Central Africa
CIM	Computer-integrated manufacturing
CMEA	Council for Mutual Economic Assistance
CPE	Centrally planned economies
DIP	Digital image processing
DIS	Draft international standard (ISO)
DP	Data processing
ECE	Economic Commission for Europe (United Nations)
ECOWAS	Economic Community of West African States
ECU	European currency unit
EEC	European Economic Community
EFTA	European Free Trade Association
EITB	Engineering Industry Training Board (United Kingdom)
ESPRIT	European Strategic Programme for Research and Development in Information Technology (EEC)
FAST	Forecasting and assessment in the field of science and technology (EEC)
FMC	Flexible manufacturing cell
FMS	Flexible manufacturing system
FMU	Flexible manufacturing unit
f.o.b.	Free on board
GATT	General Agreement on Tariffs and Trade
GDP	Gross domestic product
GEAD	General Economic Analysis Division (ECE)
GNP	Gross national product
GO	Gross output
IEC	International Electrotechnical Commission
IEEE	Institute of Electrical and Electronics Engineers
IFMBE	International Federation for Medical and Biological Engineering
IFR	International Federation of Robotics
IMEMO	Institute of World Economics and International Relations, Academy of Sciences of the USSR
ISCP	International specialization and co-production (CMEA)
ISIC	International Standard Industrial Classification of All Economic Activities
ISO	International Organization for Standardization
ISTAT	Central Statistical Office (Italy)
ITA	International Trade Administration (United States)
ITD	Industry and Technology Division (ECE)
ITU	International Telecommunication Union
JET	Joint European Torus (EEC thermonuclear research)
JIRA	Japan Industrial Robot Association
LAIA	Latin American Integration Association
LAN	Local area network
LSI	Large-scale integration
MAP	Manufacturing automation protocol
MMS	Manufacturing message specification
n.e.s.	Not elsewhere specified
NC	Numerical control
NIC	Newly industrialized country
NMP	Net material product

OECD	Organisation for Economic Co-operation and Development
OIME	Other industrial market economies
OPEC	Organization of Petroleum Exporting Countries
OSI	Open systems interconnection (ISO)
RACE	Research and Development in Advanced Communication Technologies for Europe (EEC)
R and D	Research and development
SIC	Standard Industrial Classification (United Kingdom, United States)
SITC	Standard International Trade Classification
TR	Transferable rouble (USSR)
UNIDO	United Nations Industrial Development Organization
VA	Value added
VLSI	Very-large-scale integration
WIFO	Austrian Institute for Economic Research
WIPO	World Intellectual Property Organization

SYMBOLS USED IN THE STUDY

A dash (-) indicates nil or negligible

One dot (.) indicates not applicable

Three dots (...) indicate not available

Use of a hyphen (-) between dates representing years, for example 1979-1986, signifies the full period involved, including the beginning and end years

An asterisk (*) indicates estimate (provisional data)

The term "billion" signifies a thousand million

References to dollars ($) are to United States dollars

CHAPTER 0 INTRODUCTION

"Causarum cognitio cognitionem eventorum facit ..."

(Cicero, Topika 67) [0.1]

0.1. PREFACE

The period under review was probably one of the most dynamic in the post-war industrial development, and in that of the engineering industries in particular. The completion of industrial reconstruction and the renovation of basic manufacturing facilities in the 1950s, followed by the boom in the production and application of computer technology - its proliferation in the civil sectors of the economy was one of the few positive results of the world-wide political and military tension - created advantageous conditions for further growth of the engineering industries in the late 1960s and the 1970s.

Certain events in the years immediately preceding the period reviewed in the present study negatively affected progress in the 1980s. The oil shocks and other energy problems, the further decrease of already scarce material resources, the world population explosion, increased immigration into the ECE region and grave environmental problems, amongst others, influenced industrial policies in most ECE countries. The imbalance in the growth of the developed economies of the region and that of the developing countries continued, with the "scissors" opening wider every year and the indebtedness of many developing areas becoming a serious problem.

The recently published Report on Our Common Future (Brundtland Report) stressed the need for new environmental and resource-saving policies directed towards "sustainable development". Originally intended to deal exclusively with environmental strategies, the Brundtland Report turned into a real global study covering all aspects of the "global challenge". As regards the future role of industry, the Report states, in particular, that:

> "The world manufactures seven times more goods today than it did as recently as 1950. Given population growth rates, a five- to tenfold increase in manufacturing output will be needed just to raise developing world consumption of manufactured goods to industrialized world levels by the time population growth rates level off next century." [0.2]

The ongoing "third technological revolution", characterized, *inter alia,* by the "joining of computers and telecommunications into a single ... system", will lead, at the beginning of the twenty-first century, to the implementation of the concept of the "wired society". This revolution will also result "in the elimination of geography as a controlling variable", as can already be illustrated by the rapidly changing nature of today's markets.

> "Historically, a market was a place where roads crossed, rivers crossed: merchants and caravans stopped en route, farmers brought their food, artisans their skills. In the new economy this is no longer true. Take the spot oil market. It is called 'the Rotterdam market' because tankers with capacities above their contracted deliveries would come to Rotterdam to sell extra oil 'on the spot'. There, brokers would make their deals and reroute the tankers to the new destinations. The market is no longer in Rotterdam. Where is it? Everywhere. It is a telex-radio-computer network linking brokers all around the world and redirecting ships on the high seas to new destinations.
>
> Work increasingly is becoming detached from place, operations from their central headquarters. Communications networks, with bursts of data speeding thousands of kilometers, means the breakup of old geographical habits and locations. Thus we see a change of extraordinary historical and sociological importance - the change in the nature of markets from 'place' to 'networks'." [0.3]

Similar processes of "internationalization" are also taking place in the research and development area, in investment policies and in the manufacturing sphere, in particular. In this respect, the engineering industries, producing a wide assortment of products - from needles to the most sophisticated spacecraft -, will certainly continue to occupy their leading position and to play their present decisive role in general economic development (see chapters II, IV and VI).

0.2. ACTIVITIES OF THE ECE WORKING PARTY ON ENGINEERING INDUSTRIES AND AUTOMATION

"ECE activities in the field of engineering industries date back to the late 1940s. It was only in 1969, however, that an *ad hoc* meeting of government experts on engineering industries was convened, followed by the creation of a Working Party on Automation the following year.

Since that time activities in these fields have intensified greatly, reflecting the dynamic developments occurring in member countries of the region. The establishment in 1980 of the Working Party on Engineering Industries and Automation was further recognition of the creative role which ECE could play in this field in the context of east-west economic co-operation.

The Working Party has been at the leading edge of new technologies in the region which have set the pattern not only for ECE countries but the world as a whole. In this connection, the Working Party in recent years has concentrated on topics of current and future interest for sophisticated industrial development." [0.4]

At its annual plenary meeting, the Working Party evaluates the work accomplished and decides on its programme of work for the next five-year period. [0.5] The programme of work is divided into five standard subprogrammes, as follows:

- Medium- and long-term assessment of engineering industries within national, regional and global economies;
- Current developments and prospects in the engineering industries and automation, including appropriate aspects of international trade;
- Selected issues in the engineering industries and in automation;
- Environmental and resource-saving problems; and
- Statistics on engineering industries and automation.

The fulfilment of the programme of work takes the form either of specialized studies undertaken by the secretariat or of topical seminars organized in various countries. (For a list of seminars held since 1970, see annex I and for a list of sales publications see annex II to the present study.) Regarding the preparation of standardized *Annual Reviews* and *Trade Bulletins,* this work is closely related to the regular statistical activities of the Working Party, including the development of internationally comparable statistics (see subchapter 0.3).

In 1974, ECE published a study entitled *Role and Place of Engineering Industries in National and World Economies,* covering the period 1955-1970. [0.6] The study was updated to the period 1970-1975 in two phases: in 1977, *Up-dating to 1970-1975 of the Main Statistical Tables,* in 1980, *Up-dating to 1970-1975 of the Analytical Part* [0.7] and, in 1981, *Up-dating to 1975-1978 of the Main Statistical Tables* were issued. [0.8]

As from that time, a new sales publication entitled *Annual Review of Engineering Industries and Automation* has been issued on a regular basis. The first issue of the *Annual Review,* covering 1979, was published in 1981. Subsequent issues covered 1980, 1981 and 1982. [0.9] Following the decision of the Working Party on Engineering Industries and Automation at its fifth session, in March 1985, a new, accelerated schedule for the preparation of the document was introduced, with a view to reducing the interval between the availability of data and their publication to less than one year. As an interim solution, it was decided to issue a single publication, covering both 1983 and 1984. The 1985 and 1986 issues of the *Annual Review* were published in early 1987 and early 1988, respectively. [0.10]

In parallel with the above publications reviewing general trends and recent developments in the production and use of engineering technologies and products, a special *Bulletin of Statistics on World Trade in Engineering Products* has been published annually since 1963. [0.11]

All these ECE publications complement, in aggregated form and with emphasis on the various regional aspects including east-west co-operation issues, other generally available statistical publications, both as prepared by the Statistical Office of the United Nations in New York (*Statistical Yearbook, Monthly Bulletin of Statistics, Yearbook of Industrial Statistics*) and by other international governmental (OECD, EEC, CMEA, UNIDO etc.) and non-governmental organizations. [0.12]

0.3. EDITORIAL REMARKS

At its fifth session, in February 1985, the Working Party reiterated its interest in undertaking a complete review of the role and place of engineering industries in national, regional and world economies, with due emphasis on structural changes and decided to prepare such a study mainly on the basis of the *Annual Reviews.* [0.13] The first *ad hoc* Meeting for the present study, held in June 1985, agreed on the outline for the study, as well as on the timetable for its preparation. It also invited Governments and relevant international organizations to co-operate on the study and to provide statistical data. [0.14]

Following the decision of the Working Party at its sixth session, the second *ad hoc* Meeting for the study was convened in February 1987. It considered the first draft of the study, including national contributions from four member countries only. The immediately following seventh session of the Working Party agreed on the proposal of the second *ad hoc* Meeting that ECE member countries should be urged to contribute actively by transmitting original and additional material in order to assist the secretariat in the preparation of the second consolidated draft of the study. [0.15]

The third *ad hoc* Meeting for the study, held in December 1987, considered the second draft of the study, including national contributions from eight more countries. The eighth session of the Working Party, in February 1988, endorsed all the recommendations put forward by the third *ad hoc* Meeting and entrusted the secretariat with the final editing and issuing of the study in 1988 in the form of a sales publication. [0.16]

The draft of the present study originally covered the period 1979-1985. In accordance with the recommendation of the third *ad hoc* Meeting, in December 1987, approved by the Working Party in February 1988, [0.17] its coverage was extended to include 1986 data wherever available and possible. During the final editing by the secretariat, in the second quarter of 1988, some parts of the study were even updated with 1987 data (e.g. chapters I and IV).

Chapter I of the study describes the scope and objectives of the study, provides statistical definitions for the subject and reviews the general economic situation prevailing in the world economy in 1979-1986 (1987).

Chapter II illustrates the place of engineering industries in total industry and manufacturing. It shows the contribution (share) of the sector to national and regional economies in terms of production, investments, employment, research and development. It also provides information on the structural development within the sector during the period under review.

Chapter III examines the performance of each of the five subsectors of the engineering industries with emphasis on the development in the ECE region. These five subsectors are the manufacture of: metal products, except for machinery and equipment (ISIC, Rev.2, Major Group 381); non-electrical machinery (ISIC, Rev.2, Major Group 382); electrical machinery (ISIC, Rev.2, Major Group 383); transport equipment (ISIC, Rev.2, Major Group 384) and precision instruments (ISIC, Rev.2, Major Group 385).[0.18]

Chapter IV analyses long- and medium-term trends and developments in international trade in engineering products during the period under review. It also shows the geographic and product pattern of regional trade, developments in export prices and other forms of international co-operation, including standardization in engineering industries.

Chapter V complements the preceding chapters by national contributions and country statements. It contains compiled replies as received from Austria, Belgium, Bulgaria, Czechoslovakia, the German Democratic Republic, Hungary, Italy, Poland, Sweden, the Ukrainian SSR, the USSR, the United Kingdom and the United States. Wherever in chapter V no source is indicated for a particular table, it should be understood that the information provided was received from the respective national authorities.

Chapter VI briefly reviews current trends and foreseeable developments in engineering industries. This chapter also includes the main conclusions that can be drawn from the present study (see subchapter VI.4). Overall economic perspectives to the year 2000 were studied by the Senior Economic Advisers to ECE Governments and recently published.[0.19]

Selected statistical tables, based on the data bank of the ECE/ITD Engineering Industries and Technology Section, are reproduced in a standardized computer print-out form. They have been integrated in the text (chapters II and III), instead of being presented in a separate volume, in order to make the study more compact. These tables result mainly from Government replies to ECE questionnaires for the *Annual Reviews of Engineering Industries and Automation.* Most of the derived indicators (gross output, value added etc.) are calculated on the basis of values expressed in national currencies at current prices, as provided by individual Governments. All tables (and figures) are presented, for technical reasons, at the end of individual subchapters or chapters. References (explanatory notes, bibliography, other comments) are placed at the end of individual chapters.

0.4. ACKNOWLEDGEMENTS

The ECE secretariat wishes to express its gratitude to Governments and international organizations which have provided valuable material and other support in the preparation of the present study.[0.20]

The ECE secretariat's thanks go, in particular, to the Chairman and Vice-Chairman of the three *ad hoc* Meetings for the study held at Geneva from June 1985 to December 1987, Mrs. I. Platonova (USSR) and Mr. D. Hewer (United Kingdom), for their outstanding professional contribution.

It is a pleasure to offer thanks to the Chairman of the Working Party on Engineering Industries and Automation, Mr. V. Novotny (Czechoslovakia), and to the Chairman of the Meeting on Questions of Statistics concerning Engineering Industries and Automation, Mr. E. Smith (United States), for their special efforts and enthusiasm for the subject and for their generosity in making available their extensive experience in the conduct of the study.

Amongst the many individuals who made important comments and provided useful information, special gratitude goes to two consultants to the secretariat: Mr. A.S. Elekoev from the Institute of World Economics and International Relations (IMEMO), Academy of Sciences of the USSR in Moscow (chapters II and VI), and Mr. T.C. Kopinski from the International Business Monitor, Washington D.C. (United States) (chapters IV and VI).

During the preparation of the present study, the ECE/ITD Engineering Industries and Technology Section has maintained close working contacts with the ECE/GEAD division, responsible within the ECE secretariat for the preparation of regular annual *Economic Surveys of Europe*.

REFERENCES

0.1 Translation: "To understand the outcome one must know the cause..."
Cicero, Marcus Tullius (106-43 B.C.), Roman philosopher, statesman and famous orator.

0.2 "Our Common Future" - better known as the Bruntland Report - was prepared by a 22-member Commission headed by Gro Harlem Brundtland, Prime Minister of Norway, at the request of the Secretary-General of the United Nations. United Nations document A/42/427.

0.3 D. Bell "The World in 2013" in *Dialogue,* No. 3, 1988. The author has written many books, including *The Coming of the Post-Industrial Society*.

0.4 Statement by G. Hinteregger, Executive Secretary of ECE, to the eighth session of the Working Party on Engineering Industries and Automation. Press Release ECE/GEN/6, Geneva, 22 February 1988.

0.5 The ECE Working Party on Engineering Industries and Automation was born of a merger of the activities of the former Working Party on Automation and the *ad hoc* Meeting of Government Experts on Engineering Industries. Since 1981, its annual reports have carried the following symbols:

First session 1981	ECE/ENG.AUT/2
Second session 1982	ECE/ENG.AUT/6
Third session 1983	ECE/ENG.AUT/9
Fourth session 1984	ECE/ENG.AUT/14
Fifth session 1985	ECE/ENG.AUT/21
Sixth session 1986	ECE/ENG.AUT/24
Seventh session 1987	ECE/ENG.AUT/28
Eighth session 1988	ECE/ENG.AUT/34

0.6 *Role and Place of Engineering Industries in National and World Economies,* 1955-1970, vol. I (analytical part) and vol. II (main statistical tables). UN/ECE publication, Sales No. 74.II.E.Mim.7.

0.7 *Role and Place of Engineering Industries in National and World Economies. Up-dating to 1970-1975 of the Analytical Part* (vol. I) and *Up-dating to 1970-1975 of the Main Statistical Tables* (vol. II.). UN/ECE publication, Sales No. E.81.II.E.6.

0.8 *Role and Place of Engineering Industries in National and World Economies. Up-dating to 1975-1978 of the Main Statistical Tables.* ECE document ECE/ENGIN/17.

0.9 Sales numbers of *Annual Reviews* covering the years 1979-1982:

Annual Review 1979	E.81.II.E.16
Annual Review 1980	E.82.II.E.18
Annual Review 1981	E.83.II.E.20
Annual Review 1982	E.84.II.E.12

0.10 Sales numbers of *Annual Reviews* covering the years 1983-1986:

Annual Review 1983-1984	E.85.II.E.43
Annual Review 1985	E.86.II.E.30
Annual Review 1986	E.88.II.E.16

0.11 Sales numbers of *Bulletins of Statistics on World Trade in Engineering Products* since 1979:

1979	E/F/R.81.II.E.13
1980	E/F/R.82.II.E.5
1981	E/F/R.83.II.E.8
1982	E/F/R.84.II.E.5
1983	E/F/R.85.II.E.11
1984	E/F/R.86.II.E.10
1985	E/F/R.87.II.E.10
1986	E/F/R.88.II.E.14

0.12 The ECE Working Party on Engineering Industries and Automation maintains regular working contacts with some 30 international governmental and non-governmental organizations. In the field of statistics concerning engineering industries, close co-operation has been established, for instance, with OECD Working Party No. 9 of the Industry Committee on Industrial Statistics (harmonization of questionnaires). Useful contributions to the present study were provided by the secretariats of UNIDO and CMEA. Regarding non-governmental organizations, common efforts in the improvement of statistical methodology, including the development of nomenclatures, data collection and processing, are being undertaken with, amongst others, ISO and IEC (see chapter IV), IFR (statistics on the production and use of robots), IFMBE (statistics on biomedical equipment). Joint statistical meetings were held with the ECE Conference of European Statisticians in 1983 and 1988.

0.13 Report of the fifth session of the Working Party, held from 27 February to 1 March 1985 (document ECE/ENG.AUT/21, paras. 16-19 and project 01(a).1.1. in annex I).

0.14 Report of the first *ad hoc* Meeting for the present study, held from 19 to 21 June 1985 (document ENG.AUT/ AC.10/2).

0.15 For details, see the following documents:

- Report of the sixth session of the Working Party, held from 19 to 21 March 1986 (document ECE/ENG.AUT/24, paras. 10-11 and project 01(a).1.1. in annex I);
- Report of the second *ad hoc* Meeting for the present study, held on 23 and 24 February 1987 (document ENG.AUT/AC.10/4); and
- Report of the seventh session of the Working Party, held from 25 to 27 February 1987 (document ECE/ENG.AUT/28, paras. 9-10 and project 01(a).1.1. in annex I).

0.16 For details, see the following documents:

- Report of the third *ad hoc* Meeting for the present study, held on 17 and 18 December 1987 (document ENG.AUT/AC.10/6); and
- Report of the eighth session of the Working Party, held from 22 to 24 February 1988 (document ECE/ENG.AUT/34, paras. 9-10 and project 01(a).1.1. in annex I).

0.17 See report of the third *ad hoc* Meeting for the present study (document ENG.AUT/AC.10/6, para. 20).

0.18 In order to facilitate the task of the reader, while at the same time keeping the present study reasonably detailed and giving a clear picture of the main trends and developments in the engineering industries, most of the data and information are presented only at the level of individual subsectors (major groups). This has proved sufficient in recognizing and identifying major structural changes within the whole sector (as well as in industry). Individual *Annual Reviews* and specialized techno-economic studies undertaken by the Working Party investigate the structural development within selected subsectors.

0.19 *Overall Economic Perspective to the Year 2000* (UN/ECE Publication, Sales No. E.88.II.E.4) comprises the following chapters:

(1) Trends, challenges and opportunities (ECE region in the world economy, resources and growth factors, common economic problems and policy objectives);

(2) Scenarios exploring long-term economic growth conditions in the ECE region;

(3) Human resources development in the ECE region;

(4) Energy prospects in the ECE region;

(5) Interrelationship between the environment and economic development; and

(6) Impact of science and technology in long-term economic development.

0.20 Chapter V of the present study includes edited national contributions from Austria, Belgium, Bulgaria, Czechoslovakia, the German Democratic Republic, Hungary, Italy, Poland, Sweden, the Ukrainian SSR, the USSR, the United Kingdom and the United States. The following countries responded and commented on or confirmed the draft conclusions contained in chapter VI: the Byelorussian SSR, Czechoslovakia, Finland, Switzerland, the USSR and the United Kingdom. Helpful assistance in the preparation of the study was received from the following international organizations: ISO, UNIDO, IEC and CMEA.

CHAPTER I GENERAL ECONOMIC DEVELOPMENT AND OBJECTIVES OF THE STUDY

I.1. SCOPE AND OBJECTIVES OF THE STUDY

The engineering industries play a key role in industry and in national economies in general. Besides being a major producer of consumer durables, investment goods produced by them have an extensive impact on the technological changes taking place in the economy of all countries. Investment goods supplied by the engineering industries determine the technological and, consequently, the productivity level. As engineering industries are not only the producers but also the main consumers of advanced machinery and equipment, their level of technological development is to a considerable extent indicative of the level of development of industry and the economy as a whole. [I.1]

The present study is based on material provided by Governments and on information contained in the *Annual Reviews of Engineering Industries and Automation* for the period under consideration. It also reflects the results and conclusions of studies undertaken and seminars convened in recent years under the auspices of the ECE Working Party on Engineering Industries and Automation. [I.2]

The performance of the engineering industries is studied mainly on the basis of a comprehensive data bank created by the ECE secretariat and containing time-series of relevant statistics, collected over recent years by means of annual questionnaires and extracted from national statistics and from other official United Nations sources. [I.3]

The study is intended as a general review of developments, with several objectives, e.g.

- To illustrate the dominant role of engineering industries in supplying all other industrial sectors with machinery and equipment, as well as providing the non-industrial sphere (including services and households) with products and devices of basic importance;

- To reveal the continuing leading position of engineering technologies and products in various forms of international trade; and

- To present the engineering sector as a leading-edge investor in high technologies (research and development input, capital investments, redeployment and personnel retraining, etc.), and also their major users, e.g. NC-machining centres, industrial robots, new materials, telecommunication equipment and other information technologies, based on recent developments in microelectronics. In this respect, special attention has been given to the ongoing process of restructuring the engineering industries in favour of less labour-, energy- and raw-material-intensive technologies.

The aim of the study is to examine long-term trends and developments in engineering industries as a whole as well as in each of their five subsectors. On the basis of a short review of the overall economic situation prevailing in the world economy during the period 1979 to 1987, an attempt is made to define the place of engineering industries and their subsectors in this general development. In order to asses the performance of the engineering industries in the period under review, various parameters have been analysed, such as volume and structure of output, research and development expenditure, gross fixed capital formation in the engineering industries, employment, productivity etc. The structure of the engineering industries and their subsectors has been evaluated and an attempt made to derive some general trends. Another objective of the study is the analysis of trends and developments in world trade in engineering products with a view to identifying new forms and particularities in its structure and geographical distribution.

I.2. STATISTICAL DEFINITIONS

In accordance with the United Nations International Standard Industrial Classification (ISIC), Revision 2, all economic activities are classified in 9 major divisions, ranging from Major Division 1: "Agriculture, Hunting, Forestry and Fishing" to Major Division 9: "Community, Social and Personal Services". Industrial activities are classified in Major Division 2: "Mining and Quarrying"; Major Division 3: "Manufacturing"; and Major Division 4: "Electricity, Gas and Water". [I.4] Manufacturing constitutes by far the largest category of economic activities. It includes 9 divisions, of which Division 38: "Manufacture of Fabricated Metal Products, Machinery and Equipment" represents all those economic activities which are the subject of the present study and are labelled "engineering industries". The complete structure of the engineering industries in accordance with ISIC, Rev.2 is reproduced in table I.1. [I.5]

Statistics collected on the basis of ISIC, Rev.2 are establishment-oriented. Examples of such data are statistics on output, value added, investment, prices, employment, research and development expenditure and others. Commodity-oriented statitics have been collected on the basis of the Standard International Trade Classification (SITC), Revision 2. [I.6] Products produced by the engineering industries can be defined in terms of SITC, Rev.2 as Division 69: "Manufactures of metal n.e.s"; Section 7: "Machinery and transport equipment"; Division 87: "Professional, scientific and controlling instruments and apparatus n.e.s."; and Groups 881: "Photographic apparatus and equipment n.e.s", 884: "Optical goods n.e.s." and 885: "Watches and clocks". The complete breakdown of engineering products, including metal products, in accordance with SITC, Rev.2 is reproduced in table I.2.

In order to permit the compilation of comparable statistics on production and trade of selected engineering products and product groups, it was necessary to establish a correspondence key between ISIC, Rev.2 and SITC, Rev.2. As a result of the efforts of the Working Party's statistical group, the engineering industries were broken down into five subsectors on the basis of the ISIC, Rev.2 and the following correspondence presented in table I.3 was agreed upon. [I.7] This correspondence, which has proved its practical usefulness in recent years, has been employed in the present study.

The attention of the reader is, however, drawn to the fact that certain publications which served as a source of data for the present study use a different concept. "Engineering products" are sometimes defined exclusively as machinery and transport equipment (in terms of SITC, Rev.2: Section 7), i.e. metal products (SITC, Rev.2: 69) and precision instruments (SITC, Rev.2: 87 + 881 + 884 + 885) are *not* included under this heading. Thus, to avoid any risk of error, the subject of all statistical analysis is clearly indicated throughout the study.

I.3. GENERAL ECONOMIC DEVELOPMENT IN THE PERIOD UNDER REVIEW

An overview of development during 1979-1987

In 1979, the world economy went into a phase of declining economic growth which was accompanied by a worsening of inflation and by wide changes in the foreign trade balances between regions and countries. [I.8] The slow-down in economic activity lasted for a period of roughly three years, during which economic policies were strongly influenced by the priority given to reducing external trade imbalances.

The early 1980s were characterized by wide-ranging economic disturbances which influenced the economic performance of both centrally planned and market economies in the ECE region. Following the rapid rise in oil prices (which more than doubled between the end of 1978 and the beginning of 1980 and which were superimposed on rising commodity prices), the deceleration in the growth of output became a world-wide phenomenon, while international trade slowed down even more than output. The period of faltering economic activity reached its lowest point in 1982. The economies of the ECE region contributed significantly to this process since their policies, which were aimed in the majority of cases towards adjustment and stabilization, were highly restrictive and had a strongly negative influence on the growth of output.

The period 1983-1984 witnessed an economic upswing in the ECE region. Total output growth accelerated through 1983 and well into 1984, also boosting growth in the rest of the world. The period 1985-1987 was characterized by a relatively moderate and steady rate of expansion in the ECE market economies, while the acceleration which occurred in 1986 in the centrally planned economies was followed in 1987 by a slow-down (see table I.4).

The problem of imbalances

(a) The developed market economies

In the 1980s, several major imbalances have been building up to the point that they are now a threat to the stability of the world economy. The large and growing current account deficit of the United States has perhaps been the main recent source of tension in the international economic climate. The indebtedness of the developing countries has also been the cause of major concern for half a decade already.

The combined current account balance of the developed market economies has remained in deficit for a number of years. This means that some of the industrialized countries have been absorbing resources from the rest of the world, and mostly from the developing countries. This direction of the flow of resources can hardly be reconciled with the aim of supporting the developing countries in their efforts to reduce poverty. The weak growth performance in the developing countries, unsatisfactory plan fulfilments in many centrally planned economies, and high unemployment in western and southern Europe are other symptoms of a general malaise in the world economy.

Several recent economic events, such as the persistent volatility of exchange rates, the negative effects of the stock market crash which occurred in October 1987 and the mounting pressure of protectionist tendencies, show clearly the urgency of the need to take concerted international action in order to reduce the present imbalances and the danger of a recession.

For some time now the main issue concerning adjustment in the world economy has been how to bring down, without a serious economic crisis, the United States current account deficit with a corresponding reduction in surpluses elsewhere. The process has been under way for the past three years, and as a result, there has been a fall in the United States dollar exchange rate and a corresponding improvement in the competitiveness of goods produced in the United States. Although there have been no visible signs of a sustained reduction in the current account deficit itself, the underlying trend towards improvement has been evident for some time: in 1987 the volume of United States exports increased considerably, while the growth of import volume slowed down. One important reason for the slow reaction in the current accounts is that, in 1986, exports (in US dollars) were equal to only some two thirds of imports.

In western Europe, on the other hand, the adjustment is evidently not moving with sufficient momentum to offset the weaker growth in the United States. The adverse effect of the expected deterioration in the

European external accounts should be cushioned by stronger domestic demand. The chances that sufficient growth will be generated to improve the employment situation in western Europe are declining. The prevailing objective seems to be the reduction of the size and scope of the public sector. Hence, fiscal policies for 1988 are not expected to counter the tendencies towards weak growth and rising unemployment in western Europe. The overall restrictive stance will continue. As regards monetary policy, after the stock market crisis, there was an easing to support the liquidity of the financial system as the deflationary risks increased. The general course of monetary policy, however, is uncertain since its conduct is influenced by both domestic policy commitments and external exchange-rate considerations.

From the point of view of the prospects for the world economy, it is important to note that, in Japan, following recent policy action, the rate of growth of domestic demand is expected to be maintained. Since the volume of Japanese net exports is declining, the country's economy is providing some support to developments in the rest of the world. Nevertheless, the economic performance in western Europe and the United States will exert a dominant influence on the world economy.

(b) The centrally planned economies

The period under review was marked by unprecedented challenges for most European centrally planned economies. Towards the end of the 1970s and at the beginning of the 1980s, most of these countries experienced the lowest growth rates of the post-war period. Moreover, deficits in their foreign trade with the developed market economies had built up to high levels during the 1970s. The resulting external debt, combined with unusually high interest rates in international financial markets and the virtual disappearance in the early 1980s of new credit lines, led these countries to place restoration of the foreign trade balances as a top priority.

Given the requirements of the external balances, and deliberate policies to maintain or even further increase living standards, investment inevitably bore the brunt of this strategy. Implicitly at least, rapid growth lost its ranking among economic policy priorities, although the five-year plans of all the countries clearly aimed at checking the deceleration of growth at some stage during the quinquennium. A wide range of economy measures, and also some modifications to management and planning systems, were promulgated to help implement this economic strategy.

All in all, it seems that after the unexpected slowdown in output growth in 1987, the European centrally planned economies are now poised to accelerate economic expansion in 1988. The strength of this acceleration will largely depend on agricultural production and gains in overall efficiency. It will also depend on developments in international markets, especially prices and demand for oil and refined oil products.

The recession of 1980-1982

The recovery from the recession, which, on the basis of previous cycles, should have followed the downturn in 1980 did in fact materialize, but it petered out. A new recession thus appeared which resulted in a renewed fall in output in 1982 in North American countries and in virtual stagnation in western Europe.

The policy-induced character of the recession derived from a tightening of economic policies acompanied by high and rising interest rates. Fixed investment dropped in western Europe in 1981 and 1982 and in the United States in 1980 and, after a virtual stagnation in 1981, again and more rapidly in 1982.

The monetary factors did not only contribute to the 1982 recession within national economies. The internal banking system also began to show signs of fragility which led to the emergency of new forces of contraction. The negative effects were particularly negative for non-oil developing countries, which were faced after 1979 with several external disturbances at the same time: first, higher oil prices raised directly and indirectly the cost of imports; secondly, the economic policies of the industrialized countries raised interest rates in world financial markets and, thirdly, the recession led to a reduction in the demand for imports.

In 1982, the developing countries' capacity to import was drastically affected by falling terms of trade, rising debt-service costs and, above all, by the shrinking availability of external credit. Their imports declined, adding further deflationary pressure on economic activity in the rest of the world.

For all the centrally planned economies, 1980 was the closing year of a five-year plan period. A deceleration of output growth had already been foreseen in the plans for this period, but actual developments were below envisaged plans. Factors such as the slow-down in the growth of the labour supply and increasing scarcities of energy and raw materials severely restricted the possibilities of further strong expansion. Besides, various unexpected events and developments, such as the oil price increase and, particularly, the deepening recession in the developed market economies, contributed to the weaker-than-planned output growth.

In this context, Governments opted in 1981 for a slow rate of growth in the domestic use of output, in order to provide room for improvements in foreign trade balances without endangering consumption. The deceleration of production, in this policy framework, had a severe impact on investments, which had been faltering already before 1981. Weak investment inevitably delayed the implementation of necessary structural adjustments.

As a result, the volume growth of aggregate domestic demand (NMP used for consumption and accumulation) declined in 1981 and in some countries there was even an absolute fall. In 1982, when eastern Europe had attained a surplus in the current account with the market economies, the volume of aggregate domestic demand fell. In the USSR, on the other hand, the development of aggregate domestic demand roughly followed that of production.

The tight stance of policy in eastern Europe was also necessitated by the changing situation in intra-CMEA trade. The big rise in world commodity prices was absorbed into intra-CMEA trade prices with a time lag, and was eventually reflected in changes in the terms of trade between the USSR and the east European countries. Hence, adjustment efforts were also needed in this part of their foreign trade. Thus, in 1981-1982 the volume of east European exports to the USSR increased considerably while their imports fell.

In 1982, there was near stagnation in the world economy. The east European countries (other than the USSR) again registered a poor performance. The rapid changes in international financial conditions during 1981 had a profound impact on the east European economies. As financial difficulties became serious, for Poland and Romania in particular, international lending institutions became more cautious in their lending policies and credits to most centrally planned economies began to dry up. In 1982 certain financial institutions also withdrew their deposits from east European countries and thereby added to the liquidity shortage in these countries. In these external financial conditions, some of the most indebted countries had to adopt adjustment measures which involved abrupt and sharp import cuts and strong export promotion, hence aggravating internal imbalances and inhibiting the achievement of the modest growth targets. As a result, the east European countries succeeded in turning their current accounts from deficit into surplus in 1982. The counterpart of these strong adjustment measures was, however, a low output growth in all the east European countries. In the USSR, the situation was quite different: an improvement in the terms of trade permitted imports to continue to rise substantially and the rate of output growth was higher than that of the preceding three years.

The upturn in 1983-1984

In 1983, the economies of the ECE region began to recover from the longest post-war slow-down in real growth. Output for the region as a whole increased by some 3% in 1983. While there was an upturn throughout the region, its strength was uneven: in the centrally planned economies, the output growth was stronger than in the market economies; among the market economies the recovery in the United States gathered considerable momentum, whereas output growth in western Europe lagged behind. The recovery in 1983 continued in 1984 with more strength than generally expected.

The upswing of economic activity in the ECE region has spread its influence to the rest of the world. The dominating sources for this uplift were the United States and, outside the ECE region and to a lesser extent, Japan. The developing countries' exports increased sharply and, as their balance of payments constraints eased somewhat, they were able, as a group, to raise their imports. Thus the revival in the industrialized countries provided the impetus to some modest growth in the developing countries in 1984.

The recovery had a marked effect on unemployment only in the United States, where it fell from 10% in 1982 to around 7% in the fourth quarter of 1984. In western Europe, in contrast, unemployment remained on a rising trend and, at the end of 1984, was around 10%.

The strong investment performance during the recovery in the United States restored the ratio of gross fixed investments to total output (GNP) to the level attained during the previous peak, in 1979. However, this investment boom was financed to a great extent by foreign savings through a growing current account deficit.

In western Europe, the investment ratio did not increase and remained well below the ratios of the late 1960s and early 1970s. Thus it was not only western Europe's phase of slow growth which was a matter of concern: weak investment activity compromised the possibility of attaining future rates of output growth that would provide for growing employment and a decline in unemployment.

A notable feature of the recovery was the continued fall in inflation rates. Contrary to often-expressed views, fiscal stimulus such as that experienced in the United States did not lead to accelerating inflation. There were, at any rate, several factors at work which tended to reduce inflation during the revival of economic activity: conditions of large excess supply in goods and labour markets continued to dampen inflation; inflationary expectations were on the decline; and inflationary pressures from the international commodity markets did not flare up, although there was some upward pressure in the early stage of recovery.

The upturn in the rate of growth of the centrally planned economies got under way in 1983. However, the pace and timing of development was not uniform. After stagnating in 1982, total output growth in eastern Europe rose sharply in 1983 by nearly 4% and accelerated further, to over 5% in 1984. In the USSR, on the other hand, output growth, hovering between 3 and 4%, increased relatively less from the rates of the immediately preceding years.

In the centrally planned economies as a whole, total domestic demand (NMP used) continued to expand at a slower pace than output (NMP produced), reflecting the policy targets with respect to external balance. As regards the structure of domestic demand, consumption remained relatively protected during the austerity period, but in 1983 growth in consumption was less than that in total domestic demand in several countries because of stronger fixed investments. In 1984, consumption again rose considerably, leaving correspondingly less room for expansion in fixed investment. Investment growth thus became somewhat more buoyant after the recovery, which began during 1983.

The slow-down of 1985

After gathering momentum in 1983 and 1984, world output growth declined in 1985. The slow-down affected both industrialized and developing countries, but it was more pronounced in industrialized countries and particularly in market economies belonging to the ECE region. The deceleration of economic growth in the region was mainly due to the United States.

In the centrally planned economies of the ECE region, the plans for 1985 indicated that the economic growth was expected to continue at least at the rate attained during the two previous years. However, these expectations were not fulfilled.

Recent economic trends and prospects in market economies

The slight loss of momentum in the world economy during 1987 contained considerable diversity of performance among countries and regions. The main elements in this slow-down were the fall in output growth in the developing countries and the less-than-expected expansion in the European centrally planned economies. Output growth in the developed market economies was maintained at the same rate as in 1986. This steady growth in the developed market economies was, however, modest and slightly below that of the world economy as a whole. In the ECE region, growth in 1987 was about 2.6% compared with 3.2% in 1986.

The upswing in the United States lasted for more than five years. After the boom year of 1984, when GNP rose by nearly 7 percentage points, output growth fell back and stayed close to 3%, without any further slackening during the three-year period of 1985-1987. In contrast to the development in aggregate supply, aggregate demand and its main components have fluctuated sharply in the course of the upswing.

During the first four years of the five-year upswing, real total domestic demand outran the growth of output. In the peak year of 1984, demand growth was as much as 2 percentage points higher than output growth. As seen from the demand side, the upswing still remained quite strong throughout 1985 and most of 1986. Only since 1987 has the deceleration of domestic demand been clearly evident, while growth has not faltered.

Economic activity in western Europe picked up very slowly from the recession in the early 1980s. In contrast to developments in the United States, there has been no genuine recovery in the sense of a significant narrowing of the gap between actual and potential output. Instead, output growth quickly settled down at an annual rate of 2.5% until 1987 when even this modest rate of expansion left productive capacities in western Europe underutilized. In turn, this discouraged the formation of new capacity as much of the investment was, in fact, carried out to offset the scrapping of obsolete equipment.

As regards the interaction between total demand and output in western Europe, developments, in broad terms, have been a mirror-image of those in the United States. The rate of growth of real domestic demand was continuously below the rate of output growth in 1983-1985. However, during the past two years domestic demand has strengthened but, with declining real net exports, much of it has leaked abroad leaving output growth unaffected. The weakened competitiveness of western Europe in relation to the United States is reflected in these changes in the demand pattern. Western Europe (together with Japan), in principle, is expected to take over from the United States as the leading source of growth for the world economy. In point of fact, while domestic demand in the United States has slowed down to make room for more exports, domestic demand in western Europe has accelerated, giving some impulse to growth in the rest of the world. However, from the point of view of providing a boost to the world economy, reactions in western Europe have been weak. In fact, short-term forecasts indicate that the growth of total real domestic demand threatens to fall back after two years of somewhat more rapid growth.

Developments in individual west European countries have not been uniform. In a few countries, among them the United Kingdom and Italy, growth has lately been stronger than in the rest of western Europe. However, the majority of countries, and especially those with close economic ties with the Federal Republic of Germany, have experienced very sluggish growth. Since the major part of west European trade is intra-trade, there is the risk that weak economic growth will also spread to the countries which have so far shown greater dynamism. Except in the surplus countries, widening external imbalances may limit the scope for many individual countries to take action against weakening economic growth and increasing unemployment.

After a long period, during which the effects of the turmoil of the early 1980s were felt more strongly than in most other market economies, the years 1986-1987 saw an improvement in the economies of southern Europe. Growth began to accelerate and in both 1986 and 1987 output performance was, on average, better than in the rest of western Europe.

Based on both national and international forecasts, output growth in market economies of the ECE region is likely to account in 1988 for some 3%.

The forecasts for the market economies implicitly assume that the adjustment process in the United States will proceed smoothly. In response to relative price changes and changed external conditions, resources should move from domestic, sheltered users to the sectors producing tradeable goods, fixed investment and net exports should rise at the expense of consumption, private and public, and foreign demand is assumed to be sufficiently buoyant to absorb the increased supply of United States' products.

Recent economic trends and prospects in centrally planned economies

After the very strong expansion in 1986, the centrally planned economies of the ECE region recorded a considerable slow-down in output growth in 1987. In eastern Europe the net material product (NMP produced) increased by 3.2% in 1987 as compared with 4.6% in 1986. In the Soviet Union, NMP produced grew at a rate of 2.3% compared with 4.1% in 1986. Performance varied within eastern Europe, with growth rates in individual countries ranging from 2 to 5%, but almost all countries recorded a slow-down. However, in Hungary, where output growth in the two previous years had lagged considerably behind the average for the seven-country grouping, the growth of NMP produced accelerated in 1987. Bulgaria maintained the growth of NMP produced at roughly the same rate as in 1986 - somewhat above 5%. Smaller-than-expected efficiency gains, together with weak export demand and deteriorating terms of trade in western markets, seem to have been the main reasons explaining the slow-down in 1987.

The industrial sector provides the bulk of national output in all the centrally planned economies of the ECE region. Growth rates recorded in 1987 were also below those of 1986, but the slow-down was less pronounced. In the Soviet Union, the 3.8% rate of increase of gross industrial production was 1 percentage point below that recorded in 1986 but coincided with the target for 1987. In eastern Europe, the aggregate 3.5% rate of increase of gross industrial output was also 1 percentage point below the rate attained in 1986.

Short-term considerations apart, recent economic developments point to a moderation of the more buoyant medium-term economic growth registered since 1983 in both eastern Europe and the Soviet Union. Thus the maintenance of the pace of expansion recorded in the past five years or so proved to be much more demanding than may have been assumed at the time when the five-year plan strategies for the period 1986-1990 were laid down. In fact, average output growth in 1986-1987 in eastern Europe and the Soviet Union lagged behind the growth rates recorded in 1983-1985. The annual plans for 1988 call for high growth rates of NMP produced, both in eastern Europe (4.8%) and, in particular, in the Soviet Union (5.9%).

East-west trade has been losing momentum during the past three years and developments during 1987 were no exception in this respect. Apart from the big rise in Soviet exports after the oil price decline, the tendency towards sluggishness has affected all the major components of this trade. Given that east-west trade is relatively important in the total trade of the centrally planned economies, its weakening may affect their capacity to activate and sustain a more resource-efficient pattern of economic growth. The economic reforms pertaining to foreign trade have clearly reflected the perception of such a risk. The steady decline of market shares in the west is a long-term challenge to eastern Europe to improve export competitiveness in order to sustain and widen the scope of imports from the west.

Against a background of changing domestic and international development conditions, several countries accelerated the process of economic reforms along the lines announced in the previous two years. This was particularly the case in the Soviet Union, where economic reforms have been coupled with a broader programme of economic and social reconstruction (perestroika), which was worked out in 1987. Within the second stage of reforms now under way, greater enterprise independence has been extended to many new branches and a limited, but still important, scheme of state orders, combined with contracts between enterprises based on output plans, are gradually to replace the traditional system of mandatory planning indicators. Significant changes have also taken place in external economic relations, where increasing numbers of enterprises and associations are now allowed to conduct foreign transactions on their own account, and will enjoy expanded rights with regard to the disposal of foreign currencies so earned. A subsequent, third stage of the reform is scheduled for the beginning of the next decade, when substantial changes in domestic price, income, tax and budget systems are to be implemented.

The reform process also continued in Hungary, where it started at the end of the 1960s. In addition to earlier changes concerning macro-economic planning and the management of enterprises, a value-added tax for enterprises and progressive personal income tax were introduced at the beginning of 1988. In Poland, a second - more radical - phase of economic reform is being implemented. Although fundamental economic reform is considered the only solution for the problems of the Polish economy, heavy external indebtedness continues to pose serious constraints. A reform process has also begun in Bulgaria, where a new set of rules for economic activity became effective on 1 January 1988. In Czechoslovakia, the movement towards reform strengthened at the end of 1987 and in the course of 1988, and a draft reform programme has been elaborated. In the German Democratic Republic, it appears that the national authorities consider that previously introduced modifications to the country's management and planning system have not yet exhausted their potential. In Romania, the traditional system of central management and planning is still thought to be appropriate for the country.

Table I.1

Structure of the engineering industries in accordance with the International Standard Industrial Classification of all Economic Activities (ISIC), Revision 2

Division	Major Group	Group	Type of product
38			Manufacture of fabricated metal products, machinery and equipment
	381		Manufacture of fabricated metal products, except machinery and equipment
		3811	Manufacture of cutlery, hand tools and general hardware
		3812	Manufacture of furniture and fixtures primarily of metal
		3813	Manufacture of structural metal products
		3819	Manufacture of fabricated metal products except machinery and equipment n.e.c.
	382		Manufacture of machinery except electrical
		3821	Manufacture of engines and turbines
		3822	Manufacture of agricultural machinery and equipment
		3823	Manufacture of metal-and wood-working machinery
		3824	Manufacture of special industrial machinery and equipment except metal- and wood-working machinery
		3825	Manufacture of office, computing and accounting machinery
		3829	Manufacture of machinery and equipment except electrical n.e.c.
	383		Manufacture of electrical machinery, apparatus, appliances and supplies
		3831	Manufacture of electrical industrial machinery and apparatus
		3832	Manufacture of radio, television and communication equipment and apparatus
		3833	Manufacture of electrical appliances and housewares
		3839	Manufacture of electrical apparatus and supplies n.e.c.
	384		Manufacture of transport equipment
		3841	Ship-building and repairing
		3842	Manufacture of railroad equipment
		3843	Manufacture of motor vehicles
		3844	Manufacture of motorcyles and bicycles
		3845	Manufacture of aircraft
		3849	Manufacture of transport equipment n.e.c.
	385		Manufacture of professional and scientific, and measuring and controlling equipment, n.e.c, and of photographic and optical goods
		3851	Manufacture of professional and scientific, and measuring and controlling equipment n.e.c.
		3852	Manufacture of photographic and optical goods
		3853	Manufacture of watches and clocks

Source: See I.4.

Table I.2

Structure of engineering products (including metal products) in accordance with the Standard International Trade Classification (SITC), Revison 2

Section	Division	Group	Type of product
	69		Manufactures of metal n.e.s.
		691	Structures and parts of structures n.e.s. of iron, steel or aluminimum
		692	Metal containers for storage and transport
		693	Wire products (excluding insulated electrical wiring) and fencing grills
		694	Nails, screws, nuts, bolts, rivets and the like, of iron, steel or copper
		695	Tools for use in the hand or in machines
		696	Cutlery
		697	Household equipment of base metal n.e.s.
		699	Manufactures of base metal n.e.s.
7			Machinery and transport equipment
	71		Power generating machinery and equipment
		711	Steam and other vapour generating boilers, superheated water boilers
		712	Steam and other vapour power units, not incorporating boilers; steam engines
		713	Internal combustion piston engines
		714	Engines and motors, non electric
		716	Rotating electric plant
		718	Other power generating machinery
	72		Machinery specialized for particular industries
		721	Agricultural machinery (excluding tractors)
		722	Tractors, whether or not fitted with power-take-offs, winches or pulleys
		723	Civil engineering and contractors plant and equipment
		724	Textile and leather machinery
		725	Paper mill and pulp mill machinery, paper cutting machines and other machinery for the manufacture of paper articles
		726	Printing and bookbinding machinery
		727	Food processing machines
		728	Other machinery and equipment specialized for particular industries
	73		Metalworking machinery
		736	Machine tools for working metal or metal carbides
		737	Metalworking machinery (other than machine tools)

Source: See I.6.

Table continued.

Table I.2 (continued)

Section	Division	Group	Type of product
	74		General industrial machinery and equipment n.e.s.
		741	Heating and cooling equipment
		742	Pumps (including motor and turbo pumps) for liquids; liquid elevators of bucket, chain, screw, band and similar kinds
		743	Pumps (other than pumps for liquids) and compressors; fans and blowers; centrifuges; filtering and purifying apparatus
		744	Mechanical handling equipment
		745	Other-non electrical machinery, tools and mechanical apparatus
		749	Non-electric parts and accessories of machinery
	75		Office machines and automatic data processing equipment
		751	Office machines
		752	Automatic data processing machines and units thereof; magnetic or optical readers, machines for trans cribing data onto data media in coded form and machines for processing such data
		759	Parts of and accessories suitable for use with machines of a kind falling within heading 751 and 752
	76		Telecommunications and sound recording and reproducing apparatus and equipment
		761	Television receivers
		762	Radio broadcast receivers
		763	Gramophones, dictating machines and other sound recorders and reproducers; television image and sound recorders and reproducers, magnetic
		764	Telecommunications equipment, and parts of and accessories for the apparatus and equipment falling within division 76
	77		Electrical machinery, apparatus and appliances
		771	Electric power machinery (other than the rotating electric plant of heading 716)
		772	Electrical apparatus for making and breaking electrical circuits, for the protection of electrical circuits, or for making connections to or in electrical circuits
		773	Equipment for distributing electricity
		774	Electric apparatus for medical purposes and radiological apparatus
		775	Household type electrical and non-electrical equipment
		776	Thermionic, cold cathode and photo-cathode valves and tubes; photocells; mounted piezo-electric crystals; diodes; transistors and similar semi-conductor devices; electronic microcircuits
		778	Electric machinery and apparatus n.e.s.

Table continued.

Table I.2 (concluded)

Section	Division	Group	Type of product
	78		Road vehicles
		781	Passenger motor cars
		782	Motor vehicles for the transport of goods or materials and special purpose motor vehicles
		783	Road motor vehicles n.e.s.
		784	Parts and accessories, n.e.s. of the motor vehicles falling within headings 722, 781, 782, 783
		785	Motorcycles, motor scooters and other cycles, motorized and non-motorized; invalid carriages
		786	Trailers and other vehicles, not motorized and specially designed and equipped transport containers
	79		Other transport equipment
		791	Railway vehicles (including hover trains) and associated equipment
		792	Aircraft and associated equipment
		793	Ships, boats (including hovercraft) and floating structures
8			Miscellaneous manufactured articles
	87		Professional, scientific and controlling instruments and apparatus
		871	Optical instruments and apparatus
		872	Medical instruments and appliances
		873	Meters and counters
		874	Measuring, checking, analysing and controlling instruments and apparatus n.e.s.; parts and accessories n.e.s of the instruments and apparatus of groups 873 and 874
	88		Photographic apparatus, equipment and supplies and optical goods n.e.s.; watches and clocks
		881	Photographic apparatus and equipment n.e.s.
		884	Optical goods n.e.s.
		885	Watches and clocks

Table I.3

Correspondence key ISIC, Rev.2 - SITC, Rev.2

Economic activity (ISIC, Rev.2 code)	*Type of products manufactured (Terminology used throughout the study)*	*Product classification (SITC, Rev.2 code)*
38	Engineering products, including metal products	69+7+87+884+884+885
381	Metal products	69
382	Non-electrical machinery	71(except 716)+72+73+ 74+75
383	Electrical machinery	716+76+77
384	Transport equipment	78+79
385	Precision instruments	87+881+884+885

Table I.4

Annual changes in output growth in the ECE region, 1979-1987

(Percentage)

Region	*1979*	*1980*	*1981*	*1982*	*1983*	*1984*	*1985*	*1986*	*1987*
ECE market economies [a]	3.0	0.8	1.1	-0.8	2.5	4.5	2.8	2.8	2.7
of which:									
North America	2.6	0.0	2.1	-2.6	3.5	6.7	3.1	3.0	2.9
Western Europe	3.3	1.5	0.3	0.8	1.6	2.6	2.6	2.7	2.4
ECE centrally planned economies [b]	2.1	2.7	1.7	2.8	4.1	3.6	3.5	4.2	2.6
of which:									
Eastern Europe	2.0	0.1	-1.9	0.1	3.9	5.3	3.7	4.6	3.2
USSR	2.2	3.9	3.3	3.9	4.2	2.9	3.5	4.1	2.3

Source: Economic Surveys of Europe in 1985-1986, UN/ECE publication, Sales No. E.86.II.E.1, and 1987-1988, Sales No. E.88.II.E.1.

a For North America: gross national product; for western Europe: gross domestic product.

b For eastern Europe and the USSR: net material product produced.

REFERENCES

I.1 See chapter II of the present study entitled "The place of engineering industries and their subsectors in the economy as a whole".

I.2 Since 1979, the ECE Working Party on Engineering Industries and Automation (and its predecessors) has published (in the form of UN/ECE sales publications):

- ten techno-economic studies on selected topics;
- seven issues of the *Annual Review of Engineering Industries and Automation*; and
- eight issues of the *Bulletin of Statistics on World Trade in Engineering Products*

and organized 11 seminars on various engineering industries or related issues (see also annexes I and II of the present study).

I.3. Whenever specific data and/or information are used in the study, the exact source is mentioned. The large majority of the statistical tables are based on national sources, as collected and compiled by the ECE Industry and Technology Division.

I.4 *International Standard Industrial Classification of All Economic Activities*, Statistical Papers, Series M, No.4, Rev.2. United Nations publication, Sales No. E.68.XVII.8.

I.5 Within the present study, most of the comparative production analyses distinguish between ISIC major groups (381-385) only. These five ISIC major groups, in fact, represent all the main engineering subsectors. A more detailed analysis (by individual groups of engineering products) has been started within individual *Annual Reviews* and specialized studies.

I.6 *Standard International Trade Classification*, Statistical Papers, Series M, No.32, Rev.2 (United Nations publication, Sales No. E.75.XVII.6).

I.7 The Working Party continues its statistical methodological work on a regular basis, within the scope of annual meetings on Questions of Statistics concerning Engineering Industries and Automation. The tenth statistical meeting and, at the same time, the second joint meeting with the Conference of European Statisticians, will take place in October 1988.

I.8 The present subchapter has been compiled mainly on the basis of the *Economic Survey of Europe in 1986-1987* (UN/ECE publication, Sales No. E.87.II.E.1) and in *1987-1988* (Sales No. E.88.II.E.l), as prepared by the ECE General Economic Analysis Division.

CHAPTER II THE PLACE OF ENGINEERING INDUSTRIES AND THEIR SUBSECTORS IN THE ECONOMY AS A WHOLE

II.1 THE ENGINEERING INDUSTRIES - THE KEY SECTOR FOR TECHNICAL PROGRESS

The engineering industries form the core of the economy in all industrialized countries. The engineering industries first and foremost produce capital (investment) goods which determine the technological and, consequently, the productivity level of industry. "The capital goods industry ... by its unique position in being linked directly to virtually all sectors of the economy, is a motive force in development and an engine for technological growth. It creates the conditions not only for operating the industrial system but also for the self-reproduction of this system". [II.1] In addition to capital goods (mainly machinery and equipment), the engineering industries produce and supply a wide range of intermediate products, i.e. semi-finished products, components and parts for use in all other industrial sectors, and consumer durables for non-industrial use in homes and services (see figure II.1).

The engineering industries are not only the producers and suppliers but also major users of technologically advanced equipment. The technological level of the engineering industries is, therefore, representative of the technological potential of the national economy in general. To meet the needs of all sectors of the economy in technologically advanced, highly productive equipment, the engineering industries have continuously to develop and to produce new types of capital goods which are competitive on the world market as concerns both their technical parameters and their prices. To meet these challenges, considerable efforts are needed in research and development (R and D). It is, therefore, not surprising that R and D expenditure in the engineering industries has grown from year to year in most countries (see subchapter II.3).

The late 1970s and the 1980s brought about the widespread introduction of new labour-, material- and energy-saving technologies in the engineering industries of all industrialized countries. Technological progress was centred in those sectors of the engineering industries producing electronics, computers, machine tools, telecommunication equipment, measuring and controlling instruments, and others. The major consumers of the products of those high-technology industries also belong to the engineering industries. Recent fast developments in the automotive industry (considerably reduced fuel consumption, improved safety, higher comfort, reduced noise and pollution levels, etc.), in the aircraft industry (introduction of more fuel-efficient, quieter engines, substantial weight reductions through improved design and the use of new materials, etc.) and in shipbuilding (high level of automation and, consequently, broad application of labour-saving technologies), for instance, would be unthinkable without the product and technology developments in the key engineering branches described above.

The engineering industries of the industrialized countries have developed into large and complex undertakings in which science is harnessed to production, and they are the basis for the general integration of science and production. On the one hand, the engineering industries supply the most up-to-date technology and consumer goods, in which respect they play a key role in the introduction of technical advances into, *inter alia,* the production process, personal consumption, health care, education and science. [II.2] On the other hand, the engineering industries themselves are a very large customer for machinery and equipment. Through their connections with all their customers, the engineering industries receive feedback, and are thus able to interpret needs and to decide the scale and priorities of the innovation process and structural change.

The engineering industries themselves are a major consumer of their own output, as is shown, for example, by the following figures for the United States of America: intermediate consumption 24.6%, capital investment in machinery and equipment 3.3%. [II.3] Consequently, these industries consume 25-30% of their output. The remainder goes to other sectors of the economy (40-50%), or is used to increase reserves (3%), or on personal consumption (15-20%), or for exports and the military sphere. [II.4]

It is a feature of the engineering industries that their output largely comprises short-run and custom-made items. Mass production accounts for less than 20% of the total volume of their output. The objective contradiction between the trend towards individualization of demand (with concomitant losses in the continuity of output) and the attempts of entrepreneurs to reduce production costs by economies of scale has given rise to the need to develop flexible and rapidly applicable means of automation. The technological basis for this reprogrammable equipment is the rapid progress made in microelectronics. During the last few years there have been several increases in the component density of integrated circuits, in the capacity of memory stores and in the rapidity of operation. There is basic innovation in the manufacture of microelectronic components every two or three years. These developments have resulted in the miniaturization of electronic devices, which have become more reliable and relatively cheaper. Microprocessors have brought about a real revolution in the use of computer techniques in the engineering industries. These miniature devices, which are relatively cheap but quite powerful, are widely employed in automated process-control systems. They are used in numerically controlled (NC) machine tools, in industrial robots, in automated transport within the plant, in monitoring and measuring equipment etc., and to optimize the operation of conventional equipment (electric motors, for example). The incorporation of the microprocessor into production in the engineering industries has resulted in considerable improvements in the quality and consumer characteristics of both capital goods and consumer durables (the term "mechatronics" has been coined for this marriage of mechanical and electromechanical devices with electronic controls).

Ever wider use is being made in the engineering industries of computer-aided design (CAD) and computer-aided engineering and manufacturing (CAE, CAM), many thousands of flexible manufacturing units and cells (FMU, FMC) have been set up, and engineers and scientists have been given training in the use of personal computers. The results are greatly increased flexibility in production, improved quality, higher productivity of the work-force, smaller production stocks and smaller working areas. Complex hierarchical systems and integrated computerized production have become possible. The automation of medium-batch and, sometimes, even of small-batch production has become profitable.

The rapid growth rates of the high-technology industries and their increasing share in the total volume of output are a feature of the modern engineering industries. The branches concerned include electrical engineering, electronic engineering, the aerospace industry, computers and communication equipment, automated equipment and automated production systems, instrument making etc. In 1980, these high-technology industries accounted for 53% of the output of the engineering industries in the United States, 42% in the Federal Republic of Germany, and 35% each in the United Kingdom and Japan; their share has continued to increase subsequently. [II.5]

As an example of the significant increase in the high-technology production of centrally planned economies, one can use the figures for the USSR. The production of computer technology increased from 2.7 billion roubles in 1981 to 4.8 billion in 1986 (estimated value for 1987: 5.2 billion roubles), and the production of precision instruments (mainly measuring and controlling apparatus) from 3.8 billion roubles in 1981 to 4.8 billion in 1986 (estimated value for 1987: 5.1 billion roubles). Similar trends were also registered in other east European countries. [II.6]

Scientific and technological progress in these industries is raising the overall technical level of the engineering industries, which are among the main users of high-technology output, and is modernizing the "conventional" branches of the economy. Thus, it is expected that motor vehicles, which consume half of all petroleum products in market-economy countries, will be on average 20% more effective by their use of fuel in 1990, by comparison with 1980, and 25 to 35% more effective by the year 2000. The ecological characteristics, safety and reliability of machines and mechanisms are being improved. The applications of automation to production and process control are expanding rapidly, as is the use of new materials (plastics, composites, ceramics, optical fibres, metals of exceptionally high purity and their alloys). The development of information technology, including telecommunication equipment, is at the heart of this process.

Accelerated development of the industrial infrastructure has also been registered. The process of introduction of new engineering technologies requires a variety of types of services, e.g. education and training, data processing and other information services, transportation and distribution services, management and consultation, maintenance and specialized architectural and engineering services. [II.7] Increasing importance is also being accorded to numerous financial and trading aspects of the design, investment, production and supplying processes.

Consequently, the engineering industries are playing a most important role in the social development of the industrialized countries. Their significant technological potential and highly skilled work-force make them the main driving force and the prime source of scientific and technical progress in the modernization and technical re-equipping of branches of the national economy and in increasing labour and capital productivity. At the same time, the engineering industries themselves depend for their development and structure on the demand for their products from all sectors and spheres of the national economy and on the state of the world market. Development of the engineering industries is greatly affected by the demand for consumer durables.

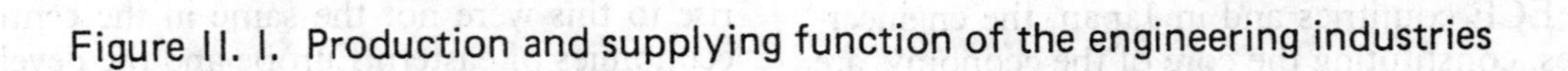

Figure II. I. Production and supplying function of the engineering industries

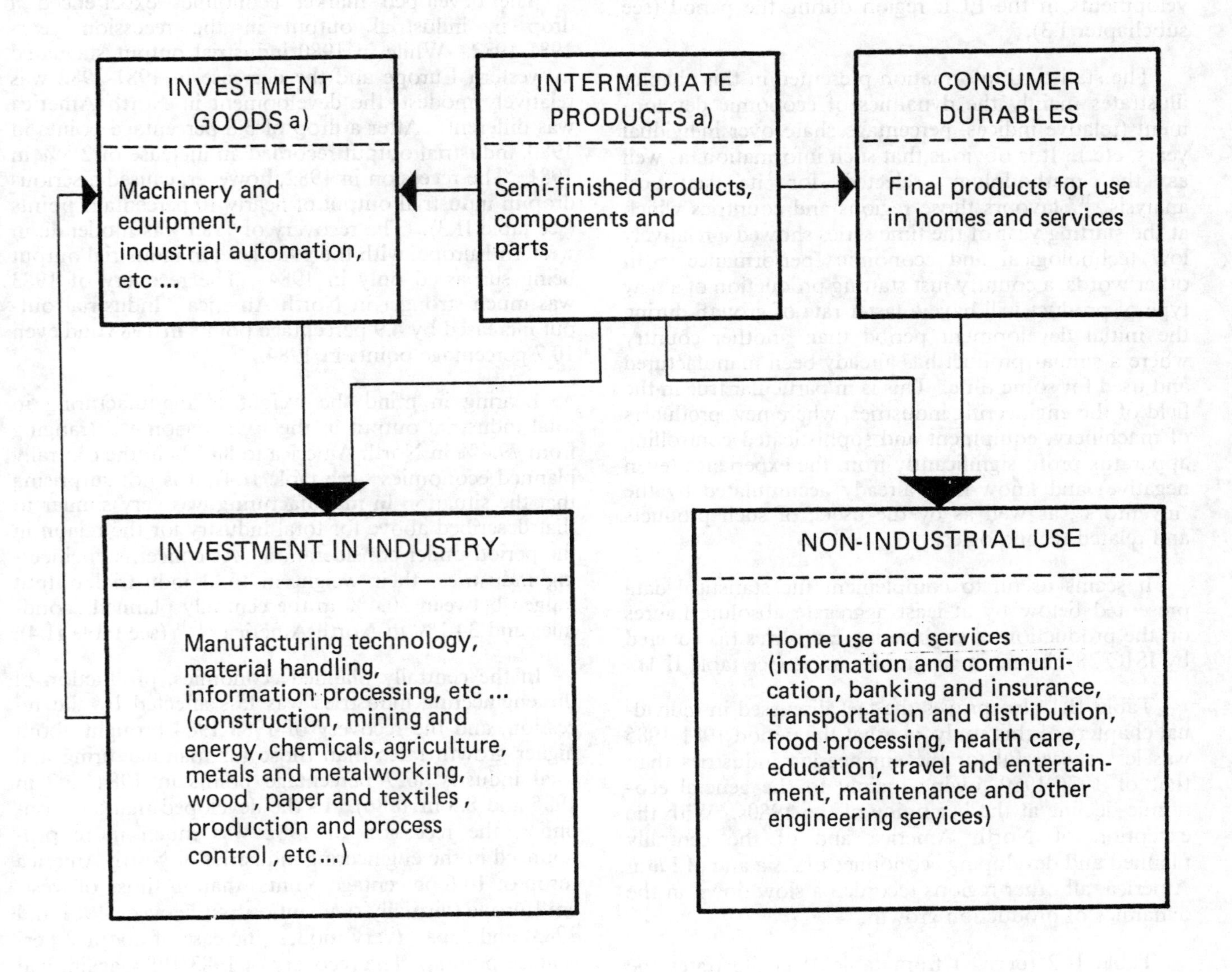

Source: ECE secretariat.

a) Investment goods and intermediate products account in developed countries for some 70–80% of total engineering output (in terms of value).

II.2. THE RELATIVE PERFORMANCE OF THE ENGINEERING INDUSTRIES IN THE ECONOMY AS A WHOLE

In most ECE countries and in Japan, the engineering industries, constituting the core of the economy, are amongst the most sensitive to the general economic climate and, owing to their high share in export performance, to the situation of the world market in particular. [II.8] These considerations are confirmed by the analysis of the development of engineering industries in recent years, bearing in mind the overall economic developments in the ECE region during the period (see subchapter I.3).

The statistical information presented in this chapter illustrates mainly the dynamics of economic development (relative indices, percentage share over individual years, etc.). It is obvious that such information, as well as the methodology selected for its statistical analysis, [II.9] favours those regions and countries which at the starting year of the time series showed a relatively low technological and economic "performance". In other words, a country just starting production of a new type of product will have a faster rate of growth during the initial development period than another country where a similar product has already been manufactured and used for some time. This is in particular true in the field of the engineering industries, where new producers of machinery, equipment and sophisticated controlling apparatus profit significantly from the experience (even negative) and know-how already accumulated by the "inventors", as well as by the users, of such products and related technologies.

It seems useful to complement the statistical data presented below by at least aggregate absolute figures on the production of engineering industries (as covered by ISIC 38), by main geographic areas (see table II.1).

Table II.1 also proves the fact discussed in individual chapters of this study, i.e. that the period 1981-1985 was less successful for the engineering industries than that of 1976-1980, owing mainly to the general economic decline at the beginning of the 1980s. With the exception of North America and of the centrally planned and developing economies of Asia and of Latin America, all other regions recorded a slow-down in the dynamics of production growth.

Table II.2 (derived from table II.1) illustrates the share of individual regions in total world engineering production (1975-1985). It shows that North America's stable position (approx. 20% share) continues, while western Europe's share decreased from 27.2% in 1975 to under the 20% level in 1985. This also affected the position of the whole ECE region: its share in total world engineering production decreased from 80.4% in 1975 to 79.1 in 1980 and 76.7 in 1985. However, the unique place and dominant role of the ECE region in the overall development of world engineering is still clearly evident.

World industrial production started to decrease in 1980, continued its decline in 1981 and reached its lowest level in 1982. The trend was reversed in 1983, with considerable growth recorded in 1984 (6.1 percentage points), in 1985 (3.3 percentage points) and in 1986 (3.7 percentage points). The developments giving rise to this were not the same in the centrally planned economies of eastern Europe and the developed market economies (see table II.3). The centrally planned economies managed to keep positive growth rates over the whole period, although the weak growth in 1981 and 1982 reflects the problems faced also by these countries at times of world-wide economic recession.

The developed market economies experienced a drop in industrial output in the recession years 1981-1982. While in 1980 industrial output stagnated in western Europe and the decrease in 1981-1982 was relatively modest, the development in North America was different. After a drop of 3.8 percentage points in 1980, industrial output recorded an increase of 2.5% in 1981. The recession in 1982, however, caused a serious drop in industrial output of nearly 10 percentage points (see table II.3). The recovery of 1983 was moderate in western Europe, with the 1980 level in industrial output being surpassed only in 1984. The recovery of 1983 was much stronger in North America. Industrial output increased by 4.9 percentage points in 1983 and even 10.7 percentage points in 1984.

Bearing in mind the weight of manufacturing in total industrial output in the ECE region [II.10] (ranging from 79.4% in North America to 88.2% in the centrally planned economies - see table II.4), it is not surprising that the situation in manufacturing was very similar to that described above for total industry for the region in the period under consideration. As concerns engineering industries, their weight in total industrial output ranged between 30.4% in the centrally planned economies and 34.1% in North America [II.11] (see table II.4).

In the centrally planned economies, production of the engineering industries was less affected by the recession, and the recovery of 1983-1984 brought about higher growth rates than those for manufacturing and total industry (7.9 percentage points in 1984, 9.2 in 1985 and 8.8 in 1986). In the developed market economies, the recession of 1982 was much more pronounced in the engineering industries of North America (drop of 10.6 percentage points) than in those of western Europe (virtually stagnant output between 1981 and 1983) and Japan (very modest increase of about 2 percentage points). The recovery of 1983-1984, again, had a much greater and immediate impact on the engineering industries in North America (increase of 5.2 percentage points in 1983 and even 7.0 in 1984) than in western Europe (virtually no change in 1983, a small increase of 2.0 percentage points in 1984 and a real recovery only in 1985, with an increase of 6.0 percentage points). After facing some difficulties in 1982, the Japanese engineering industries recovered in 1983-1984 and - as a heavily export-oriented branch of industry - profited extensively from the recovery in the United States (increases in the index of engineering industries' output of 6 percentage points in 1983, a record 21 percentage points in 1984 and 11 percentage points in 1985.

Tables II.5 and II.6 show the development (1979-1986) of industrial production indices in total industry (ISIC, Rev.2: 2-4) and in engineering industries (ISIC, Rev.2: 38) in individual ECE member countries and in Japan.

In several countries, the growth of total industrial production over the period under consideration exceeded the average annual increase of 4% (Bulgaria, Denmark, German Democratic Republic, Ireland, Luxembourg, Norway, Romania and USSR). In terms of gross output and/or value added, the performance of industry was even better.

The economic recovery of 1984 marked a turning-point in the period under review, leading to substantial growth rates in the output of engineering products, including metal goods, on a world-wide basis, although with regional variations. In some of the smaller countries, particularly Bulgaria, Ireland and Turkey, the accelerated growth began a year earlier. Only very few countries showed negative or lower growth rates in the latter part of the period compared with the 1979-1982 figures.

The above-mentioned reversal was less marked in the engineering industries of the centrally planned economies of eastern Europe, which by and large continued to record high growth rates in their output. But even here, there was a marked pick-up in some of the east European countries in the latter part of the review period, in the case of Czechoslovakia, Hungary and Poland reversing the trend from negative to positive growth rates. Almost consistently, the highest growth rates in the area were posted by Bulgaria, whose engineering output index rose from 94 in 1979 to 160 in 1986. The USSR marked the second highest overall increase in the area, from 94 to 145 in 1986.

Regarding overall industrial growth, it slowed down in 1981-1985 compared with 1976-1980 in all east European countries and the Soviet Union. On average, eastern Europe reached a 2.8%, and the Soviet Union a 3.7%, annual growth rate of gross industrial output (see table II.7).

Industrial developments differed in individual years. Overall industrial growth in 1981-1985 was strongly influenced by the marked slow-down which occurred in almost all countries concerned in 1981-1982. Worsened external conditions, as well as domestic shortages of production inputs, contributed significantly to this. The slow-down of industrial output growth was more pronounced in eastern Europe than in the Soviet Union. Industrial production rose at the relatively low rates of 2.2 and 2.4% for the seven countries combined in 1981 and 1982, respectively. In 1983 and 1984, a marked upswing took place in most of the countries concerned. The most important factors contributing to this acceleration were successes in removing supply constraints and in improving the efficiency of resource use. Though industrial performance in 1985 did not fully meet expectations, output nevertheless grew substantially faster than in the early 1980s.

The acceleration of industrial growth, which had started in 1983, continued into 1986. Gross industrial output of the six east European countries increased by 4.6% in 1986, about half a percentage point more than in 1985. An even stronger acceleration occurred in the Soviet Union, where industrial output rose by almost 5%, or one percentage point more than in 1985. Thus the year 1986 saw the best performance for eight years in eastern Europe and the Soviet Union as a whole. None the less, the variation between countries ranged between 1.9% in Hungary and 7.7% in Romania. Various factors on both the supply and demand sides contributed to these generally favourable developments.

The market economies of the ECE region fared differently. In western Europe, several countries ended 1985 at output levels below those of 1980 (France, Greece, Italy, Spain and Switzerland), although the production of the west European engineering industries as a whole tended to gain strength towards the end of 1984 and in 1985. A more detailed analysis shows that the recovery started already in 1983/84 in some smaller countries (Denmark, Ireland, Luxembourg, Sweden and Turkey) while the output of the engineering industries still stagnated in 1984 in France and Italy and showed only a moderate recovery in the United Kingdom and the Federal Republic of Germany. In the case of the latter, the nearly 10% growth in output in 1985 and 6% in 1986 is proof, however, that the engineering industries of the Federal Republic of Germany revived strongly after the difficulties of the preceding years.

In North America, the strong recovery of manufacturing in general, and the engineering industries in particular, which started in 1983, continued throughout 1984. The growth rates recorded in engineering output in 1984 and 1985 underline the unusual strength of this recovery. This contrasted with negative growth rates in output for the years 1980 and 1983, and only a 2% increase posted in 1981. Owing to the close links between the two economies, the engineering industries in Canada and the United States recorded parallel developments over the period under review.

When comparing the development of industrial production indices in total industry and in engineering industries (see tables II.5 and II.6), it may be noted that in the majority of countries (from 68% to 79% over the period 1981-1986) higher growth rates were recorded in engineering industries. The other countries (e.g. Canada, Greece, Norway, Portugal, Spain, Switzerland and the United Kingdom), for different reasons (priority development of other industrial sectors such as petroleum and chemicals and light industry including food products), recorded a relative slow-down in engineering industries.

In order to assess the relative performance of the engineering industries (by individual countries) in industry as a whole, an attempt was made to compare the development of gross output (GO) and value added (VA) indicators. Tables II.8-II.11 show the annual percentage changes over the period 1979-1986 of GO and VA, both for total industry and for the engineering industries. Tables II.12 and II.13 illustrate the share of GO and VA of the engineering industries in total industry (see also figures II.2 and II.3).

In the the majority of countries listed, the share of engineering industries was relatively stable throughout the period 1979-1986. Among the exceptions were Bulgaria, Hungary and Poland, where the time-series was interrupted in 1981, 1980 and 1982, respectively; the German Democratic Republic, with the share continuously declining between 1980 and 1983; [II.12] and Norway, with a considerable drop in 1983.

In the case of Bulgaria, Hungary and Poland, the development can be attributed to important modifications in prices in the respective years (and as concerns Poland, the general economic problems at the beginning of the 1980s). In the German Democratic Republic, the decrease in the share of engineering industries in total industrial gross output (figures should be compared with caution owing to variations in the price bases for the years under consideration) occurred at times when annual increases in the index of engineering industries' production were higher than those of total industry (pointing to changes in the price structure of industrial branches as the reason for the described trend). In the case of Norway, the engineering industries in 1982, and in particular in 1983, recorded an absolute decline in some important sectors (e.g. shipbuilding) which caused the aggregated index of production of the engineering industries to drop by 8 percentage points in 1983 without again reaching the 1980 level in 1984 or in 1985.

Data for Japan and the United States confirm the strong position of the engineering industries in both countries, placing Japan in the top position of all countries reviewed, with a continuously increasing share of engineering industries in gross output of total manufacturing (over 44% in 1984), followed by the Federal Republic of Germany, the United States and Sweden.

Figures II.2 and II.3 illustrate the share of gross output and value added of engineering industries in total industry (1979-1986). Four "standard" and four "additional" countries, [II.13] all with a long tradition and experience in engineering manufacture, have been selected in order to show that, when measured in terms of value added, the share of the engineering industries in total industry is 4-8% higher than when measured in terms of gross output. This is not surprising, since the engineering industries continue to be a typically labour-intensive sector of industry; many engineering products consist of hundreds of thousands of parts and components, thus requiring not only precise machining, but also costly assembly operations and high professional skills throughout the entire production process, i.e. from design to the final testing.

Table II.1

World production of engineering industries (ISIC, Rev.2: 38) by geographic region, 1975-1985

Region	*Value in billions of constant 1975 US dollars* [a]			*Index*	
	1975	*1980*	*1985* [b]	*1980/1975*	*1985/1980*
Total **world**	1 424	1 898	2 456	133.3	129.4
of which:					
Developed market economies	832	1 056	1 301	126.9	123.2
of which:					
North America	295	379	498	128.5	131.4
Europe	387	439	476	113.4	108.4
Japan and others	150	238	327	158.7	137.4
Centrally planned economies	516	740	982	143.4	132.7
of which:					
Europe	464	684	908	147.4	132.7
Asia	52	56	74	107.7	132.1
Developing countries	76	102	173	134.2	169.6
of which:					
Africa	4	6	6	150.0	100.0
Asia	26	35	69	134.6	197.1
Latin America	46	61	98	132.6	160.7

Source: Sectoral Studies Series, No. 15, vol. II, UNIDO, Vienna 1986. (Based on United Nations Industrial Statistics).

a The figures given in this table are an indicator of quantitative growth. Accurate comparisons between individual regions are not possible owing to the unavailability of realistic exchange rates.

b UNIDO estimate.

Table II.2

Share of geographic regions in world production of engineering industries, 1975-1985

(Percentage)

Region	*1975*	*1980*	*1985* [a]
Total **world**	100.0	100.0	100.0
of which:			
Developed market economies	58.4	55.6	53.0
of which:			
North America	20.7	20.0	20.3
Europe	27.2	23.1	19.4
Japan and others	10.5	12.5	13.3
Centrally planned economies	36.2	39.0	40.0
of which:			
Europe	32.5	36.0	37.0
Asia	3.7	3.0	3.0
Developing countries	5.4	5.4	7.0
of which:			
Africa	0.3	0.3	0.2
Asia	1.9	1.9	2.8
Latin America	3.2	3.2	4.0
ECE region [b]	80.4	79.1	76.7

Source: As for table II.1.

a UNIDO estimate.

b All European countries, United States and Canada.

Table II.3

Index of world industrial production by branches of industry and by selected regions, 1979-1986

(1980 = 100)

Branch of economy	*Year*	*World*	*Centrally planned economies* [a]	*Region: Developed market economies — Total* [b]	*Europe*	*North America*
Industry (ISIC, Rev.2: 2-4)	1979	100.5	96.3	100.7	100.1	103.8
	1980	100.0	100.0	100.0	100.0	100.0
	1981	99.4	101.9	100.4	98.6	102.5
	1982	96.8	104.6	96.5	97.4	93.2
	1983	99.6	109.2	99.2	98.6	98.1
	1984	105.7	114.1	105.9	101.3	108.8
	1985	109.0	119.1	108.9	104.8	110.8
	1986	112.7	124.0	110.3	107.0	111.8
Manufacturing (ISIC, Rev.2: 3)	1979	99.7	96.2	101.4	100.4	105.9
	1980	100.0	100.0	100.0	100.0	100.0
	1981	100.6	102.0	100.0	98.2	102.3
	1982	98.5	104.8	95.7	96.5	92.1
	1983	101.9	109.6	98.8	97.3	99.0
	1984	108.7	114.7	105.8	100.2	110.6
	1985	112.8	120.0	108.9	103.2	113.1
	1986	116.8	125.2	110.8	105.4	115.4
Engineering industries (ISIC, Rev.2: 38)	1979	98.3	94.3	99.7	98.3	106.6
	1980	100.0	100.0	100.0	100.0	100.0
	1981	101.8	104.7	101.3	98.8	102.3
	1982	99.5	109.2	97.3	98.1	91.7
	1983	102.8	115.5	100.0	98.2	96.9
	1984	112.6	123.4	110.2	100.2	113.9
	1985	119.1	132.6	116.0	106.2	117.7
	1986	123.4	141.4	117.8	109.4	118.5

Source: *Monthly Bulletin of Statistics*, vol. XLII, No. 2, Statistical Office, United Nations, New York, February 1988.

a Bulgaria, Czechoslovakia, German Democratic Republic, Hungary, Poland, Romania and USSR.

b North America, Europe (other than eastern Europe and Yugoslavia), Australia, Israel, Japan, New Zealand and South Africa.

Table II.4

Weight of the output of industry, manufacturing and engineering by selected regions, calculated on the basis of 1980 data

Region	*Industry (ISIC, Rev.2:2-4)*	*Manufacturing (ISIC, Rev.2:3)*	*Engineering (ISIC, Rev. 2:38)*
World	100	77.8	29.1
of which:			
Centrally planned economies [a]	100	88.2	30.4
Developed market economies [b]	100	83.2	34.3
of which:			
Europe	100	85.7	34.0
North America	100	79.4	34.1

Source: *Monthly Bulletin of Statistics*, vol. XL, No. 8, Statistical Office, United Nations, New York, August 1986.

For notes, see table II.3.

Table II.5 Index of industrial production in total industry (ISIC, Rev.2:2-4) in ECE member countries and Japan, 1979-1986

(1980=100)

COUNTRY	1979	1980	1981	1982	1983	1984	1985	1986
Austria	97	100	99	98	99	104	109	110
Belgium	101	100	97	98	99	102	104	105
Bulgaria a/	96	100	105	110	115	120	124	129
Cyprus	94	100	104	106	108	113	112	...
Czechoslovakia b/	97	100	102	104	107	111	115	118
Denmark c/	100	100	100	102	106	116	121	126
Finland	92	100	103	104	107	112	116	117
France	100	100	98	97	98	99	98	101
German Dem.Rep. d/	95	100	105	108	112	117	122	127
Germany,Fed.Rep. of	100	100	98	95	96	99	105	107
Greece	99	100	101	102	101	104	107	108
Hungary	102	100	103	105	106	109	110	111
Ireland e/	101	100	105	104	111	125	127	130
Italy	95	100	98	95	92	95	97	99
Luxembourg f/	103	100	94	90	90	114	121	124
Malta	84	100	102	107	110	...	...	...
Netherlands	101	100	98	94	97	102	106	106
Norway	96	100	99	99	108	118	121	126
Poland	101	100	86	85	90	95	99	103
Portugal	95	100	100	105	107	106	118	123
Romania b/	94	100	103	104	109	116	122	131
Spain	99	100	99	98	101	101	103	107
Sweden	100	100	98	97	102	107	109	110
Switzerland	95	100	99	96	95	97	103	108
Turkey	100	100	109	125	122	137	139	...
USSR g/	97	100	103	106	111	115	120	126
United Kingdom	107	100	96	98	102	103	108	110
Yugoslavia	96	100	104	104	105	111	114	119
Canada	102	100	101	91	97	111	116	119
United States	102	100	102	95	101	112	114	115
Japan	96	100	101	101	105	117	122	121

Source : ECE/ITD - Data bank of the Engineering Industries and Technology Section. Government replies to ECE questionnaires for the Annual Reviews of Engineering Industries and Automation.

a/ Including logging and fishing, excluding publishing, gas and water.
b/ Excluding publishing.
c/ Excluding electricity.
d/ Including fishing, excluding publishing.
e/ Excluding electricity and petroleum refineries.
f/ Excluding paper and paper products.
g/ Including fishing, logging, motion picture production, cleaning and dyeing.

Table II.6 Index of industrial production in engineering industries (ISIC, Rev.2:38) in ECE member countries and Japan, 1979-1986

(1980=100)

COUNTRY	1979	1980	1981	1982	1983	1984	1985	1986
Austria	95	100	97	99	101	106	118	...
Belgium	104	100	97	99	101	101	105	107
Bulgaria	94	100	107	116	128	140	154	160
Cyprus	93	100	117	134	124	127	134	124
Czechoslovakia	104	100	104	108	114	121	129	135
Denmark a/	99	100	98	101	104	120	130	134
Finland	88	100	109	116	116	121	131	133
France	100	100	100	100	99	98	98	...
German Dem.Rep.	93	100	108	113	119	125	134	142
Germany,Fed.Rep. of b/	97	100	100	99	99	102	112	118
Greece	92	100	102	93	87	89	85	88
Hungary	106	100	106	110	112	114	118	123
Ireland	97	100	114	115	132	166	171	...
Italy	90	100	99	94	92	93	96	100
Luxembourg	99	100	94	94	101	139	150	157
Malta	64	100	97	110	119	...	...	...
Netherlands c/	97	100	100	100	99	105	109	110
Norway	98	100	100	101	94	99	104	107
Poland	100	100	88	87	93	100	107	115
Portugal	85	100	104	112	100	85	103	103
Romania	91	100	102	106	111	121	131	141
Spain	97	100	97	93	96	93	96	105
Sweden	100	100	101	100	105	112	120	120
Switzerland d/	91	100	97	92	88	88	94	...
Turkey e/	...	...	100	106	129	155	171	189
USSR	94	100	106	111	118	126	135	145
United Kingdom	107	100	91	93	95	99	105	104
Yugoslavia	99	100	104	108	112	117	122	131
Canada	106	100	102	91	92	104	112	112
United States	103	100	102	93	100	116	120	120
Japan	89	100	106	108	114	135	146	147

Source : ECE/ITD - Data bank of the Engineering Industries and Technology Section.
Government replies to ECE questionnaires for the Annual Reviews of Engineering Industries and Automation.

a/ Excluding shipbuilding and repairing.
b/ Excluding manufacture of aircraft.
c/ Including foundries.
d/ Excluding precision instruments.
e/ 1981=100.

Table II.7

Growth performance of industry and engineering industries in eastern Europe, 1976-1986

(Average annual growth rate of gross output in constant prices)

Country/Region	1976-1980		1981-1985		1986	
	Industry	*Engineering industries*	*Industry*	*Engineering industries*	*Industry*	*Engineering industries*
Bulgaria	6.0	9.1	4.3	8.9	4.0	7.3
Czechoslovakia	4.7	6.7	2.7	4.9	3.2	4.9
German Democratic Republic [a]	5.0	7.0	4.1	6.2	3.8	5.5
Hungary	3.4	3.1	2.0	3.8	1.9	2.8
Poland	4.7	6.9	0.4	1.3	4.7	7.3
Romania [a]	9.5	12.6	4.0	5.5	7.7	7.7
Eastern Europe	5.6	...	2.8	...	4.6	...
USSR	4.5	8.2	3.7	6.2	4.9	7.4
Eastern Europe and the USSR	4.8	...	3.4	...	4.8	...

Source: ECE secretariat common data base, derived from national statistics.

a Marketable production (Romania since 1981).

Table II.8 Gross output in total industry (ISIC, Rev.2:2-4) in ECE member countries and Japan, 1979-1986

(Percentage change) a/

COUNTRY	1980/1979	1981/1980	1982/1981	1983/1982	1984/1983	1985/1984	1986/1985	1986/1979
Austria	12.2	9.0	4.8	3.9	7.9	...	...	...
Belgium	...	...	...	...	...	...	...	...
Bulgaria	4.2	33.6	4.6	4.3	4.2	3.2	2.9	7.7
Cyprus	26.2	21.5	8.9	8.2	11.3	5.3	-1.6	11.1
Czechoslovakia b/	4.1	4.8	7.2	2.8	9.8	4.8	2.9	5.2
Denmark c/	13.9	10.5	10.7	10.1	12.9	8.3	...	11.1 d/
Finland	22.2	12.8	6.6	8.3	9.9	6.6	-3.9	8.7
France	15.8	11.1	11.8	8.5	9.6	5.9	...	10.4 d/
German Dem.Rep.	4.7	...	...	4.2	7.6	12.5	...	...
Germany, Fed.Rep. of	7.9	5.1	2.4	2.4	6.6	7.3	...	5.3 d/
Greece	...	27.6	...	...	...	...	...	...
Hungary	12.9	8.5	6.4	5.6	7.2	6.5	3.8	7.2
Ireland	13.5	17.0	12.1	17.0	...	...	...	...
Italy	23.6	16.9	10.3	18.6	20.0	...	...	...
Luxembourg	4.1	3.2	16.6	5.9	...	...	...	...
Malta	14.8	3.6	1.9	-0.8	...	...	...	...
Netherlands	10.7	12.7	3.1	2.3	8.7	4.7	...	7.0 d/
Norway	23.2	14.0	7.3	8.6	14.8	13.4	...	13.4 d/
Poland	-0.2	-11.4	...	6.2	4.9	4.1	...	...
Portugal	35.7	24.3	23.0	29.9	28.0	...	...	...
Romania	...	...	...	...	...	...	...	...
Spain	22.8	12.8	11.0	16.6	10.3	...	...	...
Sweden	13.0	6.2	10.6	16.2	13.8	8.1	2.6	10.0
Switzerland	...	...	...	...	...	...	...	...
Turkey c/	118.5	67.7	43.8	27.8	55.9	...	...	...
USSR e/	3.5	3.1	13.4	7.0	3.6	3.2	...	5.6 d/
United Kingdom c/	1.9	0.5	6.2	7.4	10.4	8.8	5.3	5.7
Yugoslavia	40.0	46.3	30.1	47.2	68.0	84.5	...	51.6 d/
Canada c/	9.0	13.4	-3.7	8.7	15.1	...	...	...
United States c/	7.3	8.9	-2.8	4.8	9.7	1.1	...	4.7 d/
Japan c/	17.4	3.6	2.2	1.1	7.4	5.1	...	6.0 d/

Source : ECE/ITD - Data bank of the Engineering Industries and Technology Section.
Year to year percentage changes and annual average increase 1986/1979 are calculated from Government replies to ECE questionnaires for the Annual Reviews of Engineering Industries and Automation.

a/ Based on values expressed in national currency at current prices.
b/ State national industry only.
c/ Manufacturing industries only (ISIC, Rev.2 : 3).
d/ 1985/1979.
e/ Including logging, motion picture production, cleaning and dyeing, excluding gas.

Table II.9 Gross output in engineering industries (ISIC, Rev.2:38) in ECE member countries and Japan, 1979-1986

(Percentage change) a/

COUNTRY	1980/1979	1981/1980	1982/1981	1983/1982	1984/1983	1985/1984	1986/1985	1986/1979
Austria	11.8	7.2	9.5	3.2	3.5	...	...	...
Belgium	4.5	-0.1	17.4	7.1	3.7	5.2	...	6.2 b/
Bulgaria	6.1	-1.0	8.5	9.9	9.1	10.6	3.7	6.6
Cyprus	17.8	25.9	18.6	1.5	9.5	21.4	5.3	14.0
Czechoslovakia c/	5.2	5.0	6.8	4.3	3.6	3.9	5.4	4.9
Denmark	13.4	7.8	15.9	11.4	11.5	15.8	...	12.6 b/
Finland	23.5	16.2	14.5	4.3	7.9	13.3	3.2	11.6
France	15.0	10.5	13.5	7.8	7.2	5.7	...	9.9 b/
German Dem.Rep.	7.6	...	...	3.5	1.8	10.4	...	...
Germany,Fed.Rep. of	6.1	4.5	4.2	2.7	5.8	...	...	...
Greece	...	29.1	...	...	...	...	...	...
Hungary	-5.5	8.2	8.0	6.0	7.6	11.3	9.6	6.3
Ireland	22.7	29.2	19.6	27.9	...	...	...	...
Italy	25.6	11.1	10.5	23.0	14.8	...	...	...
Luxembourg	19.5	8.0	12.2	13.7	...	...	...	...
Malta	16.0	7.5	5.9	5.2	...	...	...	...
Netherlands	6.7	4.9	4.0	0.3	12.8	8.9	2.0	5.6
Norway	19.7	13.2	6.9	-5.0	11.0	28.9	...	12.0 b/
Poland	0.3	-12.0	...	6.9	7.1	7.1	...	...
Portugal	32.2	25.5	26.8	18.4	16.7	...	...	...
Romania	...	...	...	...	...	...	...	...
Spain	17.2	6.5	12.0	15.2	7.6	...	...	...
Sweden	8.7	8.2	10.7	17.7	14.1	11.3	7.6	11.1
Switzerland	...	...	...	...	...	...	...	...
Turkey	84.5	54.5	43.2	40.4	62.5	...	...	...
USSR	5.5	5.0	3.7	6.1	5.9	5.8	...	5.3 b/
United Kingdom	7.1	-0.2	7.4	6.1	11.4	11.1	4.4	6.7
Yugoslavia	32.8	44.3	33.7	40.2	63.7	98.2	...	50.6 b/
Canada	3.3	11.2	-4.9	9.9	25.6	...	...	...
United States	3.3	9.6	-1.5	7.0	15.3	4.0	...	6.1 b/
Japan	15.5	11.5	3.2	4.8	13.3	8.7	...	9.4 b/

Source : ECE/ITD - Data bank of the Engineering Industries and Technology Section. Year to year percentage changes and annual average increase 1986/1979 are calculated from Government replies to ECE questionnaires for the Annual Reviews of Engineering Industries and Automation.

a/ Based on values expressed in national currency at current prices.
b/ 1985/1979.
c/ State national industry only.

Table II.10 Value added in total industry (ISIC, Rev.2:2-4) in ECE member countries and Japan, 1979-1986

(Percentage change) a/

COUNTRY	1980/1979	1981/1980	1982/1981	1983/1982	1984/1983	1985/1984	1986/1985	1986/1979
Austria	6.8	7.0	5.0	4.8	6.5	...	...	...
Belgium	2.8	-1.2	9.9	9.1	7.8	4.6	...	5.4 b/
Bulgaria	...	...	...	...	...	...	...	...
Cyprus	19.6	16.4	11.6	8.7	15.4	8.0	...	13.2 b/
Czechoslovakia c/	4.9	...	2.6	5.9	4.1	4.6	3.6	0.0
Denmark d/	11.0	7.3	12.5	11.4	11.6	9.3	...	10.5 b/
Finland	15.6	12.4	9.2	10.1	12.0	5.8	0.7	9.3
France	11.3	9.6	14.5	11.4	9.5	8.5	...	10.8 b/
German Dem.Rep.	...	...	...	...	...	...	...	...
Germany,Fed.Rep. of	5.5	2.4	2.8	2.6	5.4	6.3	...	4.2 b/
Greece	...	24.2	...	...	...	...	...	...
Hungary	-9.2	9.9	7.2	3.6	8.2	8.2	3.0	4.2
Ireland	17.2	21.9	15.1	17.8	...	...	...	...
Italy	22.5	14.8	8.1	-0.7	18.0	...	...	...
Luxembourg	2.6	4.1	17.6	3.0	...	...	...	...
Malta	15.5	3.4	4.7	-2.0	...	...	...	...
Netherlands d/	4.1	4.3	0.0	10.8	11.7	4.3	...	5.8 b/
Norway	29.5	13.2	8.7	13.7	16.8	7.7	...	14.7 b/
Poland	4.3	-12.5	...	21.9	20.5	17.5	...	...
Portugal	33.6	15.2	15.8	26.9	33.1	...	...	...
Romania	...	...	...	...	...	...	...	...
Spain	16.0	10.1	11.5	15.7	10.0	...	...	...
Sweden	11.6	5.2	7.9	16.0	14.5	9.8	8.4	10.4
Switzerland	7.8	7.2	...	...	...	...	...	...
Turkey d/	111.1	76.7	39.2	18.3	39.9	...	...	...
USSR	...	...	...	...	...	...	...	...
United Kingdom d/	6.1	2.8	5.3	8.2	10.4	8.8	5.3	6.7
Yugoslavia e/	38.4	44.8	32.5	41.2	67.0	88.3	...	50.9 b/
Canada d/	8.7	12.2	-6.5	11.4	15.1	...	...	...
United States d/	3.5	8.2	-1.6	7.0	11.5	1.6	...	5.0 b/
Japan d/	10.0	3.2	4.3	-3.9	8.6	14.6	...	6.0 b/

Source : ECE/ITD - Data bank of the Engineering Industries and Technology Section.
Year to year percentage changes and annual average increase 1986/1979 are calculated from Government replies to ECE questionnaires for the Annual Reviews of Engineering Industries and Automation.

a/ Based on values expressed in national currency at current prices.
b/ 1985/1979.
c/ State national industry only.
d/ Manufacturing industries only (ISIC, Rev.2 : 3).
e/ Social product including turnover taxes.

Table II.11 Value added in engineering industries (ISIC, Rev.2:38) in ECE member countries and Japan, 1979-1986

(Percentage change) a/

COUNTRY	1980/1979	1981/1980	1982/1981	1983/1982	1984/1983	1985/1984	1986/1985	1986/1979
Austria	7.9	9.3	5.9	3.4	4.9	...	...	...
Belgium	4.3	-5.9	12.1	7.5	1.9	5.1	...	4.0 b/
Bulgaria	...	...	...	...	...	...	...	...
Cyprus	23.4	24.7	15.6	1.7	12.7	12.0	...	14.7 b/
Czechoslovakia c/	5.1	...	2.9	7.9	0.7	3.9	-2.4	0.1
Denmark	13.2	5.7	15.3	11.9	10.8	12.5	...	11.5 b/
Finland	18.0	17.7	19.7	4.9	9.4	12.8	1.0	11.7
France	12.5	13.2	13.9	7.9	8.0	6.6	...	10.3 b/
German Dem.Rep.	...	...	...	...	...	...	...	...
Germany,Fed.Rep. of	10.1	4.4	4.5	1.9	4.6	10.0	...	5.9 b/
Greece	...	28.8	...	...	...	...	...	...
Hungary	-23.4	13.2	12.2	6.7	9.0	14.8	10.4	5.3
Ireland	27.2	24.1	20.9	29.3	...	...	...	...
Italy	22.7	18.1	11.0	-4.6	15.8	...	...	...
Luxembourg	31.4	12.5	6.6	0.9	...	...	...	...
Malta	0.1	14.3	16.9	-0.9	...	...	...	...
Netherlands d/	7.6	2.0	5.4	3.3	3.4	4.0	...	4.3 b/
Norway	8.1	16.0	6.9	-3.7	7.4	13.6	...	7.9 b/
Poland	1.9	1.0	...	15.5	23.1	22.5	...	...
Portugal	34.3	17.5	17.6	20.9	14.2	...	...	...
Romania	...	...	...	...	...	...	...	...
Spain	15.1	4.8	11.4	10.2	2.0	...	...	...
Sweden	8.0	9.3	4.3	16.8	14.7	12.6	8.8	10.6
Switzerland	...	...	...	...	...	...	...	...
Turkey	75.2	54.8	45.6	28.0	54.6	...	...	...
USSR	...	...	...	...	...	...	...	...
United Kingdom	6.5	1.4	7.1	5.5	10.4	11.7	4.4	6.6
Yugoslavia e/	33.7	47.7	38.0	38.9	59.5	103.8	...	52.0 b/
Canada	6.3	14.0	-6.6	8.1	21.4	...	...	...
United States	4.4	9.6	-0.7	5.2	16.0	1.3	...	5.8 b/
Japan	11.1	11.6	5.6	-2.7	11.6	19.0	...	9.1 b/

Source : ECE/ITD - Data bank of the Engineering Industries and Technology Section.
Year to year percentage changes and annual average increase 1986/1979 are calculated from Government replies to ECE questionnaires for the Annual Reviews of Engineering Industries and Automation.

a/ Based on values expressed in national currency at current prices.
b/ 1985/1979.
c/ State national industry only.
d/ In 1979, excluding precision instruments.
e/ Social product including turnover taxes.

Table II.12 Share of gross output of engineering industries (ISIC, Rev.2:38) in total industry (ISIC, Rev.2:2-4) in ECE member countries and Japan, 1979-1986

(Percentage) a/

COUNTRY	1979	1980	1981	1982	1983	1984	1985	1986
Austria	26.1	26.0	25.5	26.7	26.5	25.4	...	...
Belgium	...	...	...	...	...	...	...	...
Bulgaria	27.1	27.6	20.4	21.2	22.3	23.4	25.1	25.2
Cyprus	9.6	8.9	9.2	10.1	9.4	9.3	10.7	11.5
Czechoslovakia b/	30.0	30.3	30.4	30.3	30.7	29.0	28.8	29.5
Denmark c/	26.1	26.0	25.4	26.6	26.9	26.6	28.4	...
Finland	17.4	17.6	18.1	19.5	18.8	18.4	19.6	21.0
France	29.8	29.6	29.4	29.9	29.7	29.0	29.0	...
German Dem.Rep.	35.0	35.9	32.4	29.7	29.5	27.9	27.4	...
Germany,Fed.Rep. of	35.5	34.9	34.7	35.3	35.4	35.2	...	...
Greece	...	13.3	13.5	...	...	...	...	...
Hungary	27.2	22.8	22.7	23.1	23.2	23.3	24.3	25.7
Ireland	17.4	18.8	20.8	22.2	24.3	...	...	...
Italy	29.2	29.7	28.2	28.2	29.3	28.0	...	...
Luxembourg	9.4	10.8	11.3	10.9	11.7	...	...	...
Malta	19.5	19.8	20.5	21.3	22.6	...	...	...
Netherlands	21.1	20.4	19.0	19.1	18.8	19.5	20.3	...
Norway	20.6	20.0	19.9	19.8	17.3	16.7	19.0	...
Poland	31.1	31.3	31.0	23.1	23.2	23.7	24.4	25.8
Portugal	16.4	16.0	16.1	16.6	15.2	13.8	...	...
Romania	...	...	...	...	...	...	...	...
Spain	23.8	22.7	21.4	21.6	21.4	20.8	...	...
Sweden	34.0	32.7	33.3	33.4	33.8	33.9	34.9	36.6
Switzerland	...	...	...	...	...	...	...	...
Turkey c/	19.2	16.2	15.0	14.9	16.4	17.0	...	...
USSR d/	27.3	27.9	28.4	26.0	25.7	26.3	27.0	...
United Kingdom c/	33.1	34.8	34.5	34.9	34.5	34.8	35.6	35.2
Yugoslavia	25.9	24.5	24.2	24.9	23.7	23.1	24.8	...
Canada c/	32.5	30.8	30.2	29.9	30.2	33.0	...	...
United States c/	36.8	35.4	35.6	36.1	36.9	38.8	39.9	...
Japan c/	37.8	37.2	40.0	40.4	41.9	44.1	45.6	...

Source : ECE/ITD - Data bank of the Engineering Industries and Technology Section.
Government replies to ECE questionnaires for the Annual Reviews of Engineering Industries and Automation.

a/ Based on values expressed in national currency at current prices (see also ref. II.12).
b/ State national industry only.
c/ Figures represent share in manufacturing industries only (ISIC, Rev.2 : 3).
d/ Total industry includes logging, motion picture production, cleaning and dyeing, but excludes gas.

Table II.13 Share of value added of engineering industries (ISIC. Rev.2:38) in total industry (ISIC, Rev.2:2-4) in ECE member countries and Japan, 1979-1986

(Percentage) a/

COUNTRY	1979	1980	1981	1982	1983	1984	1985	1986
Austria	27.5	27.7	28.3	28.6	28.2	27.8	...	...
Belgium	26.2	26.5	25.3	25.8	25.4	24.0	24.2	...
Bulgaria	...	...	...	...	...	...	...	...
Cyprus	9.8	10.1	10.8	11.2	10.5	10.2	10.6	...
Czechoslovakia b/	34.5	34.6	37.6	37.7	38.4	37.2	36.9	34.8
Denmark c/	33.4	34.0	33.5	34.3	34.5	34.3	35.3	...
Finland	23.5	24.0	25.2	27.6	26.3	25.7	27.4	27.5
France	35.0	35.4	36.6	36.4	35.2	34.7	34.1	...
German Dem.Rep.	...	...	...	...	...	...	...	...
Germany,Fed.Rep. of	39.1	40.8	41.6	42.3	42.0	41.7	43.1	...
Greece	...	17.6	18.2	...	...	...	...	...
Hungary	32.7	27.6	28.4	29.7	30.6	30.8	32.7	35.0
Ireland	20.6	22.4	22.7	23.9	26.2	...	...	...
Italy	33.1	33.1	34.1	35.0	33.6	33.0	...	...
Luxembourg	9.7	12.5	13.5	12.2	12.0	...	...	...
Malta	22.2	19.3	21.3	23.8	24.0	...	...	...
Netherlands cd/	35.5	36.7	35.9	37.8	35.3	32.7	32.6	...
Norway	20.6	17.2	17.6	17.3	14.7	13.5	14.2	...
Poland	35.1	34.3	39.7	31.2	29.6	30.2	31.5	31.8
Portugal	21.0	21.1	21.5	21.8	20.8	17.8	...	...
Romania	...	...	...	...	...	...	...	...
Spain	28.0	27.8	26.5	26.5	25.2	23.4	...	...
Sweden	38.6	37.3	38.8	37.5	37.8	37.8	38.8	38.9
Switzerland	...	...	...	...	...	...	...	...
Turkey c/	21.3	17.7	15.5	16.2	17.5	19.3	...	...
USSR	...	...	...	...	...	...	...	...
United Kingdom c/	40.4	40.5	40.0	40.7	39.7	39.7	40.7	40.4
Yugoslavia e/	28.6	27.6	28.1	29.3	28.8	27.5	29.8	...
Canada c/	32.1	31.4	31.9	31.9	31.0	32.6	...	...
United States c/	42.9	43.3	43.8	44.2	43.5	45.2	45.1	...
Japan c/	39.6	40.0	43.3	43.8	44.3	45.6	47.3	...

Source : ECE/ITD - Data bank of the Engineering Industries and Technology Section.
Government replies to ECE questionnaires for the Annual Reviews of Engineering Industries and Automation.

a/ Based on values expressed in national currency at current prices.
b/ State national industry only.
c/ Figures represent share in manufacturing industries only (ISIC, Rev.2 : 3).
d/ In 1979, engineering industries exclude precision instruments.
e/ Figures are based on social product including turnover taxes.

Figure II. 2

Share of gross output (GO) and value added (VA) of engineering industries in total industry (manufacturing) of four (standard) countries, 1979-1986

(Percentage)

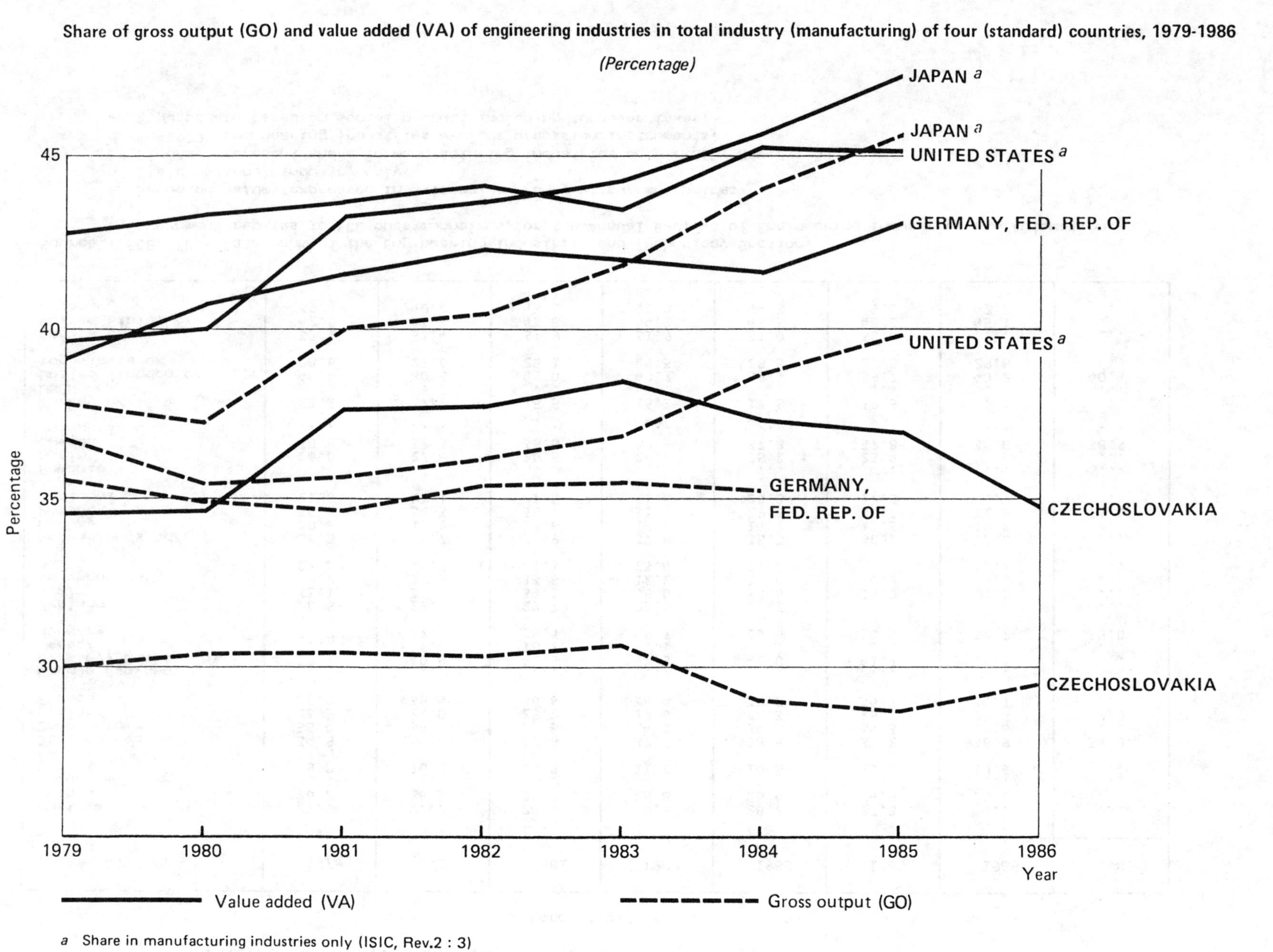

[a] Share in manufacturing industries only (ISIC, Rev.2 : 3)

Figure II. 3

Share of gross output (GO) and value added (VA) of engineering industries in total industry (manufacturing) of four (additional) countries, 1979-1986

(Percentage)

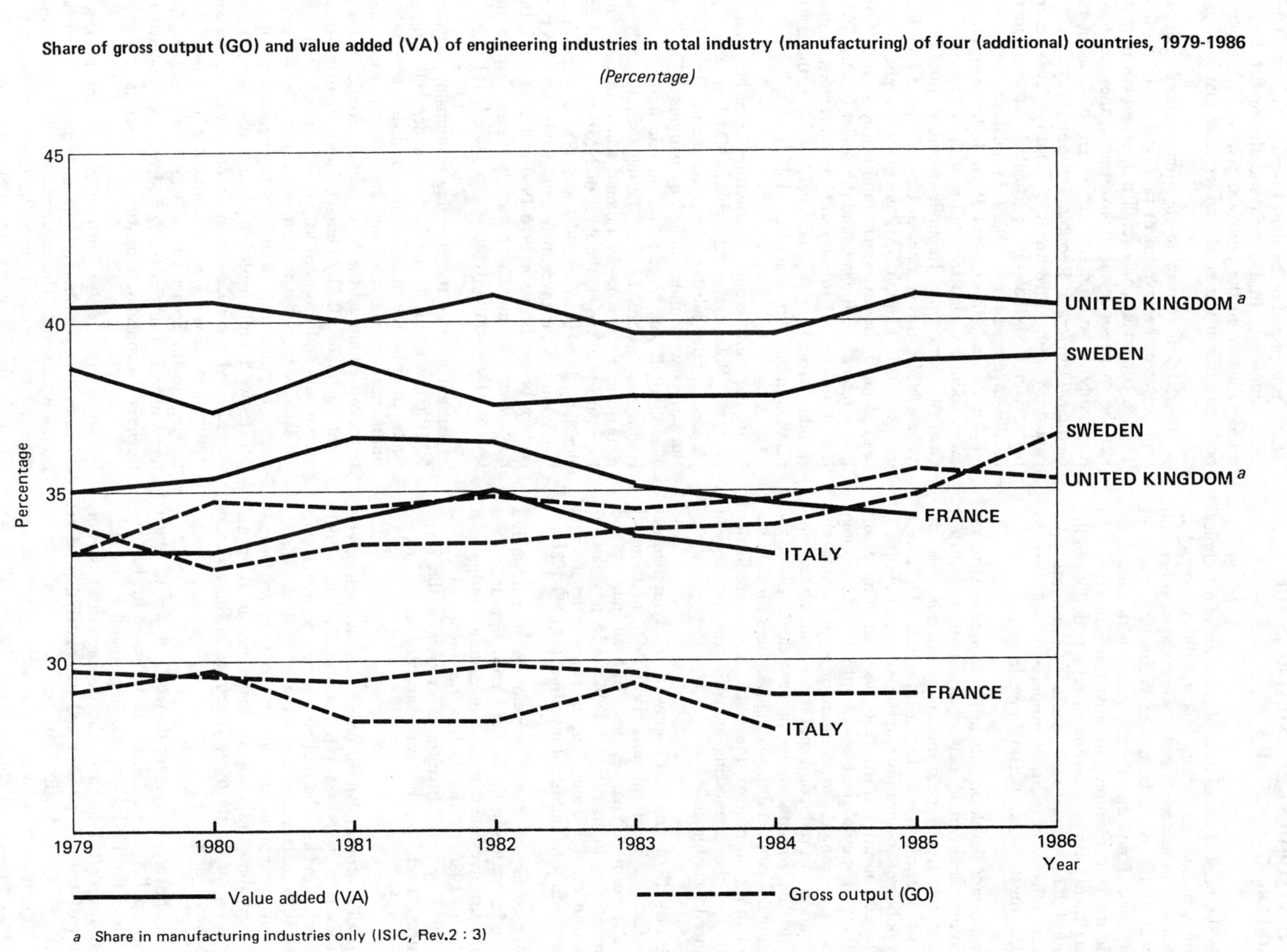

a Share in manufacturing industries only (ISIC, Rev.2 : 3)

II.3. GROSS CAPITAL FORMATION AND RESEARCH AND DEVELOPMENT EXPENDITURE IN THE ENGINEERING INDUSTRIES

In the present conditions of scientific and technical advance, almost all factors linked with economies in total labour inputs are linked in some way or other with renewal of the basic factors of production. Any of them, even if not directly connected with capital expenditure (for example, improved organization and management), become obsolete over a period when the technical basis remains the same. Continuously increasing effectiveness in all production factors is dependent on the flow of capital investment.

Important developments in the economic structure of the main industrially developed countries, and their development trends are being largely determined by the scale of investment and investment trends. The investment policy of engineering companies in the industrially developed countries in the 1980s was one of increasing the flow of capital investment into high-technology sectors, increasing the expenditure on equipment for the automation of production processes, reducing expenditure on the commissioning of new capacities in the traditional sectors and increasing resources for the integrated modernization of the whole of the production facilities of the engineering industries on the basis of advanced technologies.

Tables II.14 and II.15 show the development of gross fixed capital formation (1979-1986) in terms of annual percentage changes, both in total industry and in engineering industries. [II.14] Table II.16 provides information on the share of investments in engineering industries in total industrial investments (1979-1986).

A recovery in east European investment, and a slight acceleration in the Soviet Union, appear to have pushed up investment growth in eastern Europe as a whole in 1986 to the highest rate achieved in the 1981-1986 period. Investment growth in the seven countries taken together was estimated at 6.5% (see table II.17).

In all centrally planned economies, average investment levels in 1981-1985 were lower than in 1976-1980. In Czechoslovakia, the decline was only marginal, but it was 10% in Hungary and some 37-38% in Poland. A barely perceptible increase in the German Democratic Republic and a slightly bigger one in Romania contrasts with the relatively large rises in Bulgaria and the Soviet Union.

The above development shows the degree to which a major policy reorientation, incorporated in the five-year plans for 1981-1985, was implemented. In fact, plans for 1976-1980 had already prescribed slower growth of both investment and net material product (NMP). [II.15] Investment was expected to absorb a roughly similar share of total resources. Only in the German Democratic Republic was a planned fall in the ratio of investment to NMP (the investment ratio) actually achieved. In contrast, the 1981-1985 five-year plans of all countries foresaw substantial declines in the investment ratio - by 1.5 percentage points. Moreover, the declines which actually occurred were similar to what was expected or greater (4-6 percentage points) in most countries. In Bulgaria and the Soviet Union, some reduction in the ratio was achieved - though in both countries it was small and less than planned.

There is, of course, no short-term causal relationship between the investment ratio and NMP growth. Though constraints on overall output growth may affect the share (or level) of investment, they need not necessarily do so, and fast output growth - as in the German Democratic Republic - is not always accompanied by rising investment. Country development patterns in 1981-1985, and particularly the movement of the investment ratio itself in relation to output growth, illustrate this in 1981-1985. In this respect, the seven countries concerned fall into three groups. The German Democratic Republic recorded one of the highest NMP growth rates of east European countries in 1981-1985. Though NMP growth was slightly lower than planned, the investment ratio fell by slightly more than foreseen. In Bulgaria and the Soviet Union, NMP growth rates were also buoyant. However, investment spending rose at approximately double the rates planned in both countries (see table II.18).

In the third group of countries (Czechoslovakia, Hungary, Poland and Romania), NMP growth was planned at rates substantially lower than in 1976-1980. But targets were underfulfilled by margins of 25-40%. The shortfall was biggest in Romania (although that country also recorded one of the highest rates of NMP increase in the area). In Poland, which in fact published no plan target, the level of NMP produced declined by 15% between 1980 and 1985. While in all four countries the investment ratio fell by up to 6 percentage points between 1976-1980 and 1981-1985, the downward movement was scarcely bigger than in the German Democratic Republic (see table II.18).

Examination of the changing relationship between NMP growth and the main domestic demand components confirm the movements in investment ratios summarized above, and indicate that, in most east European countries, the burden of economic readjustment sparked off by unfavourable terms-of-trade movements and the need to pay off accumulated debt fell largely upon the investment sector. In Bulgaria, where imports and balance-of-payments adjustment policies were implemented at an early stage, investment was affected much less than in other east European countries in 1981-1985. Investment growth was also buoyant in the Soviet Union, which made substantial terms-of-trade gains in the period reviewed and where foreign debt considerations were less significant.

These patterns also illustrate the different incidence on individual countries of the changes in the world economic climate. The experience of the three groups suggests that, in the case of the German Democratic Republic, these changes - combined with much more vigorous action to improve the efficiency with which investment and fixed assets were used - affected primarily plans for investment rather than output growth in 1981-1985. In the Soviet Union and Bulgaria, they did not greatly affect either. In the other four countries, output and investment plans were both modified.

There can be no doubt that changes in trade policies considerably reduced the scope for investment activity in all countries of eastern Europe. As noted above, all the six countries concerned have attempted to expand exports and cut imports as a direct response to balance-of-payments constraints. These policies had a significant impact, in particular, upon the investment-goods sector of trade: its share in total exports rose and its share in total imports fell in all east European countries. Investment goods thus proved to be one of the most easily exportable classes of east European exports, and the import category most readily dispensed with in the short term.

The extent to which east European investment was squeezed by lower availabilities of investment goods can be measured by changes in the balance of trade in investment goods after 1975. The net east European deficit on investment-goods trade relative to 1975 declined in current dollar terms by about $6 billion up to 1980, and by $15 billion up to 1983. These figures were equivalent to 25 and 60% of actual east European investment-goods imports from all suppliers in the two years considered. The Soviet Union's deficit, on the contrary, rose after 1975 by over $5 billion up to 1980, and by nearly $14 billion up to 1983 - equivalent to 68% of its actual level of investment-goods imports in the latter year.

Besides quantitative rises in the supply of engineering goods (see subchapter II.2), the intensification process also entails qualitative improvements. The new machinery and equipment produced is expected to provide new functions, to be more productive and also, in the light of input shortages, to be more economical in the use of labour, fuel and other resources. The direct contribution of the engineering industries to this aspect of the intensification policy cannot be assessed directly.

However, an important determinant of the engineering industries capacity to fulfil these requirements is the speed at which new technology has been provided for the modernization of engineering itself. It is perhaps significant that rises in the engineering industries' share of industrial investment since 1975 have occurred only in Bulgaria, the German Democratic Republic and Romania. Sustained rises, as shown by changes in five-year averages between 1976-1980 and 1981-1985, were only achieved in Bulgaria.

With one exception, the stagnating or decreasing share of engineering in total industrial investment has closely paralleled the deceleration in economic growth in 1981-1985 compared with 1976-1980. The exception is the Soviet Union, where NMP growth objectives were satisfactory, but the engineering industries' share in total industrial investment stagnated. This reflected the big increase in the share of the fuel industry in total investments since 1975.

The relationship between the absolute levels of employment, fixed assets and output in the engineering branches of the centrally planned economies has been studied in the past by ECE. [II.17] The research carried out concluded that, while output per unit of fixed assets in engineering is higher than the average of all branches, both capital intensity (fixed assets per worker) and labour productivity (output per worker) are lower. These indicators are not easy to interpret, though their apparent uniformity throughout the region is worthy of note (see also subchapter II.4).

Engineering compares favourably with the industrial average in other respects also. Capital productivity has declined less than in industry as a whole. One reason for this is that the age structure of fixed assets in the engineering industry - as judged by the share of fixed assets under five years old in the branch total - is considerably more favourable than the average for industry in all countries but Hungary. The retirement of obsolete fixed assets in engineering seems to be generally higher than in industry as a whole (except in the Soviet Union).

In western market economies, gross fixed capital formation slowed down considerably after 1973. The investment ratio, which had reached 23% in 1972-1973, passed to an average of 21.5% in 1980-1981 and was just under 20% in 1984, in spite of a rapid recovery in investment in this year (of about 8%) which continued at a slower pace in 1985.

Manufacturing investment has in general followed a permanent gradual deceleration since 1973 but investment in machinery and equipment continued to increase during the late 1970s and, after an earlier and stronger downturn in the United States in 1980-1982 and in western Europe in 1981-1983, regained dynamism in 1984 and 1985.

The overall conditions for business investment in western Europe have been relatively favourable over the past years. Increasing demand, higher rates of capacity utilization, together with improved profitability and rising business confidence, have led to a significant rise in expenditure on machinery and equipment. Evidence from business surveys also suggests that investment in 1985 was increasingly motivated by the desire to augment production capacities.

There was a marked acceleration in 1985 in expenditure on machinery and equipment in those countries where the increase had been relatively small in 1984 (Austria, Belgium, Finland, the Federal Republic of Germany, the Netherlands, Sweden and Switzerland), while in other countries the high rates of 1984 were more or less maintained. There was a sharp drop in investment growth in Norway after a vigorous expansion in 1984: this largely reflects the volatility of investment in the oil sector. The slow-down in the United Kingdom had been expected and mainly reflects the bringing forward of investments into 1984 and 1985 in anticipation of the phasing out of tax allowances on investments (see table II.19).

It appears that in nearly all western market economies most of the surge in investment expenditure was in the manufacturing sector. Investment behaviour in the non-manufacturing sectors varied much more between countries. It should also be noted here that the underlying strength of investment in industry may be difficult to assess in so far as an increasing part of the capital equipment used is leased from other companies, usually financial companies in the service sector. However, in the national accounts statistics, capital assets are allocated to the various sectors on the basis of ownership rather than use.

Investment in structures was in striking contrast to that in machinery and equipment. The general tendency was for a further decline or only modest growth. This reflects the continuing weakness of public investment, the sluggishness in industrial building and, especially, the deep-rooted structural problems in the housing sector in many countries. The latter are partly due to the saturation of demand in some countries, the negative impact on the demand for housing of low growth rates of real disposable incomes, and the high real interest rates of recent years.

In western Europe, the net effect of these generally opposing changes in machinery and equipment, on the one hand, and structures, on the other, was an increase in total investment in 1985 which was somewhat lower than in 1984.

In the United States, there was a sharp acceleration in investment growth in 1984 and 1985, after two years of decline (1982-1983). The weaker growth of investment in 1986 was not unexpected. It reflects the normal cyclical downturn accompanying the weakening of final demand after the vigorous expansion in 1983/84.

Data available on the share of engineering in total manufacturing investment in market economies are rather inconclusive and obviously affected by the cyclical behaviour of individual countries. Data for several countries show, however, that the growth in investment in the engineering industries during recent years has been strongly concentrated in several relatively new research-intensive branches, such as office and computing machinery, and in general in branches producing electrical equipment. An extreme case of this tendency is represented by the United States, where equipment spending within manufacturing investment has been concentrated to the extent of 93% (in the period 1979-1984) in office equipment and automobiles, leading to a higher rate of investment expenditure in the branches producing these goods. Investment in these branches in most market economies during the period 1980-1983 either decreased much less than for the manufacturing average or continued to expand (Japan, France, Belgium and United Kingdom) - several small countries have also redirected successfully an increasing share of their investment towards these areas, where demand is growing favourably.

The capital investment in the engineering industries is a small proportion of the GNP of a country: in 1983 it was 1.2% in the United States, 1.4% in Japan, 1.5% in the Federal Republic of Germany, 1.1% in France, 0.7% in the United Kingdom and 2.2% in the USSR. [II.18] For all that, the share of the engineering industries in the total investments of industry was between 22 and 49%, depending on the country. A steady increase in the proportion of new high-technology sectors in the total volume of capital investment in the engineering industries has become a decisive feature in the development of investments during the 1980s in the engineering industries of the leading market economies. The average annual growth rates of capital investment in the period 1980-1984 in electrical engineering, which incorporates the output of modern industrial electronics and up-to-date telecommunication equipment, were more than twice the rates for the engineering industries as a whole, reaching 24.5% in Japan (as against 12.1% for the engineering industries as a whole), 11.1% in the United States (4.05%), 8.6% in France (4.7%), 6.7% in the United Kingdom (2.7%) and 5.7% in the Federal Republic of Germany (1.9%).

The difference between the two indicators was not so marked in countries with a centrally planned economy. In Poland, for example, the similar indicators over the same period were 18.0% (18.2%), and in Bulgaria 4.7% (6.8%). Only in Hungary was there a more appreciable disparity, the figures being 3.6% and 4.5%. [II.19]

The increase in the volume of capital investment in the high-technology industries of the United States is confirmed by the increase in investment as a proportion of total costings on output: for example, this indicator was 14.7% in the semiconductor industry in 1984 and 15.2% in 1980, as against 8.6% in 1975 (table II.20).

The bulk of capital investment in the engineering sectors in market economies is devoted to the modernization of production capacities. The resources devoted to modernization are a continuously increasing proportion of total capital investment. Thus, the figure reached was between 75 and 98% in the engineering sectors in the United States by 1985 (table II.21). The only exception was electrical engineering, where the level was 65%, precisely because the bulk of the high-technology production capacity is concentrated in this sector and is expanding. Forecasts for 1985-1987 predicted an increase of 14.5%.

The pattern of reproduction of capital investment is also changing appreciably in centrally planned economies. In the years 1980-1985 alone, expenditure on the technical re-equipping and reconstruction of existing enterprises in the USSR, as a proportion of total capital investment in industry, increased from 31.6% to 37.5%. Over the period 1981-1985, modernization extended to 729,000 units of productive equipment, including 159,000 (22%) in enterprises of the engineering and metalworking industries. Total expenditure on the introduction of new techniques and technologies over the period 1981-1985 was 29% up on the period 1976-1980. As a result, the proportion of mechanized and automated operations in Soviet industry reached 51%, as against 45.7% in 1975 (data on the structure of the work force). [II.20]

The major part of the capital investment of engineering companies in market economies goes to the purchase of machinery and equipment. They allocated 79-89% of total investment to those ends in 1984. The coresponding figure in centrally planned economies is slightly lower, 52-62%. [II.21]

The devoting of such a high proportion of expenditure to the acquisition of new equipment by the engineering industries is connected with the extremely extensive technical re-equipping of production in the traditional sectors - the automotive industry, electrical engineering and non-electrical engineering. There is a very great need for the renewal and updating of production equipment. The current United States Administration has put forward a programme for the re-industrialization of the United States, a particular feature of which is an increase in the technical level of industry to a point at which the active and large-scale introduction of electronic technology and automated systems would largely erase the boundary between the new and the traditional industries.

Expenditure on the acquisition of automation equipment forms a constantly increasing proportion of the total capital investment of the engineering industries. The development in the United States in 1970-1983 is shown in table II.22.

Computerized and automated technologies in general are now gradually invading all stages of the production process from design of the product to its dispatch from the enterprise. The scale of the use of automation equipment in the modern engineering industries is being even further expanded by the possibilities available to firms of the most varied sizes, right down to very small firms, to acquire automated equipment through leasing, with preferential credit terms from government bodies, a favourable and flexible depreciation policy and reliable after-sales servicing (table II.23).

In the United Kingdom, more than 60% of the computers installed in engineering enterprises in 1984 were implemented in factories with fewer than 500 employees. Global expenditure on computers alone in the engineering industries was some 500 million pounds sterling in 1984-1985, and in some enterprises investment in computers was greater than expenditure on metalworking machinery. [II.22] The expanded introduction of automation equipment has modified the structure of capital investment. According to a study conducted in British companies employing flexible manufacturing systems (FMS), only 17% of total capital investment went on FMS, while 30% was devoted to equipment for the automation of transportation and storage operations, 30% to computer-aided design (CAD), 10% to flow control of goods and materials, 10% to the assembly and testing system, and 4% to the automated stock control system. [II.23]

The nature and scale of investment activity in the engineering industries reflects general structural changes in the industrial output of the country as a whole. Official policy may considerably affect fixed capital formation. In market economies, taxation measures and credit-finance measures serve, as a rule, to create a healthy investment climate in a given industrial sector, a group of sectors or individual geographic regions.

Great significance is also attached to new ways and means of encouraging engineering enterprises to increase capital investment in new equipment and technology in centrally planned economies. In the USSR, in particular, wider use is being made of profit-and-loss accounting, enterprises are being given more independence, and conditions are being created for the effective introduction of technological achievements. The integrated programme of CMEA member States provides for the establishment of direct relations between companies, industrial enterprises and the scientific and research organizations of the countries on the basis of specific proposals embodied in bilateral and multilateral agreements and contracts.

The countries concerned will establish joint scientific, technical and production groupings, international engineering and technological centres for the development and promotion of new processes, technologies and materials, centres for study, training and improvement of qualifications, and other joint undertakings for the solution in common of problems of importance.

The measures agreed for implementation of the programme will be carried out by the use of national resources, credits from the International Investment Bank and the International Bank for Economic Co-operation, and common funds to be established by the countries concerned to finance the most important individual measures.

There is a great deal of very wide-ranging R and D in the industrialized market economies. More than 70% of R and D expenditure in United States industry in 1983 was in the engineering industries, which accounted for some 57% of total R and D expenditure in the country. The figures for Japan were 59 and 37%, for the Federal Republic of Germany 67 and 47%, and for the United Kingdom 64 and 42%.

The laboratories and scientific departments of engineering companies largely devote themselves to applied research and development. They provide a constant stream of innovation to production; only 1.5 to 3.0% of the total R and D expenditure of engineering companies goes to basic research, which is, as a rule, carried out at their request in institutes of higher education, government laboratories and other non-commercial organizations on a contractual basis.

The United States is a leader among the countries with a market economy as regards the scale of R and D in the engineering industries. According to 1981 data, R and D expenditure in the United States aerospace industry was 4.4 times the total expenditure of Japan, the Federal Republic of Germany, the United Kingdom, France and Italy in this sphere; the figure for the office equipment and computer industry was 4.3 times; that for precision instruments 3.3 times. [II.24]

Expenditure on scientific development in the engineering industries increased by 60% between 1975 and 1984 in the United States (in current prices), by 160% in Japan, 130% in the Federal Republic of Germany, and 220% in the United Kingdom (table II.24). The countries of western Europe and Japan, in which scientific research facilities developed slightly later than in the United States, are constantly striving to expand the volume of R and D. Japan thus succeeded in overtaking the United States in the mean annual rate of increase in R and D expenditure over the period 1979-1983, when its average was 9.7% as against 4.6% for the United States; the figure for France was 5.5%. [II.25] In consequence, the country indicators for R and D expenditure as a proportion of GNP may be seen to be evening out by 1984-1985 (table II.25).

The number of scientists and engineers employed on R and D per 10,000 of the working population is an indicator that is evening out rapidly. The figure for Japan had already reached 96% of the United States level by 1984 and that for the Federal Republic of Germany 73% (table II.26).

The efforts of Japan to increase the numbers of scientific research staff in electrical engineering are particularly noteworthy. By 1984, the laboratories and research divisions of this sector were employing more than 130,000 people, as against 116,000 in the United States (table II.27).

The electrical engineering and transport equipment sectors employ 65 to 77% of the research work-force, and these workers carry out 70 to 80% of all R and D in the engineering industries of the leading market-economy countries (tables II.24 and II.27).

Considerably increased appropriations for R and D have been allocated during the past decade to the sectors producing machinery and equipment, apparatus and instruments, whose products embody the most recent advances in science and technology and may themselves be embodied in new technologies throughout the engineering industries. These high-technology industries include, first and foremost, those that manufacture office machines and computers, which are included under non-electrical engineering in the United States, those that manufacture communication equipment, which are included in electrical engineering, the aerospace industry, which is included in transport equipment, and precision engineering. These sectors account for up to 70% of total R and D expenditure in the engineering industries in, for example, the United States and the United Kingdom.

R and D expenditure is increasing most remarkably in precision engineering. During the period 1975-1984 it increased 4.5 times in Japan, 2.3 times in the United States, and 1.7 times in the United Kingdom and the Federal Republic of Germany. This testifies to an appreciable growth in demand for new precision instruments, measuring instruments and control equipment, the need for which is increasingly determined by the extent of automation in modern engineering industries. In the United Kingdom, for example, the average annual rate of increase of R and D expenditure in precision engineering was around 23% in the period 1981-1983, as against 4.8% for the engineering industries as a whole. The figures for Japan for the years 1980-1984 were respectively 20 and 10%.

The increased R and D expenditure is connected with the development of modern products that are complicated both to design and to manufacture. R and D expenditure as a proportion of production costs is far higher in the high-technology industries than the average for the engineering industries as a whole. For example, the figure was 17.2% in the aerospace industry of the United States in 1985. There is also a significant proportion of R and D expenditure in the sales of firms producing communication equipment. Thus, the figure was 21.4% for the British Plessey Company, 8.8% for the German Federal Republic company Siemens, 8.6% for the Swedish company Ericsson, and 8.5% for the Japanese company Fujitsu. [II.26] The similar statistic for the engineering industries as a whole was 6.4% for the United States, 4.1% for the United Kingdom, 3.8% for the Federal Republic of Germany and only 3.0% for Japan (table II.28).

Applied research and development is proceeding along three main lines in the engineering industries of the countries considered: the development of new types of products, the development of new technologies and the improvement of products; the order of priority varies in relation to the sector and over time. The bulk of the resources allocated by engineering companies for R and D in the United States in 1975 was mainly for the improvement of traditional production, but by 1985 the development of new forms of production had taken first place (table II.29).

The reason for such an appreciable change in the targeting structure of R and D expenditure in the United States engineering industries during the years 1975-1985 is a considerable expansion in the range of products made in small and medium batches, and also considerably accelerated innovation. Similar trends of change in the structure of R and D expenditure are to be noted in the engineering industries of almost all the industrialized countries.

Expansion of the scale of R and D in the development of new types of production is speeding change in the range of products manufactured, a particular feature of industries whose output is in small and medium batches. According to previous forecasts, as much as 42% of United States output in 1987 may be of products not manufactured in the preceding five years (as against 25% in 1980) in all branches of non-electrical engineering, 27% (21%) in precision engineering, 28% (17%) in the aerospace industry, 25% (18%) in the automotive industry, and 18% (16%) in electrical engineering. [II.27]

The development of the modern sphere of research and development in the engineering industries is closely linked to government policy on science and technology, which is based on national priorities and includes a targeting system of measures concerning, in particular, the financing of R and D by government establishments and organizations, the indirect stimulation of R and D through preferential credit, tax and depreciation policies, use of the contract system for major projects and programmes, and other measures.

Although there are national differences of scale and type in the implementation of these measures, they all have the same aim, namely to speed up the rate of development and commercial application of programme results. In the United States, for example, some 40% of R and D expenditure in the engineering industries in 1984 was financed by government bodies. Some 80% of the public resources were allocated to research on communication equipment, electronics and aerospace technology, i.e. the high-technology sectors of production (table II.30).

In Japan a number of major projects are being carried out under the direction and with the financial support of the Office for Science and Technology, the Ministry of Foreign Trade and Industry, the Ministry of Education and other State bodies. These projects are concerned with the development and introduction of the most recent technologies to be used in building up future output. Both in the United States and in Japan, vast importance is attached to R and D in the sphere of equipment for integrated automation, flexible technologies, and new-generation computer technology (supercomputers), in which many engineering companies are participating.

The level of public expenditure in the overall R and D expenditure of the engineering industries varies greatly in the countries of western Europe. In the Federal Republic of Germany, for example, the Government finances only 11% of R and D in the engineering industries, but far more in individual sectors - 15% for the electrical industry and 58% for the aerospace industry. [II.28]. Public funding of R and D in the engineering industries of the Federal Republic of Germany is channelled mainly through Fraunhofer Gesellschaft, which manages 25 research institutes. In the United Kingdom, the functions of directing and planning government-funded R and D in the engineering industries devolve on ministries and their departments, and on various advisory committees. Direct measures are widely used by government bodies to stimulate R and D. A number of highly important systems are currently used to subsidize R and D and to promote its results. Most of these systems were introduced back in the 1970s; in particular, a system that subsidizes the development of new products and technologies in almost all spheres of industrial production has been in operation since 1977.

R and D in the field of microelectronics and informatics is subsidized in a more selective way. Thus, the British Government is subsidizing 57% of expenditure under the £350 million Alvey programme that relates to developments in informatics, fifth-generation computers and artificial intelligence. [II.29]

The practice of providing direct government subsidies to industry is of a narrowly specialized nature in France, where it tends to be applied to developments in advanced technologies in which there is a risk factor. The bulk of the public expenditure on R and D in the engineering industries is applied to the aerospace industry (62%) and electronic engineering (23%). [II.30] The forms of public subsidizing for industrial R and D are constantly being made more flexible. Thus, the direct awards for assistance to new inventions, which have been operating successfully since the late 1970s, are being expanded. Such awards now cover all stages of the process from the formulation of the idea to the completion of the finished industrial prototype. Engineering R and D has been actively stimulated since 1962 by the introduction of tax concessions for companies that engage in research and development.

Thanks to the intensified activity of engineering companies in R and D, the west European countries were able to hold their own in the world market for high-technology products by the middle of the 1980s. Their total share of the world export market was practically 42% in aerospace technology, 29% in precision engineering, 22% in communication equipment and 19.3% in computer technology (table II.31).

Scientific and technical co-operation at the most varied levels is now becoming increasingly important in bringing new engineering techniques and technological processes on to the market more rapidly. Increasing international co-operation has been noted in the 1980s in the development of equipment for integrated automation: robotics, NC-machine tools, computer technology, software, flexible production systems, modern means of communication, etc. Much valuable experience in the organization of joint research centres and pilot plants, in the training and exchange of skilled workers, the exchange of information, and trade in patents and licences had already been accumulated by the middle of the 1980s. All of this provided engineering companies with the conditions for their participation in a number of international programmes, especially those carried out within the European region. They include the ESPRIT, FAST, RACE and BRITE programmes, and the recently adopted EUREKA programme. [II.31]

Table II.14 Gross fixed capital formation in total industry (ISIC, Rev.2:2-4) in ECE member countries and Japan, 1979-1986

(Percentage change) a/

COUNTRY	1980/1979	1981/1980	1982/1981	1983/1982	1984/1983	1985/1984	1986/1985	1986/1979
Austria	12.5	13.0	-2.4	-7.4	13.3	...	...	...
Belgium	18.1	-5.9	17.6	3.1	9.6	8.7	...	8.2 b/
Bulgaria c/	49.6	-17.6	27.9	-12.4	-17.8	19.6	...	5.2 b/
Cyprus	17.6	16.7	6.5	-24.4	14.7	10.1	-3.9	4.3
Czechoslovakia d/	3.5	-0.7	1.1	-0.5	1.6	6.1	-7.4	0.4
Denmark e/	15.2	-5.0	3.1	8.1	35.9	31.1	...	13.8 b/
Finland	37.6	3.7	18.8	-6.1	7.7	14.3	3.9	10.7
France	20.9	5.1	8.4	4.7	11.6	12.2	...	10.4 b/
German Dem.Rep.	4.2	3.2	-1.5	4.0	-7.6	2.4	...	0.7 b/
Germany,Fed.Rep. of	-0.3	-0.2	1.7	3.2	3.9	11.3	...	3.2 b/
Greece	...	35.3	...	...	...	...	...	...
Hungary	-12.7	-6.1	5.0	4.0	2.2	4.4	-1.6	-0.9
Ireland	5.7	11.7	19.6	-20.4	...	...	...	...
Italy	29.0	19.5	9.7	14.5	19.1	...	...	...
Luxembourg	9.7	8.1	-13.2	1.7	-4.2	12.3	...	2.0 b/
Malta	-13.0	12.2	9.6	4.9	...	...	...	...
Netherlands	17.4	-8.0	-4.8	1.9	27.6	21.6	9.5	8.6
Norway	11.1	63.5	-22.4	10.3	46.6	-13.6	...	12.0 b/
Poland	-14.0	-26.4	...	6.0	13.8	10.0	...	...
Portugal e/	19.4	32.9	39.3	2.5	-17.4	...	...	...
Romania	2.6	...	-3.7	10.0	11.1	-4.0	...	...
Spain e/	1.4	28.7	-15.3	14.1	20.8	...	...	...
Sweden	33.2	6.3	3.9	14.5	17.6	16.6	-3.3	12.2
Switzerland	...	...	...	...	...	...	...	...
Turkey e/	-30.8	107.1	43.2	36.3	79.0	...	...	...
USSR	16.6	3.8	2.7	6.5	3.7	4.4	...	6.2 b/
United Kingdom	8.0	-0.8	7.3	-1.0	10.0	3.4	3.0	4.2
Yugoslavia	27.4	28.6	19.8	23.2	44.5	59.3	...	33.1 b/
Canada	25.2	24.0	1.0	-13.1	-3.4	...	...	...
United States	15.0	13.9	-1.6	-5.3	15.2	7.4	-7.2	5.0
Japan e/	21.4	13.9	5.7	-2.1	12.8	11.8	...	10.4 b/

Source : ECE/ITD - Data bank of the Engineering Industries and Technology Section. Year to year percentage changes and annual average increase 1986/1979 are calculated from Government replies to ECE questionnaires for the Annual Reviews of Engineering Industries and Automation.

a/ Based on values expressed in national currency at current prices.
b/ 1985/1979.
c/ Excluding land.
d/ State national industry only.
e/ Manufacturing industries only (ISIC, Rev.2 : 3).

Table II.15 Gross fixed capital formation in engineering industries (ISIC, Rev.2:38) in ECE member countries and Japan, 1979-1986

(Percentage change) a/

COUNTRY	1980/1979	1981/1980	1982/1981	1983/1982	1984/1983	1985/1984	1986/1985	1986/1979
Austria	25.2	57.4	-19.1	-26.1	0.1	...	...	...
Belgium	40.9	-0.7	17.3	-7.8	24.6	12.8	...	13.4 b/
Bulgaria c/	139.8	-42.7	31.4	63.0	-13.8	18.7	...	20.2 b/
Cyprus	-22.2	36.9	-22.4	-16.0	56.0	64.1	-10.9	6.8
Czechoslovakia d/	6.7	2.9	-19.7	5.2	1.4	18.1	-4.2	0.9
Denmark	25.5	-6.7	13.8	7.2	29.6	46.3	...	18.1 b/
Finland	55.3	7.1	24.9	2.9	8.9	19.0	-10.6	13.8
France	22.2	-1.0	6.1	3.1	24.3	19.8	...	12.0 b/
German Dem.Rep.	9.4	15.9	-8.4	0.7	-1.5	0.8	...	2.5 b/
Germany,Fed.Rep. of	17.2	-0.3	5.7	3.5	-1.0	24.2	...	7.8 b/
Greece	...	20.9	...	...	...	...	...	...
Hungary	-15.5	-12.7	5.3	2.1	-3.5	-14.5	3.1	-5.5
Ireland	38.3	10.5	24.5	-24.0	...	...	...	...
Italy	19.2	13.4	22.1	21.0	11.3	...	...	...
Luxembourg	22.2	107.6	-9.6	-49.1	36.4	-4.9	...	7.1 b/
Malta	-69.2	87.9	59.4	85.0	...	...	...	...
Netherlands	19.0	-11.3	0.7	-5.2	38.1	34.5	12.3	11.2
Norway	-0.3	16.0	0.3	-17.5	23.9	34.4	...	8.1 b/
Poland	-20.1	-20.8	...	16.9	12.0	3.1	...	...
Portugal	38.8	40.9	52.6	-21.3	4.4	...	...	...
Romania	12.2	...	-21.8	8.8	10.4	-7.0	...	...
Spain	-8.3	12.7	-4.3	21.8	63.1	...	...	...
Sweden	28.9	11.7	-4.5	13.4	36.8	21.1	15.2	16.9
Switzerland	...	...	...	...	...	...	...	...
Turkey	-24.4	135.1	42.6	47.1	85.0	...	...	...
USSR	16.7	4.3	0.7	5.9	5.0	4.4	...	6.0 b/
United Kingdom	-3.2	-11.2	-2.6	6.8	24.1	5.7	6.3	3.2
Yugoslavia	29.3	20.7	22.4	23.9	41.7	81.6	...	35.1 b/
Canada	66.4	1.1	-29.0	3.8	9.3	...	...	...
United States	16.9	6.1	-8.1	-5.4	27.1	9.8	-5.7	5.1
Japan	43.1	23.6	8.4	-5.0	23.9	18.8	...	17.9 b/

Source : ECE/ITD - Data bank of the Engineering Industries and Technology Section.
Year to year percentage changes and annual average increase 1986/1979 are calculated from Government replies to ECE questionnaires for the Annual Reviews of Engineering Industries and Automation.

a/ Based on values expressed in national currency at current prices.
b/ 1985/1979.
c/ Excluding land.
d/ State national industry only.

Table II.16 Share of gross fixed capital formation of engineering industries (ISIC, Rev.2:38) in total industry (ISIC, Rev.2:2-4) in ECE member countries and Japan, 1979-1986

(Percentage) a/

COUNTRY	1979	1980	1981	1982	1983	1984	1985	1986
Austria	18.2	20.2	28.2	23.4	18.6	16.5	...	...
Belgium	16.5	19.7	20.8	20.8	18.6	21.1	21.9	...
Bulgaria b/	16.3	26.1	18.1	18.6	34.7	36.4	36.1	...
Cyprus	11.5	7.6	8.9	6.5	7.2	9.8	14.6	13.6
Czechoslovakia c/	17.2	17.8	18.4	14.6	15.4	15.4	17.2	17.7
Denmark d/	25.4	27.7	27.2	30.0	29.8	28.4	31.7	...
Finland	12.7	14.3	14.8	15.6	17.0	17.2	17.9	15.4
France	26.2	26.5	25.0	24.4	24.1	26.8	28.6	...
German Dem.Rep.	23.7	24.8	27.9	25.9	25.1	26.8	26.3	...
Germany,Fed.Rep. of	27.7	32.6	32.5	33.8	33.9	32.3	36.1	...
Greece	...	15.1	13.5	...	...	...	...	...
Hungary	17.9	17.3	16.1	16.2	15.9	15.0	12.3	12.9
Ireland	12.9	16.9	16.7	17.3	16.6	...	...	...
Italy	24.3	22.5	21.3	23.8	25.1	23.5	...	...
Luxembourg	6.0	6.7	12.9	13.5	6.7	9.6	8.1	...
Malta	40.4	14.3	23.9	34.8	61.4	...	...	...
Netherlands	16.6	16.9	16.3	17.2	16.0	17.3	19.1	19.6
Norway	8.3	7.5	5.3	6.9	5.1	4.3	6.8	...
Poland	24.8	23.1	24.9	21.1	23.2	22.9	21.4	22.3
Portugal d/	14.1	16.4	17.4	19.1	14.6	18.5	...	...
Romania	26.4	28.9	25.9	21.0	20.8	20.7	20.1	...
Spain	27.8 d/	25.1 d/	22.0 d/	24.9 d/	11.9	16.9	...	...
Sweden	22.4	21.7	22.8	21.0	20.8	24.2	25.1	29.9
Switzerland	...	...	...	...	...	...	...	...
Turkey d/	15.4	16.8	19.1	19.0	20.5	21.2	...	...
USSR	24.6	24.6	24.7	24.2	24.1	24.4	24.4	...
United Kingdom	21.6	19.3	17.3	15.7	17.0	19.1	19.6	20.2
Yugoslavia	16.4	16.7	15.6	16.0	16.1	15.7	18.0	...
Canada	7.2	9.6	7.8	5.5	6.6	7.4	...	...
United States	25.1	25.5	23.7	22.2	22.1	24.4	25.0	25.4
Japan d/	34.9	41.1	44.6	45.8	44.4	48.7	51.8	...

Source : ECE/ITD - Data bank of the Engineering Industries and Technology Section.
Government replies to ECE questionnaires for the Annual Reviews of Engineering Industries and Automation.

a/ Based on values expressed in national currency at current prices.
b/ Gross fixed capital formation excluding land.
c/ State national industry only.
d/ Figures represent share in manufacturing industries only (ISIC, Rev.2 : 3).

Table II.17

Development of gross fixed capital formation in industry in eastern Europe, 1976-1986

(Annual percentage changes)

Country-region	*1976-1980*	*1981-1985*	*1986*
Bulgaria	6.9	4.5	2.0
Czechoslovakia	5.8	-0.4	2.8
German Democratic Republic	5.0	0.1	5.0
Hungary	5.5	-2.0	2.5
Poland	6.0	-7.5	3.0
Romania	10.9	1.3	1.2
Total **eastern Europe**	6.9	-1.3	2.6
USSR	5.0	3.3	8.0
Total **eastern Europe and the Soviet Union**	5.6	1.9	6.5

Source: National statistics, plan documents and plan fulfilment reports. II.16

Note: Data for Bulgaria and the German Democratic Republic are at 1980 prices, for Czechoslovakia at 1977 prices, for Hungary and Romania at 1981 prices, for Poland at 1982 prices, and for the Soviet Union at (assumed) 1976 prices. These price bases are adhered to in table II.18 except where otherwise stated.

Table II.18

Investment ratios of the engineering industries in eastern Europe, 1976-1986

(Gross fixed capital formation as a percentage of NMP produced)

Country	*1976-1980*	*1981-1985*	*1983*	*1984*	*1985*	*1986*
Bulgaria	35.4	34.9	34.7	33.2	35.5	34.3
Czechoslovakia	33.7	29.9	30.3	28.1	28.7	28.6
German Democratic Republic	30.6	24.9	25.2	22.7	22.4	22.6
Hungary	37.2	31.0	31.1	29.2	28.8	29.3
Poland	30.7	24.6	23.8	25.1	25.8	25.3
Romania	41.3	36.4	36.4	35.9	34.4	32.4
USSR	30.3	29.8	30.1	29.8	29.6	30.7

Source: Data bank of the ECE General Economic Analysis Division.

Table II.19

Changes in the components of gross fixed capital formation in selected ECE member countries, 1984-1986

(Percentage change over previous year)

	Machinery and equipment			*Structures, buildings*		
Country	*1984*	*1985*	*1986*	*1984*	*1985*	*1986*
France	2.0	5.1	7.0	-4.0	-0.6	1.0
Germany, Federal Republic of	-0.5	9.4	4.6	1.6	-6.2	1.9
Italy	14.1	9.9	4.5	-0.5	-1.7	0.7
United Kingdom	10.1	7.8	1.0	8.3	-3.1	3.0
Austria	4.3	9.4	5.0	0.5	2.6	3.6
Belgium	-1.8	3.6	9.5	3.8	-0.4	2.1
Denmark	14.1	16.5	14.0	6.8	13.1	6.0
Finland	1.8	8.9	5.5	-3.9	-0.5	-1.3
Ireland	8.1	5.5	-1.3	-13.5	-7.5	-5.5
Netherlands	8.3	14.0	10.0	2.8	-3.0	4.0
Norway	-0.4	-41.0	12.5	12.7	-10.7	27.0
Sweden	8.5	14.7	0.7	3.0	0.7	0.4
Switzerland	4.3	10.4	12.0	4.1	2.9	3.0
United States	20.1	10.1	3.6	12.8	5.6	-0.2
Canada	3.1	4.4	...	-0.2	6.9	...

Source: National statistics. II.16

Table II.20

Share of capital investment in costings on the output of selected industries in the United States, 1975-1984

(Percentage)

Industry	*1975*	*1980*	*1984*
Electronic computing equipment (SIC 3573)	3.4	6.4	5.7
Radio and television equipment (SIC 3662)	2.1	4.1	5.6
Telephone and telegraphic apparatus (SIC 3661)	3.0	4.2	5.5
Electronic components (SIC 367)	5.3	9.5	7.8 [a]
Semiconductors and related devices (SIC 3674)	8.6	15.2	14.7
Measuring and controlling instruments (SIC 3822, 3823, 3824, 3829)	2.1	3.5	3.3
X-ray and electromedical apparatus (SIC 3693)	3.4	6.6	3.8

Source: *U.S. Industrial Outlook, 1987, 1988.* Department of Commerce, Washington, D.C.

a 1983.

Table II.21

Changes in the reproductive pattern of capital investment in the engineering industries of the United States, 1970-1985

(Percentage)

Branch	*1970*		*1980*		*1985*	
	I	*II*	*I*	*II*	*I*	*II*
Electrical engineering	65	35	60	40	35	65
Non-electrical engineering	43	57	47	53	13	87
Automotive industry	27	73	13	87	2	98
Aerospace industry	47	53	29	71	22	78
Precision instruments	69	31	43	57	25	75
Metal products	63	37	47	53	14	86

Source: *Annual McGraw-Hill Spring Survey of Business Plans for New Plants and Equipment*, Issues 24/1971, 34/1981 and 38/1985. McGraw-Hill Publications Co., New York.

I-expansion; II-modernization (I + II = 100 %).

Table II.22

Changes in expenditure on automation means as a percentage of total capital investment in the engineering industries of the United States, 1970-1983

Branch	*1970*	*1980*	*1983* [a]
Non-electrical engineering	28	37	41
Electrical engineering	37	27	53
Automotive industry	14	16	6
Aerospace industry	9	32	...
Precision instruments	15	17	56
Metal products	10	13 [b]	...

Source: As for table II.21.

a Data from the McGraw-Hill survey corrected in the light of data in the ECE secretariat common data base.
b 1981.

Table II.23

Proportion of enterprises in the engineering industries of the United States employing automated technology [a]

(Percentage)

Equipment	*Agricultural engineering*	*Constructional engineering*	*Machine tools*	*Electrical engineering*	*Aerospace industry*
Equipment with programmable control	20	43	58	36	68
Equipment with CNC control	23	37	58	44	70
Data processing computers	63	69	61	62	82
CAD systems	10	21	19	28	31
Computers for process control	34	49	46	40	55
Progressive equipment for transport operations	4	6	5	7	18

Source: New technology in the American Machinery Industry. A study prepared for the use of the Joint Economic Committee, Congress of the United States, Washington, D.C., 2 March 1984.

a Taken from an investigation of 628 enterprises in five different sectors of the engineering industries.

Table II.24

Changes in R and D expenditure [a] and its breakdown by sector in the industry of the United States, the Federal Republic of Germany, the United Kingdom and Japan, 1975-1984

	United States		United Kingdom		Fed.Rep. of Germany		Japan	
	Million dollars	%	Million pounds	%	Million mark	%	Billion yen	%
1	2	3	4	5	6	7	8	9
Industry								
1975	40 779		1 352		14 450		1 589	
1980	51 940		3 793		23 826		2 665	
1984	65 816		4 163 [b]		33 070 [b]		4 560	
Manufacturing industry								
1975	39 540		1 302		13 664		1 459	
1980	49 819		3 512		22 207		2 447	
1984	63 380		3 870 [b]		30 380 [b]		4 257	
Engineering industries								
1975	30 248	100.0	890	100.0	8 974	100.0	1 027	100.0
1980	38 826	100.0	2 626	100.0	15 154	100.0	1 828	100.0
1984	48 838	100.0	2 881	100.0	20 255	100.0	2 685	100.0
Non-electrical engineering								
1975	5 388	17.8	104	11.7	17 110	19.1	146	14.2
1980	6 886	18.7	437 [b]	16.6	3 060 [c]	20.7	186	10.1
1984	8 949	17.2	508 [d]	17.6	3 707 [d]	18.3	312	11.6
Electrical engineering								
1975	8 607	28.5	352	39.6	3 671	40.9	397	38.7
1980	10 707	28.7	1 181 [b]	45.0	6 021 [c]	39.7	694	38.0
1984	13 991	26.8	1 354 [d]	46.3	7 787 [d]	38.4	1 416	52.8
Transport equipment								
1975	13 729	45.4	397	44.6	2 928	32.6	426	41.5
1980	17 066	43.2	953 [b]	36.3	4 572 [c]	30.2	819	44.9
1984	20 656	43.7	968	33.6	6 582	32.8	715	26.6
Precision instruments								
1975	1 978	6.5	26.1	2.9	285	3.2	35	3.4
1980	3 535	7.6	33 [b]	1.2	428 [c]	2.8	77	4.2
1984	4 543	11.3	49 [d]	1.7	483 [d]	2.4	159	5.9
Metal products								
1975	546	1.8	11	1.2	380 [e]	4.2	23	2.2
1980	642	1.8	23 [b]	0.9	1 079 [c]	7.1	52	2.8
1984	699	1.0	23 [d]	0.8	1 696 [d]	8.4	83	3.1

Sources: *Science Indicators*, 1985; *National Patterns of Science and Technology Resources*, 1986; *Japan Statistical Yearbook*, 1985; *Annual Abstract of Statistics*, 1986; *Statistisches Jahrbuch*, 1986.

a In current prices.
b 1979.
c 1981.
d 1983.
e Including office machines and computers.

Table II.25

Changes in the level of R and D expenditure as percentage of GNP in selected ECE member countries and Japan, 1970-1985

Year	*France*	*Fed. Rep. of Germany*	*Japan*	*United Kingdom*	*United States*	*USSR*
1970	1.92	2.06	1.85	...	2.57	3.28
1975	1.80	2.22	1.96	2.19	2.20	3.78
1980	1.84	2.42	2.22	...	2.29	3.76
1984	2.24	2.52	2.60	...	2.58	3.95
1985 [a]	2.27	2.61	...	...	2.72	...

Source: Science and Technology Data Book, National Science Foundation, Washington, D.C. 1987.

Table II.26

Changes in the number of scientists and engineers employed on R and D per 10,000 of the working population in selected ECE member countries and Japan, 1970-1985

Year	*France*	*Fed. Rep. of Germany*	*Japan*	*United Kingdom*	*United States*	*USSR*
1970	27.3	30.8	33.4	...	64.1	58.1 - 64.0
1975	29.4	38.6	47.9	31.1	55.3	78.2 - 87.2
1980	32.4	...	53.6	...	60.0	86.4 - 97.9
1984	39.1	47.1	58.1	35.1	63.8	92.9 - 106.1
1985 [a]	...	47.3	62.4	34.8	65.1	...

Source: As for table II.25.

Table II.27

Changes in the numbers of the R and D work-force in the engineering industries of the United States, Japan and the Federal Republic of Germany, 1975-1984

(Thousands of employees)

Sector/subsector	*United States*	*Japan*	*Fed. Rep. of Germany*
Engineering industries			
1975	256.1	209.1	115.5
1980	322.8	223.4	155.9 [a]
1984	371.8	243.7	162.5 [b]
Non-electrical engineering			
1975	52.8	33.9	24.2
1980	62.1	26.3	33.9 [a]
1984	75.0	31.7	34.5 [b]
Electrical engineering			
1975	82.6	85.3	52.1
1980	94.5	93.3	66.5 [ab]
1984	116.1	130.7	66.7 [ab]
Transport equipment			
1975	95.4	76.1	30.4
1980	125.6	85.3	39.7 [ab]
1984	125.1	56.1	43.1 [ab]
Precision instruments			
1975	17.9	7.9	4.4
1980	32.8	10.1	5.2 [a]
1984	42.1 [d]	16.4	4.8 [bc]
Metal products			
1975	7.4	5.9	4.4 [c]
1980	7.8	8.4	10.7 [ac]
1984	12.9	8.8	14.0 [bc]

Sources: *National Patterns of Science and Technology Resources*, 1986; *Japan Statistical Yearbook*, 1985; *Statistisches Jahrbuch*, 1986.

a 1979.
b 1983.
c Including output of office machines and computers.
d 1982.

Table II.28

Changes in the level of R and D expenditure in the engineering industries of the United States, the United Kingdom, the Federal Republic of Germany and Japan, 1975-1984

(Percentage)

Sector/subsector	*United States*	*United Kingdom*	*Fed. Rep. of Germany*	*Japan*
Industry				
1975	3.1	1.2	1.5	1.5
1980	3.0	1.9 [a]	1.8 [b]	1.6
1984	3.8	1.8 [c]	2.1 [c]	2.0
Manufacturing industry				
1975	2.2	1.4	1.7	1.7
1980	2.3	2.1 [a]	2.0 [a]	1.7
1984	3.2	2.0 [c]	2.3 [c]	2.3
Engineering industries				
1975	4.9	2.7	2.8	2.5
1980	4.7	4.3 [a]	3.5 [b]	2.6
1984	6.4	4.1 [c]	3.8 [c]	3.0
Non-electrical engineering				
1975	4.8	2.0	1.9	1.9
1980	5.0	2.3 [a]	2.6 [b]	1.9
1984	6.0	2.6 [c]	2.8 [c]	2.6
Electrical engineering				
1975	6.5	5.9	4.4	3.7
1980	6.6	8.0 [a]	5.7 [b]	3.6
1984	7.2	7.3 [c]	6.9 [c]	4.7
Transport equipment				
1975	7.2	5.1	3.7	2.2
1980	7.2	5.2 [a]	3.5 [b]	2.4
1984	8.3	4.7 [c]	3.8 [c]	2.3
Precision instruments				
1975	5.9	2.9	3.2	2.7
1980	7.5	1.9 [b]	3.2 [b]	3.0
1984	9.3	2.5 [c]	3.2 [c]	4.0
Metal products				
1975	1.2	...	0.7	1.0
1980	1.4	0.3 [a d]	1.5 [b d]	1.3
1984	1.7	0.3 [c d]	2.0 [c d]	1.3

Sources: *National Patterns of Science and Technology Resources*, 1986; *Survey of Current Business*, U.S. Department of Commerce, January 1987; *Annual Abstract of Statistics*, 1985; *Japan Statistical Yearbook*, 1985; *Statistisches Jahrbuch*, 1986.

a 1981.
b 1979.
c 1983.
d Including output of office equipment and computers.

Table II.29

Changes in the structure of R and D expenditure in the engineering industries of the United States, 1975-1985

(Percentage)

Sector/subsector	*Development of new types of products*	*Development of new technologies*	*Improvement of existing types of products*
Electrical engineering			
1975	19	10	71
1980	35	29	36
1985 *	56	18	26
Non-electrical engineering			
1975	43	11	46
1980	24	5	71
1985 *	56	7	38
Aerospace industry			
1975	47	8	45
1980	51	12	37
1985 *	35	31	34
Automotive industry			
1975	21	29	50
1980	27	11	62
1985 *	51	13	36
Precision instruments			
1975	47	16	37
1980	73	8	19
1985 *	54	8	38
Metal products			
1975	24	26	50
1980	48	19	33
1985 *	47	15	37

Source: 21st Annual McGraw-Hill Survey of Business Plans for Research and Development Expenditures, May 1976; *ibid., 26th Annual Survey,* June 1981; *ibid., 29th Annual Survey,* May 1984.

Table II.30

Changes in the public financing of R and D in United States engineering industries, 1975-1984

(Percentage)

Sector/subsector	*1975*	*1980*	*1984*
Industry	35.5	31.5	32.4
Manufacturing industry	35.3	29.9	31.2
Engineering industries	43.4	37.4	39.3
Metal products	8.3	8.9	8.1
Non-electrical engineering	15.9	11.0	12.6
Manufacture of office machines and computers	21.9	...	...
Electrical engineering	45.4	40.8	39.4
Communication equipment	44.3	41.1	40.5
Transport equipment	58.2	50.8	57.7
Aerospace industry	77.5	72.3	75.8
Precision instruments	14.7	18.9	13.4

Source: *National Patterns of Science and Technology Resources*, National Science Foundation, Washington D.C., 1986.

Table II.31

Country structure of the world export trade in high-technology engineering production in 1984

(Percentage)

Type of production	*United States*	*Japan*	*Fed. Rep. of Germany*	*France*	*United Kingdom*
Aerospace technology	45.1	0.5	15.2	11.8	14.5
Radio and television equipment	0.5	7.9	8.2	1.0	2.2
Office machines and computers	35.5	19.1	9.2	5.6	9.5
Communication equipment	26.5	35.5	10.4	6.1	6.4
Precision instruments	13.7	31.2	15.3	5.7	7.4

Source: *International Science and Technology Data Update*, 1986. National Science Foundation, *DRI Special Tabulations of International Trade*, Washington D.C., 1986.

II.4. EMPLOYMENT AND PRODUCTIVITY TRENDS IN THE ENGINEERING INDUSTRIES RELATIVE TO THE ECONOMY AS A WHOLE

The engineering industries have remained one of the highly labour-intensive sectors of the economy. This regards not only their quantitative development, both in absolute and in relative terms (total employment, share of employment in total industry, etc.), but also the continuing need for further restructuring of the work-force. The introduction of new technologies and the manufacture of new engineering products requires new professional skills and retraining of engineers, operators, system and program analysts, as well as of managers at all levels. [II.32] The improvement of the qualification structure has become the main challenge for all engineering companies. While in many countries the increased application of automated technologies has led to important reductions in the engineering work-force (and contributed to relatively high unemployment), in some other countries, for several reasons (including social security), employment in engineering industries has remained high and the expected growth of labour productivity has not been realized .

Tables II.32-II.35 show the indices of employment and the derived annual changes in employment for the period 1979-1986, for both total industry and the engineering industries. Table II.36 illustrates the development of the share of engineering employment in total industry.

The level of employment in the engineering industries is primarily determined by the rate of increase in the volume of output and by the rate at which and the extent to which automation and other scientific and technical developments of both a technological and an organizational nature are incorporated into production processes. Over the period 1980-1985 as a whole, industrial employment in the majority of market economies declined. Especially large reductions of the work-force relative to 1980 (more than 10 percentage points) were noted in 1984-1986 in the industry of Belgium, Finland, France, the Federal Republic of Germany, Ireland, the Netherlands, Norway, and the United States, while in the United Kingdom the reduction was 21%. The contraction of employment in the engineering industries of these countries was on approximately the same scale, slightly less than in industry as a whole in the Federal Republic of Germany and in the United States, and slightly greater in Belgium and the United Kingdom.

During the period under consideration, employment increased by 2% on average in the industry of centrally planned economies, and by 3% in their engineering industries, except in Hungary and Poland, where there was a perceptible reduction in employment. The proportional nature of the contraction of employment in industry and in the engineering industries may be seen from tables II.32 and II.33. The share of the engineering industries in the industry of almost all the developed countries remained at a stable high level in the period 1979-1986 (ranging from 30% in Spain to nearly 53% in the Federal Republic of Germany and from 28% in Bulgaria to 44% in the German Democratic Republic (see table II.36).

A number of factors were obstacles to increased employment in the engineering industries of market economies, among them the low level of utilization of production capacity (table II.37); investment predominantly for the modernization of existing capacity, rather than for expansion; and structural shifts in favour of high-technology industries in which the high level of automation, the trend towards miniaturization and the high demands made on the qualifications of personnel [II.33] afford fewer opportunities for the use of a large work-force.

Women form an ever-increasing share of the work-force. Thus, the proportion of women in the engineering industries of the United States, 21.9% of the work-force in 1970, had already reached 28.0% by 1983. The proportion is greatest (more than 42%) in electrical engineering and the production of precision instruments.

Employment is increasing most rapidly in electrical and electronic engineering and in precision-instrument manufacture, in both market economies and centrally planned economies.

Labour productivity may be assessed from the relationship between the volume of output (in fixed prices) and the size of the work-force. When other conditions are equal, there is a reciprocal relationship between these two measures (productivity and employment levels), although there are further factors affecting the level of labour productivity measured as the ratio of the volume of output to the size of the work-force (or the number of production workers). [II.34] Having regard to the shortening of the working week, which is a widespread current practice, and to the desire of employers to retain their most highly qualified workers when production is in recession, the statistic of output per production-worker man-hour provides the most reliable indication of labour productivity.

Tables II.38 and II.39 show the labour productivity indices, both for total industry and for engineering industries, by individual countries (1979-1986).

At the same time, the increasing technical level of the goods produced and the growing complexity and costliness of R and D and of production management are creating a situation in which the proportion of industrial production workers is tending to decline, while that of researchers, engineers and managerial staff is tending to rise. On average, throughout the market economies, production workers account for 65 to 70% of the total work-force in the engineering industries, although they do not exceed 40% in such high-technology sectors as the aerospace industry.

The overall average rate of increase in the volume of engineering output in the centrally planned economies was 5.5% in the years 1980-1985. Employment rose over the same period at the moderate rate of 0.75% annually. Consequently, the average annual rate of increase in labour productivity in the engineering industries of these countries was 4.8%.

Table II.40 shows recent trends in individual centrally planned economies in labour productivity in industry as a whole and in engineering, as well as the rise in fixed assets per unit of increase in labour productivity. As can be seen in the table, the growth in industrial labour productivity in 1980-1985 was - with the important exception of the USSR - lower than during the previous five-year period, but in that country, as in every other listed except Hungary, the productivity growth of the engineering branches was lower in 1980-1985 than in the previous period considered. The labour productivity in relation to the use of capital, however, shows a somewhat less regular pattern, with improvements for both total industry and engineering in the USSR, Hungary and Poland and deterioration in Bulgaria and Czechoslovakia. Although somewhat broader than in the recent past, the range of labour productivity increases in engineering within the group of centrally planned economies remained relatively narrow (3.1 to 6.5% - see table II.40).

Table II.41 illustrates the dynamics of the growth of labour productivity in selected CMEA member countries in 1970-1984. It is proof of the leading position of the engineering industries, followed by the chemical industry.

Noteworthy among the factors that slowed the growth of labour productivity in centrally planned economies were tardiness in making practical use of scientific and technological developments; obsolescence of the basic installations of the engineering industries and low rates of replacement of machinery and equipment; disproportions in the sectors of industry; relative lag in the development of high-technology sectors and types of production with enhanced labour productivity; and the retention of a predominantly extensive type of reproduction (more capital investment devoted to new construction than to the modernization of existing capacity).

The trend of labour productivity in the engineering industries of centrally planned economies was on the whole in line with general productivity trends in the economies of these countries. Growth rates for the volume of goods produced (calculated from the produced national income) were only slightly lower during the first half of the 1970s than the high rates typical of the 1960s. However, the significant slowing that occurred in the second half of the decade continued in the first half of the 1980s.

This pattern of labour productivity was conditioned by reduced opportunities for making productive use of relatively cheap additional resources of materials and manpower (extensive growth), exhaustion of possibilities for the improvement of traditional production techniques, technologies and organization, and low rates of introduction of scientific and technical developments. Structural crises at the international level (the energy, raw-material, ecological and currency and financial crises) also had an effect, as did the downturn in international trade.

Individual centrally planned economies failed to reach their planned targets for intensified development and increased labour productivity in the periods 1976-1980 and 1981-1985. Attempts to make good the reduced work-force intake by increasing the volume of fixed capital resources in use did not yield the expected results for reasons connected with the reduced scope for increasing the volume of capital investment when profitability was declining, the failure of efficient new machinery and equipment to keep pace with its rising cost, and continuing disproportions of structure. Capital investment in the engineering industries was clearly inadequate, with the result that the criteria for the withdrawal of fixed capital resources remained extremely low (about 2% annually in the engineering industries). Machinery and equipment became increasingly dated, which made production more capital intensive, impeded improvements in the quality of production and helped to perpetuate the labour shortage.

As has already been noted, labour productivity is a general indication of the effective use of all the factors of production. In the USSR the average annual growth rates for global effectiveness in the use of productive resources declined from 2% in 1971-1975 to 1% in 1976-1980 and 0.6% in 1981-1985. [II.35]

Consequently, there was no perceptible increase in the effective use of productive resources or in the rate of scientific and technical development in centrally planned economies in the period 1980-1985. At the same time, it should be pointed out that major economic reforms aimed at the most effective distribution of productive resources, accelerated scientific and technical development, and generally increased efficiency in the economy, had been carried through in most of these countries by 1985.

While the development of labour productivity indices may well serve to illustrate the overall effectiveness of industrial production, one should not underestimate certain other comparative indicators, such as the material and energy intensity or capital productivity growth (see table II.42).

Great differences were to be observed in the pattern of labour productivity in the years 1980-1985 in the market economies of North America and western Europe (with the United States playing a key role in the North American group). Labour productivity increased by some 5% per annum in the engineering industries of North America, and by less than 4% in the countries of western Europe as a whole. The more appreciable rates of contraction of employment in the west European region (nearly 2.5% per annum, as against 1.8% in North America) did not suffice to offset the inadequate increase, by comparison with North America, in the volume of production (average annual growth rates 1.2% as against 3.3% in North America). The following may be indicated as some of the causes for such falling behind: the decline in R and D expenditure as a proportion of the GDP in the countries of western Europe, which began in the second half of the 1960s, and relative slowness in making productive use of what science has to offer; the later development of high-technology sectors and products than in the United States and Japan, and the excessively high capital and energy inputs of production; declining profitability, leaving less financial scope for the modernization of production capacity etc. The advantage of the United States in the size of the market and, consequently, its superiority as regards economies of scale in production may be distinguished as another quite separate factor. Nevertheless, both the volume of production and labour productivity are rising more rapidly in the engineering industries than in manufacturing as a whole in the overwhelming majority of west European countries (see table II.43).

During the period following 1973, signs of stagnation continued to be apparent in total capital investment in western Europe. Investments did not increase even in 1983, when they rose sharply in the United States. The investment process in the United States during the past 10 years has in the main been more active than in western Europe. The adverse consequences of the stress laid by west European countries on the extensive factors of growth, which had built up over a long period, were clearly much aggravated by the breakdown of established price structures after 1983, the intensification of competition, inflationary pressures, and the contraction of consumer and investment demand.

It was towards 1980, later than in the United States and Japan, that a shift occurred in the economic policy of the west European countries towards anti-inflationary measures, encouragement for scientific and technical developments, the priority development of high-technology sectors, and extensive use of resource-conserving technologies ("low-waste", energy-saving and labour-saving technologies). Transition to new technologies that considerably increase labour productivity is, however, a process that takes time and requires vast capital inputs. The cost of creating a single workplace in the engineering industries is now estimated at 35,000 to 40,000 dollars, and the cost of a work unit in a flexible manufacturing system (FMS) at up to 1 million dollars. Moreover, new technology does not immediately yield a proportionate increase in labour productivity. It is obvious that rates of increase of labour productivity will rise in west European manufacturing industries during the next few years, but that this process will be accompanied for some time to come by an increase in the capital intensiveness of production.

The functional relationships between indices of production, employment and labour productivity in total industry and in the engineering industries are presented in graphic form in figures II.4 (a-d) and II.5 (a-d). Four standard, traditionally engineering-oriented countries for which the relevant data were available were selected, i.e. the Federal Republic of Germany (as representative of the west European market economies), Czechoslovakia (east European centrally planned economies), the United States (North American market economy) and Japan (non-ECE region market economy), in order to illustrate the development of labour productivity in total industry (ISIC, Rev.2: 2-4) and in the engineering industries (ISIC, Rev.2: 38). This selection of "standard" countries will be continued throughout chapter III with a view to achieving the necessary compatibility in analysing the development of individual ISIC major groups 381-385 (see figure II.4). In addition, four more countries, characterized by a significant growth of labour productivity in 1979-1986, were also selected (see figure II.5). It is interesting to note that, while in some countries productivity growth was achieved owing primarily to the substantial increase of production volume (Bulgaria), in other countries it was mainly the reduction of the work-force that affected productivity development (United Kingdom).

Table II.32 Index of total number of persons engaged in total industry (ISIC, Rev.2:2-4) in ECE member countries and Japan, 1979-1986

(1980 = 100)

COUNTRY	1979	1980	1981	1982	1983	1984	1985	1986
Austria a/b/	100	100	97	94	93	93	...	...
Belgium	102	100	95	92	89	89	88	...
Bulgaria	99	100	102	103	103	103	103	103
Cyprus	95	100	104	105	105	108	110	106
Czechoslovakia a/	100	100	101	102	102	103	103	104
Denmark c/	102	100	95	95	94	99	106	103
Finland	95	100	100	98	96	95	94	90
France	100	100	97	96	94	91	89	...
German Dem.Rep. a/	100	100	101	101	102	103	104	...
Germany,Fed.Rep. of c/	99	100	98	94	90	90	91	92
Greece	...	100	100	...	...	...	...	...
Hungary	103	100	98	96	92	92	93	93
Ireland d/	100	100	98	95	89	87	83	...
Italy	101	100	96	90	93	93	...	...
Luxembourg	103	100	98	96	93	...	...	...
Malta	103	100	98	87	83	...	...	...
Netherlands	101	100	97	93	89	88	90	...
Norway	100	100	99	97	90	90	91	...
Poland	100	100	99	94	93	93	90	90
Portugal	98	100	100	99	98	95	...	...
Romania	97	100	102	104	106	106	108	...
Spain	101	100	94	89	86	82	...	...
Sweden	100	100	97	94	91	92	92	92
Switzerland d/	98	100	100	103	99	94	96	...
Turkey d/	99	100	101	105	100	103	...	...
USSR	99	100	101	102	103	103	103	...
United Kingdom e/	106	100	90	84	79	82	81	79
Yugoslavia	98	100	104	107	110	113	117	...
Canada d/	100	100	100	92	90	93	...	...
United States	94	100	97	96	90	88	93	94
Japan d/	99	100	96	96	97	98	99	...

Source : ECE/ITD - Data bank of the Engineering Industries and Technology Section.
Government replies to ECE questionnaires for the Annual Reviews of Engineering Industries and Automation.

a/ Including homeworkers.
b/ Up to 1982, excluding sawmills.
c/ Excluding electricity, gas and water (ISIC, Rev.2 : 4).
d/ Manufacturing industries only (ISIC, Rev.2 : 3).
e/ At June of each year.

Table II.33 Index of total number of persons engaged in engineering industries (ISIC, Rev.2:38) in ECE member countries and Japan, 1979-1986

(1980 = 100)

COUNTRY	1979	1980	1981	1982	1983	1984	1985	1986
Austria a/	99	100	97	94	91	92	...	...
Belgium	101	100	94	90	88	86	85	...
Bulgaria	98	100	100	102	104	107	109	106
Cyprus	93	100	106	108	106	110	119	116
Czechoslovakia a/	99	100	101	102	103	103	104	106
Denmark	100	100	94	95	94	100	110	108
Finland	93	100	101	102	100	99	99	94
France	101	100	97	96	94	91	88	...
German Dem.Rep. a/	99	100	101	102	103	105	106	...
Germany, Fed.Rep. of	99	100	98	96	92	92	95	98
Greece	...	100	102	...	...	...	...	...
Hungary	103	100	98	96	92	90	93	93
Ireland	96	100	100	100	95	92	90	...
Italy	101	100	96	90	92	91	...	...
Luxembourg	94	100	100	97	97	...	...	...
Malta	104	100	101	96	93	...	...	...
Netherlands	100	100	97	93	88	88	91	...
Norway	100	100	100	98	90	89	92	...
Poland	100	100	98	92	89	88	87	87
Portugal	98	100	102	104	98	92	...	...
Romania	96	100	102	105	107	107	111	...
Spain	102	100	93	90	88	82	...	...
Sweden	101	100	98	95	92	93	95	95
Switzerland	97	100	100	100	95	91	94	...
Turkey	100	100	105	110	103	107	...	...
USSR	99	100	101	102	103	104	...	...
United Kingdom b/	106	100	89	82	77	79	78	75
Yugoslavia	97	100	105	109	111	113	116	...
Canada	100	100	101	90	86	91	...	...
United States	103	100	100	92	89	96	97	95
Japan	101	100	105	105	109	113	116	...

Source : ECE/ITD - Data bank of the Engineering Industries and Technology Section.
Government replies to ECE questionnaires for the Annual Reviews of Engineering Industries and Automation.

a/ Including homeworkers.
b/ At June of each year.

Table II.34 Total number of persons engaged in total industry (ISIC, Rev.2:2-4) in ECE member countries and Japan, 1979-1986

(Percentage change)

COUNTRY	1980/1979	1981/1980	1982/1981	1983/1982	1984/1983	1985/1984	1986/1985	1986/1979
Austria ab/	-0.0	-2.6	-3.0	-1.1	-0.5	...	...	...
Belgium	-2.3	-4.8	-3.6	-3.4	-0.1	-0.8	...	-2.5 c/
Bulgaria	1.3	1.7	0.9	0.7	-0.2	0.2	-0.6	0.6
Cyprus	4.9	3.7	1.1	0.0	2.9	1.7	-3.6	1.5
Czechoslovakia ad/	0.1	1.2	0.6	0.4	0.4	0.5	0.6	0.5
Denmark e/	-2.0	-4.9	-0.4	-0.3	5.0	6.9	-2.5	0.2
Finland	5.2	-0.3	-1.7	-2.0	-1.2	-1.3	-3.7	-0.8
France	-0.3	-3.0	-1.2	-2.0	-2.9	-2.9	...	-2.1 c/
German Dem.Rep. a/	0.2	0.7	0.6	0.7	0.9	0.6	...	0.6 c/
Germany,Fed.Rep. of e/	0.7	-2.2	-3.5	-4.1	-1.0	1.4	1.8	-1.0
Greece	...	0.3	...	...	...	...	...	...
Hungary	-2.5	-2.2	-2.3	-3.2	-0.5	1.4	-0.5	-1.4
Ireland f/	-0.4	-1.7	-3.8	-5.6	-3.0	-3.9	...	-3.1 c/
Italy	-1.4	-4.0	-6.0	3.0	0.3	...	...	...
Luxembourg	-2.6	-1.6	-2.5	-2.6	...	...	...	...
Malta	-3.3	-2.5	-10.3	-5.1	...	...	...	...
Netherlands	-0.7	-3.4	-3.7	-4.8	-0.6	2.1	...	-1.9 c/
Norway	-0.2	-0.9	-2.6	-6.3	-0.6	1.2	...	-1.6 c/
Poland	-0.3	-0.5	-5.2	-1.0	-0.5	-3.4	0.2	-1.5
Portugal	1.5	0.4	-1.0	-1.7	-3.0	...	...	...
Romania	3.2	2.0	2.1	1.7	-0.2	1.7	...	1.8 c/
Spain	-1.5	-5.9	-5.9	-2.7	-4.4	...	...	...
Sweden	-0.1	-2.8	-3.6	-2.3	0.6	0.0	-0.3	-1.2
Switzerland f/	2.0	0.2	2.3	-3.8	-4.7	1.8	...	-0.4 c/
Turkey f/	1.2	1.4	3.8	-4.6	3.1	...	...	...
USSR	1.1	0.9	1.0	0.6	0.3	0.4	...	0.7 c/
United Kingdom g/	-5.5	-10.4	-6.7	-5.3	4.1	-1.2	-2.9	-4.1
Yugoslavia	2.5	4.0	3.3	2.7	2.8	3.5	...	3.1 c/
Canada f/	-0.2	0.2	-8.3	-1.8	3.1	...	...	...
United States	6.1	-3.0	-0.6	-6.6	-2.0	6.0	0.8	0.0
Japan f/	0.7	-4.0	-0.0	1.6	0.8	1.1	...	0.0 c/

Source : ECE/ITD - Data bank of the Engineering Industries and Technology Section.
Year to year percentage changes and annual average increase 1986/1979 are calculated from Government replies to ECE questionnaires for the Annual Reviews of Engineering Industries and Automation.

a/ Including homeworkers.
b/ Up to 1982, excluding sawmills.
c/ 1985/1979.
d/ State national industry only.
e/ Excluding electricity, gas and water (ISIC, Rev.2 : 4).
f/ Manufacturing industries only (ISIC, Rev.2 : 3).
g/ At June of each year.

Table II.35 Total number of persons engaged in engineering industries (ISIC, Rev.2:38) in ECE member countries and Japan, 1979-1986

(Percentage change)

COUNTRY	1980/1979	1981/1980	1982/1981	1983/1982	1984/1983	1985/1984	1986/1985	1986/1979
Austria a/	1.5	-2.6	-3.4	-2.9	1.2	...	...	...
Belgium	-1.0	-5.7	-4.8	-2.3	-2.0	-0.7	...	-2.8 b/
Bulgaria	2.0	0.4	1.5	2.2	2.6	1.9	-3.1	1.1
Cyprus	7.9	6.2	1.8	-1.5	3.2	8.7	-3.2	3.2
Czechoslovakia ac/	0.6	1.3	0.8	1.1	0.2	1.1	1.4	0.9
Denmark	0.4	-5.6	0.3	-0.6	5.9	10.4	-1.9	1.2
Finland	7.3	1.2	0.9	-1.8	-1.7	0.1	-4.8	0.1
France	-0.9	-2.8	-0.9	-2.1	-3.6	-3.4	...	-2.3 b/
German Dem.Rep. a/	0.5	1.0	1.0	1.3	1.2	1.3	...	1.1 b/
Germany,Fed.Rep. of	1.2	-1.6	-2.6	-3.8	-0.6	3.3	3.8	-0.1
Greece	...	2.4	...	...	...	...		...
Hungary	-2.8	-2.4	-1.3	-4.6	-1.5	3.1	-0.4	-1.5
Ireland	4.4	0.5	0.0	-5.0	-3.1	-2.6	...	-1.0 b/
Italy	-1.2	-4.1	-5.7	1.6	-0.8	...	...	...
Luxembourg	6.3	-0.1	-2.6	-0.7	...	...	...	...
Malta	-3.5	1.5	-4.9	-3.2	...	...	...	...
Netherlands	-0.2	-3.5	-3.2	-6.0	0.1	3.3	...	-1.6 b/
Norway	-0.1	-0.2	-1.5	-8.3	-0.8	2.6	...	-1.4 b/
Poland	-0.4	-1.9	-6.7	-2.4	-1.1	-1.5	-0.6	-2.1
Portugal	2.5	2.3	1.6	-5.7	-6.6	...	...	...
Romania	4.2	2.2	2.9	2.0	0.2	3.1	...	2.4 b/
Spain	-2.0	-6.6	-4.1	-2.1	-6.3	...	...	...
Sweden	-0.8	-1.6	-3.6	-3.2	1.9	1.2	0.2	-0.9
Switzerland	2.6	0.4	-0.9	-4.6	-4.0	2.5	...	-0.7 b/
Turkey	0.0	5.4	4.4	-6.1	3.3	...	...	...
USSR	1.0	0.9	0.9	0.8	1.5	...	...	...
United Kingdom d/	-6.0	-11.0	-7.9	-5.8	2.1	-1.3	-3.2	-4.8
Yugoslavia	3.0	5.2	3.6	2.3	1.4	2.8	...	3.0 b/
Canada	-0.3	1.0	-10.6	-4.8	5.8	...	...	...
United States	-3.1	0.0	-7.7	-3.4	8.1	0.7	-1.8	-1.1
Japan	-0.9	5.3	-0.2	3.4	3.8	3.0	...	2.4 b/

Source : ECE/ITD - Data bank of the Engineering Industries and Technology Section.
Year to year percentage changes and annual average increase 1986/1979 are calculated from Government replies to ECE questionnaires for the Annual Reviews of Engineering Industries and Automation.

a/ Including homeworkers.
b/ 1985/1979.
c/ State national industry only.
d/ At June of each year.

Table II.36 Share of total number of persons engaged in engineering industries (ISIC, Rev.2:38) in total industry (ISIC, Rev.2:2-4) in ECE member countries and Japan, 1979-1986

(Percentage)

COUNTRY	1980/1979	1981/1980	1982/1981	1983/1982	1984/1983	1985/1984	1986/1985	1986/1979
Austria a/	33.1	33.6	33.6	33.5	32.8	33.4	...	...
Belgium	32.5	33.0	32.6	32.2	32.6	32.0	32.0	...
Bulgaria	26.6	26.8	26.4	26.6	27.0	27.7	28.2	27.5
Cyprus	10.7	11.0	11.3	11.4	11.2	11.2	12.0	12.1
Czechoslovakia ab/	39.4	39.6	39.7	39.8	40.0	39.9	40.2	40.5
Denmark c/	38.3	39.3	39.0	39.3	39.2	39.5	40.8	41.1
Finland	29.7	30.2	30.7	31.5	31.6	31.4	31.9	31.5
France	38.8	38.5	38.6	38.7	38.7	38.4	38.2	...
German Dem.Rep. a/	43.0	43.1	43.3	43.4	43.7	43.9	44.2	...
Germany,Fed.Rep. of c/	49.4	49.6	50.0	50.4	50.6	50.8	51.8	52.8
Greece	...	19.0	19.4	...	...	...	...	...
Hungary	33.2	33.1	33.0	33.3	32.9	32.5	33.1	33.1
Ireland d/	26.6	27.9	28.5	29.7	29.8	29.8	30.2	...
Italy	37.1	37.2	37.2	37.3	36.8	36.4	...	...
Luxembourg	14.3	15.6	15.8	15.8	16.1	...	...	...
Malta	18.6	18.5	19.3	20.4	20.8	...	...	...
Netherlands	37.9	38.1	38.0	38.2	37.7	38.0	38.4	...
Norway	32.2	32.2	32.5	32.8	32.1	32.0	32.5	...
Poland	34.2	34.2	33.7	33.2	32.7	32.5	33.1	32.9
Portugal	19.5	19.7	20.1	20.6	19.8	19.0	...	...
Romania	35.3	35.6	35.7	36.0	36.1	36.2	36.7	...
Spain	29.3	29.1	28.9	29.5	29.6	29.1	...	...
Sweden	42.4	42.1	42.6	42.6	42.2	42.8	43.3	43.5
Switzerland d/	49.4	49.7	49.8	48.2	47.9	48.2	48.6	...
Turkey d/	20.8	20.5	21.4	21.5	21.2	21.2	...	...
USSR	40.0	40.0	40.0	39.9	40.0	40.4	...	...
United Kingdom e/	40.6	40.4	40.1	39.6	39.4	38.6	38.6	38.5
Yugoslavia	29.5	29.6	30.0	30.0	29.9	29.5	29.3	...
Canada d/	32.3	32.2	32.5	31.7	30.7	31.6	...	...
United States	36.5	33.3	34.3	31.9	33.0	36.4	34.6	33.6
Japan d/	40.2	39.5	43.4	43.3	44.1	45.4	46.3	...

Source : ECE/ITD - Data bank of the Engineering Industries and Technology Section.
Government replies to ECE questionnaires for the Annual Reviews of Engineering Industries and Automation.

a/ Including homeworkers.
b/ State national industry only.
c/ Total industry excludes electricity, gas and water (ISIC, Rev.2 : 4).
d/ Figures represent share in manufacturing industries only (ISIC, Rev.2 : 3).
e/ At June of each year.

Table II.37

Utilization of production capacity in the manufacturing industry of selected market economies, 1976-1985

(Percentage)

Country	*Average for*				
	1976-1980	*1981-1985*	*1979*	*1982*	*1985*
United States	81.4	76.6	86.4	70.3	80.1
Germany, Federal Republic of [a]	81.7	79.7	84.7	76.3	84.3
France	80.1	78.0	81.6	77.6	79.4
United Kingdom [b]	32.2	42.1	42.3	23.3	47.8
Italy [c]	74.3	72.1	76.3	71.5	73.8

a Excluding the tobacco and food industries.

b Returns from quarterly questionnaires; the proportion of enterprises working at full capacity.

c Including the extractive industries. Calculation based on OECD, *Main Economic Indicators*; OECD, *Historical Statistics, 1964-1983* (Supplement to Zhurnal IMEMO, Moscow, 1987, p.14).

Table II.38 Index of productivity in total industry (ISIC, Rev.2:2-4) in ECE member countries and Japan, 1979-1986 a/

(1980 = 100)

COUNTRY	1979	1980	1981	1982	1983	1984	1985	1986
Austria	97	100	102	104	106	112	...	...
Belgium	99	100	102	107	112	115	118	...
Bulgaria	97	100	103	107	111	116	120	126
Cyprus	99	100	100	101	103	105	102	...
Czechoslovakia	97	100	101	102	105	108	112	114
Denmark	98	100	105	108	112	117	114	122
Finland	97	100	103	106	111	118	124	130
France	100	100	101	101	104	109	111	...
German Dem.Rep.	95	100	104	107	110	114	118	...
Germany,Fed.Rep. of	101	100	100	101	106	111	116	116
Greece	...	100	101	...	...	...	...	...
Hungary	99	100	105	110	115	118	118	120
Ireland a/	101	100	108	111	127	147	156	...
Italy	94	100	102	105	99	102	...	...
Luxembourg	100	100	95	94	96	...	...	...
Malta	81	100	105	122	132	...	...	...
Netherlands	100	100	101	101	109	116	118	...
Norway	96	100	100	103	119	131	133	...
Poland	101	100	86	90	96	102	110	115
Portugal	96	100	100	106	110	112	...	...
Romania	97	100	101	100	103	110	113	...
Spain	98	100	105	111	117	123	...	...
Sweden	100	100	101	104	112	116	118	120
Switzerland a/	97	100	99	92	95	103	108	...
Turkey a/	102	100	108	124	127	137	...	...
USSR	98	100	102	104	108	112	116	...
United Kingdom	101	100	107	117	129	125	133	139
Yugoslavia	98	100	100	97	95	98	97	...
Canada a/	103	100	101	96	104	116	...	...
United States	108	100	105	99	112	127	122	122
Japan a/	97	100	105	105	108	119	123	...

Source : ECE/ITD - Data bank of the Engineering Industries and Technology Section. Calculated on the basis of indices of industrial production and indices of total number of persons engaged.

a/ Productivity expressed in terms of labour productivity.
b/ Manufacturing industries only (ISIC, Rev.2 : 3).

Table II.39 Index of productivity in engineering industries (ISIC, Rev.2:38) in ECE member countries and Japan, 1979-1986 a/

(1980 = 100)

COUNTRY	1979	1980	1981	1982	1983	1984	1985	1986
Austria	96	100	100	105	110	115	...	...
Belgium	103	100	103	110	116	118	123	...
Bulgaria	96	100	107	114	123	131	141	152
Cyprus	101	100	111	124	116	116	112	107
Czechoslovakia	105	100	103	106	110	117	123	127
Denmark	99	100	104	107	111	120	118	124
Finland	94	100	107	114	115	122	133	142
France	99	100	103	104	105	108	112	...
German Dem.Rep.	93	100	107	111	115	120	126	...
Germany,Fed.Rep. of	98	100	102	103	107	111	119	120
Greece	...	100	100	...	...	...	...	...
Hungary	103	100	108	115	122	126	127	132
Ireland	101	100	113	114	138	180	190	...
Italy	89	100	103	103	100	102	...	...
Luxembourg	105	100	94	97	104	...	...	...
Malta	62	100	96	114	127	...	...	...
Netherlands	97	100	104	107	113	119	120	...
Norway	98	100	100	103	104	111	113	...
Poland	99	100	90	95	104	113	123	133
Portugal	87	100	102	107	101	93	...	...
Romania	95	100	100	101	104	113	118	...
Spain	95	100	104	104	109	113	...	...
Sweden	99	100	103	106	114	120	127	126
Switzerland	93	100	97	92	93	96	101	...
Turkey a/	...	...	100	102	132	153	...	...
USSR	95	100	105	109	115	121	...	...
United Kingdom	100	100	102	113	123	126	135	138
Yugoslavia	102	100	99	99	100	104	105	...
Canada	106	100	101	101	107	114	...	...
United States	100	100	102	101	112	120	124	126
Japan	88	100	101	103	105	120	126	...

Source : ECE/ITD - Data bank of the Engineering Industries and Technology Section. Calculated on the basis of indices of industrial production and indices of total number of persons engaged.

a/ Productivity expressed in terms of labour productivity.
b/ 1981 = 100.

Table II.40

Labour and capital productivity trends in the ECE centrally planned economies, 1975-1980 and 1980-1985

(Average annual increases)

Country		*Labour productivity*		*Rise in fixed assets per unit rise in labour productivity*	
		1975-1980	*1980-1985*	*1975-1980*	*1980-1985*
Bulgaria	I	4.8	3.5 [a]	2.7	3.5 [a]
	E	7.0	6.5	-0.4	3.7
Czechoslovakia	I	3.9	2.2	2.1	3.1
	E	5.5	4.0	0.9	1.8
German Democratic Republic	I	4.4	3.6	...	...
	E	(7)	(6)	...	...
Hungary	I	5.0	3.6	2.4	0.5
	E	4.5	5.2	3.6	-1.6
Poland	I	3.9	1.7	4.5	1.2
	E	5.8	4.2	4.3	-1.3
Romania	I	5.2	2.2 [a]	...	...
	E	6.8	3.1	...	...
USSR	I	2.7	3.2	4.7	3.2
	E	5.8	5.2	3.5	1.9

I = Total industry.
E = Engineering industries.

Source: Data bank of the ECE General Economic Analysis Division. ECE secretariat estimates.

a 1980-1984.

Table II.41

Growth of labour productivity in selected sectors in selected ECE centrally planned economies, 1970-1984

(Average annual increase in per cent)

Country	*Total industry*	*Engineering industries*	*Chemical industry*	*Steel industry*	*Energy*
Bulgaria	5.34	8.27	6.45	4.51	1.72
Czechoslovakia	4.20	5.85	5.32	2.84	2.31
Hungary	4.97	5.83	7.56	3.31	4.99
Poland	4.58	6.52	5.76	3.89	3.11
USSR	4.22	6.76	5.58	2.89	5.56

Source: A. Nesporova: "Analysis of Factors of Labour Productivity Growth in Industry of CMEA Member Countries - International Comparison, 1970-1984" (in Russian); in *Review of Econometrics* (Ekonomicko-matematicky obzor), vol. 24 (1988), No.1, Czechoslovak Academy of Sciences, Prague.

Table II.42

Growth of capital productivity and material and energy intensity in the ECE centrally planned economies, 1981-1986

(Annual change in per cent)

Country	Capital productivity [a]		Material intensity [b]		Energy intensity [c]	
	1981-1985	1986	1981-1985	1986	1981-1985	1986
Bulgaria	-3.1	-1.9	-2.5	-1.6	-1.6	...
Czechoslovakia	-2.5	-2.1	-0.1	0.5	-1.2	-0.8
German Democratic Republic	-0.8	-1.0	-0.9	-1.3	-3.1	-2.5
Hungary	-2.5	...	-0.9	3.1	-2.3	-1.3
Poland	-2.6	2.3	1.5	0.1	1.6	-2.7
Romania	-4.6	-0.9	-1.8	-0.4	-3.1	...
USSR	-2.7	-0.7	0.2	0.7	-0.9	-2.4

Source: ECE secretariat common data base, ECE energy data bank, National statistics.

a The ratio of industrial gross output to fixed assets.
b Calculated from the ratio of gross output index to net output index.
c Ratio of gross primary energy consumption to NMP produced in constant prices.

Table II.43

Rates of increase in labour productivity in engineering industries in selected market economies, 1975-1985

(Average annual increase in per cent)

Country	1975-1980	1980-1985
Austria	5.2	4.8
Denmark	5.3	4.1
Finland	...	6.6
France	4.6	2.6
Germany, Federal Republic of	7.7	2.7
Ireland	...	13.8
Italy	...	5.2
Luxembourg	2.7	7.5
Netherlands	5.9	4.7
Norway	0.0	1.9
Sweden	1.8	5.6
United Kingdom	0.2	5.4
OECD Europe	4.8	3.8
Canada	2.5	2.1
United States	2.7	5.6
OECD North America	2.7	5.2
Japan	12.0	6.2

Note: Data are not strictly comparable owing to changes in weights and base year (1975 for the first period and 1980 for the second).

Sources: OECD Paris: Indicators of Industrial Activity. 1979-1986. Data bank of the ECE General Economic Analaysis Division.

Figure II.4 (a-d)

Development of production, employment and productivity indices in the engineering industries and in total industry of four (standard) countries, 1979-1986 [a]

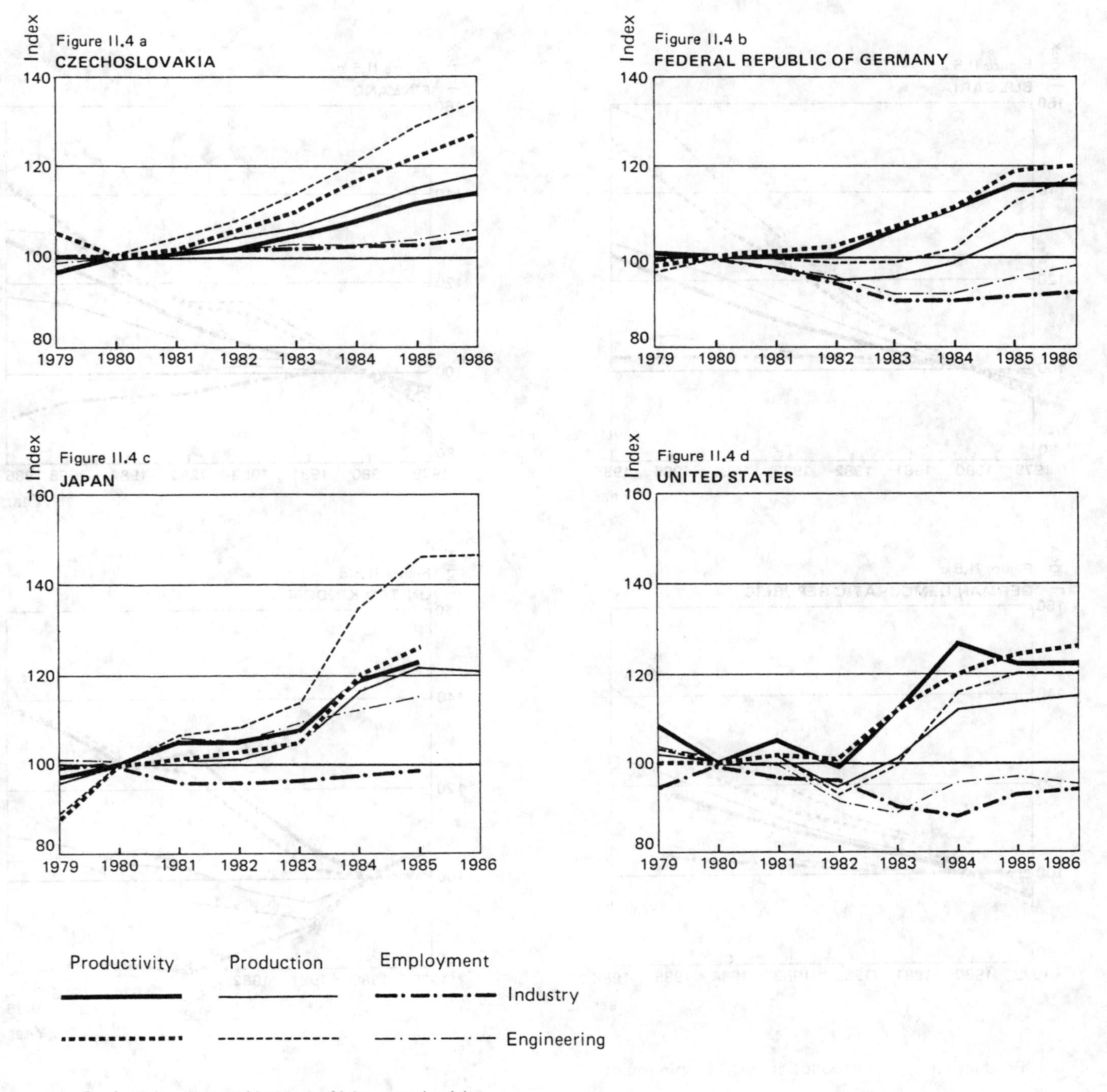

[a] Productivity expressed in terms of labour productivity.
1980 = 100.

Figure II.5 (a-d)

Development of production, employment and productivity indices in the engineering industries and in total industry of four (additional) countries, 1979-1986 [a]

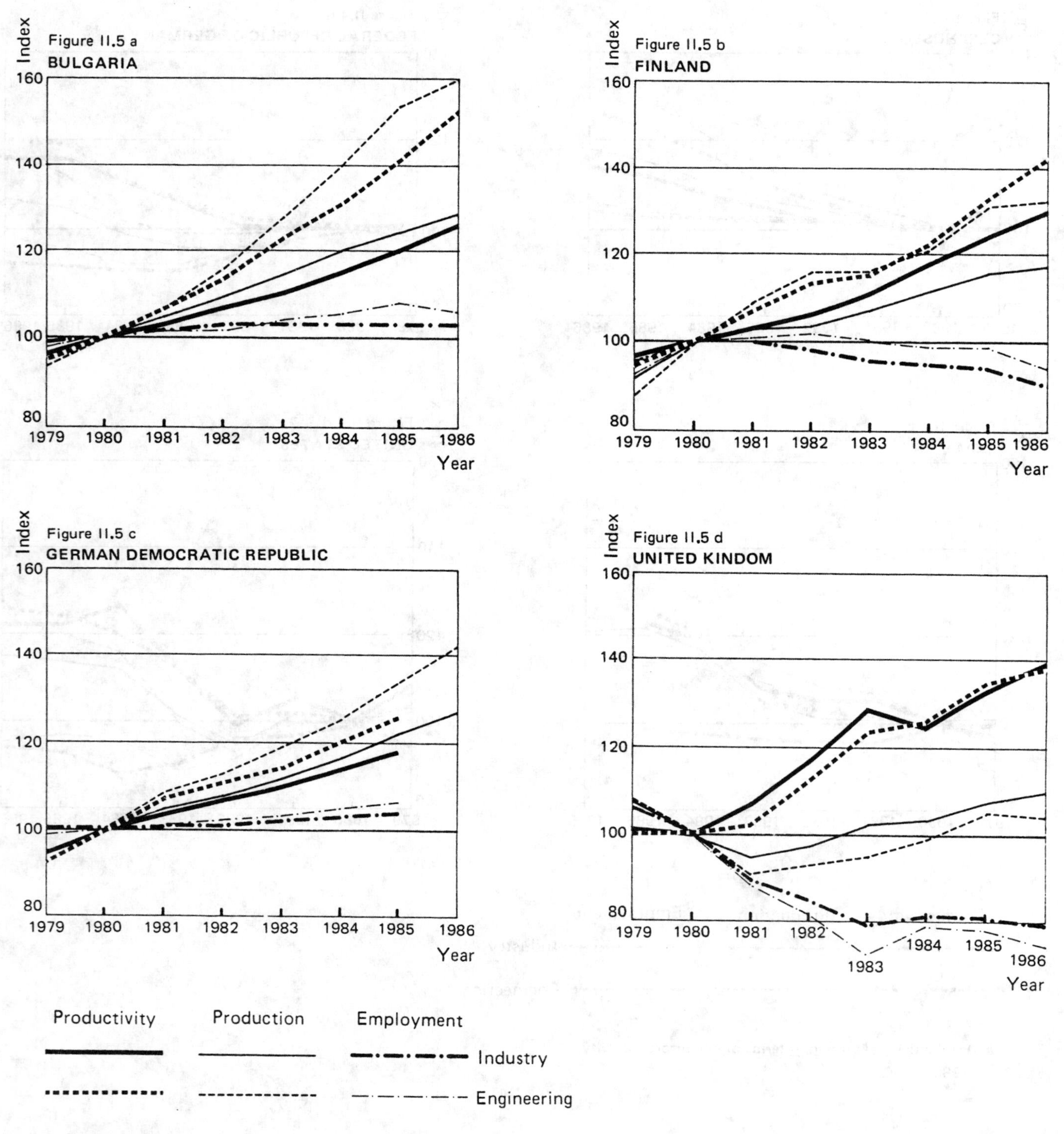

a Productivity expressed in terms of labour productivity.
1980 = 100.

II.5. THE STRUCTURE OF THE ENGINEERING INDUSTRIES AND THEIR SUBSECTORS

The structure of the engineering industries and their subsectors is analysed below in accordance with the United Nations International Standard Industrial Classification (ISIC), Revision 2. A detailed description of all industrial activities classified under "engineering industries" and a classification of relevant products manufactured by those industries - in accordance with the United Nations Standard International Trade Classification (SITC), Revision 2 - may be found in subchapter I.2.

The engineering industries, with their five subsectors, namely the manufacture of metal products (ISIC, Rev.2 : 381), of non-electrical machinery (ISIC, Rev.2 : 382), of electrical machinery (ISIC, Rev.2 : 383), of transport equipment (ISIC, Rev.2 : 384) and of precision instruments (ISIC, Rev.2 : 385), make up a very interdependent and complex industrial activity.

In the developed countries of the ECE region and in Japan, the engineering industries play a key role in the national economy. They have a long tradition and their present structure has been shaped over a long period. The engineering industries are connected through numerous economic links to virtually all other sectors of the economy. They represent increasingly integrated systems, and structural changes are therefore difficult to implement easily. Consequently, long time-series are usually necessary to see them reflected statistically.

The development (1979-1986) of basic indicators characterizing individual engineering subsectors (i.e. ISIC 381-385) is presented in chapter III. The present subchapter, which is oriented more towards the illustration of the structural development within the engineering industries, analyses the contribution of individual subsectors to overall gross output. It was thought that, before analysing the development in individual ECE member countries, it would be useful to show the production growth and share of the ECE region in total world engineering production (see table II.44). With the exception of the metal products industry (ISIC 381), all subsectors saw a slight decline in 1980-1985, compared with the preceding period of 1975-1980. This was mainly due to the overall slowdown in production dynamics in 1981-1982, which has not been fully compensated by the recovery of 1984-1985.

In terms of the ECE region's share in total world production, a relatively successful subsector remained that of precision instruments, characterized by high sophistication of products and decisive intellectual input. The second position, occupied in 1975 by electrical engineering, with an 80.7% share, was taken over during the 1980-1985 period surprisingly enough by the metal products industry - 83.0% share in 1980 and 81.2% in 1985. The third position, held in 1975 and 1980 by the transport equipment industry - with 80.1 and 79.9% share respectively -, belong in 1985 to electrical engineering with an 76.2% share. The total share of the ECE region's engineering industries has continued to be well over 75% of world production - 80.4% share in 1975, 79.1 in 1980 and 76.7 in 1985. It seems that one of the ways of reversing this decline in the share might be the introduction of new and improved forms of industrial co-operation between the three subregions of ECE, i.e. North America, western Europe and the centrally planned economies of eastern Europe.

Tables II.45-II.52 show the development of the share of individual subsectors within total engineering industries - by country, in terms of gross output (1979-1986).

In the manufacture of metal products (ISIC 381), a significant majority of countries, characterized by the well-established and balanced structure of their engineering production, accounted for a 10-20% share (in 1983 e.g. France 10.0%, Federal Republic of Germany 11.3, United Kingdom 15.6, United States 16.1, Sweden 17.6). The metal products industry recorded the highest share in relatively smaller countries, such as Cyprus (from 51.1% in 1981 to 58.4% in 1986), Austria (28.6% in 1983), Yugoslavia (27.6% in 1981), Malta (27% in 1981), Portugal (24.8% in 1979).

The countries recording the most pronounced downward tendency included Turkey (from 21.3% in 1979 to 15.1 in 1984), Ireland (from 22.8% in 1979 to 13 in 1983) and Austria (record drop in one year - from 28.6 in 1983 to 18.8 in 1984).

The traditional subsector of non-electrical machinery (ISIC 382) remained dominant in many smaller and medium-size countries such as Austria (33.8% share in 1982), Bulgaria (41% in 1984 and 1986), Czechoslovakia (45.8% in 1985), Denmark (38.4% in 1981), Finland (38.8% in 1981), Luxembourg (66.9% in 1982), Ireland (42.7% in 1983), Norway (50.3% in 1985) and Poland (37.3% in 1986). The relatively larger producers of capital goods included in this subsector accounted for some 25-30% share of total engineering production. For most countries, this share was relatively stable over the period under review, with the exception of Austria (changes in 1979-1980 and 1982-1983 to some extent point to statistical inconsistencies), Ireland and Norway (both recording considerable increases).

The manufacture of electrical machinery (ISIC 383) covers a product range which includes electric motors; power generating and transforming machinery; radio, television and telecommunication equipment; electrical appliances and housewares; and electronic components. The products of the electronics industry are particularly important for technological progress and for the penetration of information technology into all branches of national economies. The share of electrical machinery in the total output of the engineering industries varies considerably in the ECE region and Japan. The largest share was recorded throughout the period in Austria (31.9% in 1984), Bulgaria (26.7% in 1985), Hungary (30.6% in 1985), Malta (44.9% in 1983), Netherlands (40.7% in 1986), Portugal (34.6% in 1984), Yugoslavia (26.7% in 1985) and Japan (33.2% in 1984). In the United States, the share increased from 18.3% in 1979 to 21.5% in 1984; several other major engineering countries also recorded significant growth in this subsector.

The manufacture of transport equipment (ISIC 384) includes such major branches as the automotive industry (passenger cars and commercial vehicles, including buses), the railway equipment industry (locomotives, passenger carriages, goods wagons and vans), shipbuilding and the aircraft and space industry. As all the industries mentioned above depend heavily on numerous supply industries in various branches, the situation in the transport equipment sector is in many countries decisive for the health of a large segment of the national economy (e.g. the automotive industry in the United States and in the Federal Republic of Germany; the automotive industry and shipbuilding in Japan).

In most countries of the region, with a few exceptions (Cyprus and Malta), the share of transport equipment in the gross output of total engineering industries is relatively high, in the range of 25 to 35%. Transport equipment occupies a top position in the output of the engineering industries in countries such as France (36.7% share in 1983), the Federal Republic of Germany (33.2% in 1983), Portugal (36.9% in 1981), Spain (40.3% in 1983), Sweden (36.4% in 1986), Canada (56.7% in 1984) and the United States (33.1% in 1985). In Japan the share decreased from 30.8% in 1981 to 28.6% in 1984.

The manufacture of precision instruments (ISIC 385) comprises engineering branches producing professional and scientific and measuring and controlling equipment; photographic and optical goods; and watches and clocks. The products of these branches find widespread application in research and development, in manufacturing processes, in equipment used in the certification and testing of goods, in the service and maintenance sectors, as professional equipment and also as consumer goods.

Precision instruments form the smallest sector of the engineering industries. With the exception of Malta (around 20%), the output of this sector covers less than 10% of total output of the engineering industries in all countries concerned. In the following countries, however, (in addition to Malta) this sector occupies a leading position when compared with other countries of the ECE region and Japan: Denmark (6.6% share in 1984), Hungary (9.6% in 1986), Ireland (10.5% in 1983) and the United States (7.3% in 1982).

The above illustration of the structural development within the engineering industries should be considered an attempt to assess the overall trends. Some major engineering countries do not provide disaggregated data, e.g. Belgium (important automotive industry), Switzerland (leading producer of precision instruments) or USSR (significant shift from traditional subsectors to electrical and electronics industries). The statistical classification so far adopted in the German Democratic Republic does not allow serious comparison.

A more detailed analysis (see chapter III) points to the manufacture of electrical machinery (including electronics) as the most rapidly developing subsector in the region, as well as in Japan. This also influenced the significant growth in the broad field of precision instruments, in particular in the United States and in several smaller European countries. The manufacture of metal products showed a declining tendency in larger countries while increasing in relatively less-developed countries of the region. The highly capital-intensive subsectors of non-electrical machinery and transport equipment recorded a relatively stable share characterized by increased concentration and specialization of production into large companies, even transnationals.

Table II.44

Production growth and share of the ECE region in world production of engineering products, 1975-1985

Subsector	*ECE region share (percentage)*			*Production growth indices based on value*	
	1975	*1980*	*1985*	*1980/1975*	*1985/1980*
Metal products (ISIC 381)..............	78.4	83.0	81.2	125.0	128.6
Non-electrical machinery (382)	78.4	76.9	74.7	125.9	121.4
Electrical machinery 383)..............	80.7	76.2	76.2	138.0	133.1
Transport equipment (384).............	80.1	79.9	73.8	127.4	117.3
Precision instruments (385).............	91.5	86.8	88.2	139.9	132.3
Total engineering (ISIC 38)..............	80.4	79.1	76.7	131.1	125.3

Source: *United Nations Industrial Statistics Yearbooks*, 1975-1985.

Table II.45 Share of gross output of the five engineering subsectors within total engineering industries in ECE member countries and Japan in 1979

SUBSECTORS / COUNTRY	Percentage distribution a/					TOTAL ISIC 38 (in millions of national currency at current prices)
	ISIC 381	ISIC 382	ISIC 383	ISIC 384	ISIC 385	
Austria	27.3	24.0	31.6	14.3	2.7	168900
Belgium	...	...	...	...	...	629407
Bulgaria	...	...	...	...	...	7426
Cyprus	55.4	21.4	9.5	13.6	0.0	34
Czechoslovakia b/	13.9	44.6	13.8	26.5	1.2	185440
Denmark	20.5	37.5	17.8	18.3	5.9	37929
Finland	18.5	36.7	17.8	24.7	2.3	24366
France	10.7	31.1	19.2	36.3	2.6	505700
German Dem.Rep. c/	69.2	.	30.8	.	.	89800 d/
Germany,F.R.of e/	12.1	28.9	23.3	31.8	3.9	478480
Greece	...	...	...	...	...	...
Hungary	10.7	22.1	30.3	27.9	9.0	222800
Ireland	22.8	28.8	18.9	19.6	9.9	1309
Italy	15.4	25.6	22.4	31.6	5.0	55618 f/
Luxembourg g/	14.0	65.5	9.7	10.7	.	9091
Malta	24.3	5.6	30.8	18.0	21.3	42
Netherlands	19.7	21.9	34.6	20.9	2.9	46090
Norway	15.6	34.1	15.9	33.5	0.9	36589
Poland	15.8	33.9	19.8	28.5	2.1	832700 h/
Portugal	24.8	12.5	24.9	36.9	0.9	112477
Romania	...	...	...	...	...	...
Spain	24.0	17.0	22.0	35.8	1.2	2229000
Sweden	17.9	28.7	17.0	34.5	1.9	105960
Switzerland	...	...	...	...	...	...
Turkey	21.3	23.2	24.0	31.1	0.3	196859
USSR	...	...	...	...	...	158700
United Kingdom	18.3	28.1	19.2	30.2	4.3	56000
Yugoslavia	25.6	22.3	24.1	27.0	1.0	367200
Canada	18.8	14.9	14.7	49.3	2.4	58360
United States	17.9	26.2	18.3	31.7	5.9	635464
Japan	14.8	24.5	26.8	30.2	3.7	70773 f/

Source : ECE/ITD - Data bank of the Engineering Industries and Technology Section. Government replies to ECE questionnaires for the Annual Reviews of Engineering Industries and Automation.

a/ Based on values expressed in national currency at current prices.
b/ State national industry only.
c/ Metal products (ISIC, Rev.2 : 381) include non-electrical machinery (ISIC, Rev.2 : 382) and transport equipment (ISIC, Rev.2 : 384). Electrical machinery (ISIC, Rev.2 : 383) includes precision instruments (ISIC, Rev.2 : 385) and manufacture of office, computing and accounting machinery.
d/ At 1975 prices.
e/ Metal products (ISIC, Rev.2 : 381) include railroad equipment and structural steel erection. Transport equipment (ISIC, Rev.2 : 384) excludes railroad equipment.
f/ In billions of national currency.
g/ Transport equipment (ISIC, Rev.2 : 384) includes precision instruments (ISIC, Rev.2 : 385).
h/ At 1979 prices.

Table II.46 Share of gross output of the five engineering subsectors within total engineering industries in ECE member countries and Japan in 1980

SUBSECTORS / COUNTRY	Percentage distribution a/					TOTAL ISIC 38 (in millions of national currency at current prices)
	ISIC 381	ISIC 382	ISIC 383	ISIC 384	ISIC 385	
Austria	21.8	31.5	29.5	15.1	2.1	188760
Belgium	...	...	...	...	...	657918
Bulgaria	...	...	...	...	...	7878
Cyprus	54.2	19.8	12.0	14.0	0.0	40
Czechoslovakia b/	13.5	44.7	14.1	26.6	1.1	195000
Denmark	21.5	38.4	16.7	17.7	5.8	43023
Finland	18.8	37.0	18.3	23.7	2.1	30081
France	10.9	31.1	19.4	36.2	2.4	581500
German Dem.Rep. c/	68.8	.	31.2	.	.	96600 d/
Germany,F.R.of e/	12.6	28.3	24.2	30.7	4.2	507650
Greece	33.6	10.0	28.6	27.4	0.5	128923
Hungary	11.1	21.7	30.1	28.3	8.8	210563
Ireland	21.9	32.5	20.0	16.8	8.8	1606
Italy	15.5	26.2	21.7	31.4	5.1	69869 f/
Luxembourg g/	13.6	66.2	9.8	10.4	.	10866
Malta	25.1	6.2	38.7	13.4	16.6	48
Netherlands	20.3	21.7	35.4	20.7	1.8	49200
Norway	15.6	36.5	14.2	32.9	0.7	43805
Poland	15.6	33.8	20.0	28.5	2.1	834800 h/
Portugal	23.7	12.3	27.3	36.0	0.8	148728
Romania	...	...	...	...	...	....
Spain	24.3	17.2	20.5	37.0	1.0	2612000
Sweden	19.6	28.9	17.5	31.7	2.3	115145
Switzerland	...	...	...	...	...	...
Turkey	19.1	27.1	23.5	30.0	0.4	363178
USSR	...	...	...	...	...	167500
United Kingdom	15.6	30.1	21.6	29.8	3.0	59950
Yugoslavia	26.9	21.8	24.4	25.8	1.0	487500
Canada	20.4	16.8	15.8	44.3	2.7	60270
United States	17.7	27.5	19.6	28.4	6.7	656163
Japan	13.7	24.4	27.6	30.5	3.7	81758 f/

Source : ECE/ITD - Data bank of the Engineering Industries and Technology Section. Government replies to ECE questionnaires for the Annual Reviews of Engineering Industries and Automation.

a/ Based on values expressed in national currency at current prices.
b/ State national industry only.
c/ Metal products (ISIC, Rev.2 : 381) include non-electrical machinery (ISIC, Rev.2 : 382) and transport equipment (ISIC, Rev.2 : 384). Electrical machinery (ISIC, Rev.2 : 383) includes precision instruments (ISIC, Rev.2 : 385) and manufacture of office, computing and accounting machinery.
d/ At 1975 prices.
e/ Metal products (ISIC, Rev.2 : 381) include railroad equipment and structural steel erection. Transport equipment (ISIC, Rev.2 : 384) excludes railroad equipment.
f/ In billions of national currency.
g/ Transport equipment (ISIC, Rev.2 : 384) includes precision instruments (ISIC, Rev.2 : 385).
h/ At 1979 prices.

Table II.47 Share of gross output of the five engineering subsectors within total engineering industries in ECE member countries and Japan in 1981

SUBSECTORS / COUNTRY	Percentage distribution a/					TOTAL ISIC 38 (in millions of national currency at current prices)
	ISIC 381	ISIC 382	ISIC 383	ISIC 384	ISIC 385	
Austria	21.2	32.5	30.2	14.3	1.8	202270
Belgium	...	...	...	...	...	657155
Bulgaria	...	...	...	...	...	7796
Cyprus	51.1	23.7	14.3	10.9	0.0	51
Czechoslovakia b/	13.2	44.7	14.6	26.4	1.1	204760
Denmark	19.2	37.2	15.9	21.5	6.1	46392
Finland	18.1	38.8	17.2	23.5	2.4	34959
France	10.3	32.0	19.5	36.0	2.3	642700
German Dem.Rep. c/	70.4	.	29.6	.		111100 d/
Germany,F.R.of e/	12.0	28.5	23.8	31.5	4.1	530430
Greece	33.7	9.1	27.2	29.5	0.4	166445
Hungary	10.8	22.4	30.0	27.6	9.1	227870
Ireland	18.5	35.1	21.6	15.2	9.6	2075
Italy	15.7	26.9	21.4	30.6	5.5	77596 f/
Luxembourg g/	13.0	66.6	9.9	10.5	.	11740
Malta	27.0	5.6	36.8	13.9	16.7	52
Netherlands	19.1	22.0	34.6	22.1	2.1	51620
Norway	15.0	38.0	14.9	31.4	0.7	49594
Poland	15.1	34.9	20.0	27.8	2.2	734300 h/
Portugal	23.4	12.7	26.2	36.9	0.8	186702
Romania	...	...	...	...	...	...
Spain	25.0	17.2	20.2	36.7	1.0	2783000
Sweden	18.8	28.4	18.9	31.7	2.3	124640
Switzerland	...	...	...	...	...	...
Turkey	18.2	29.8	21.1	30.4	0.5	561034
USSR	...	...	...	...	...	175900
United Kingdom	15.3	28.7	22.2	30.8	2.9	59810
Yugoslavia	27.6	21.8	24.0	25.6	1.1	703600
Canada	19.3	16.7	16.5	44.7	2.7	67010
United States	17.2	28.0	19.5	28.5	6.7	718908
Japan	13.1	23.8	28.6	30.8	3.8	91188 f/

Source : ECE/ITD - Data bank of the Engineering Industries and Technology Section. Government replies to ECE questionnaires for the Annual Reviews of Engineering Industries and Automation.

a/ Based on values expressed in national currency at current prices.
b/ State national industry only.
c/ Metal products (ISIC, Rev.2 : 381) include non-electrical machinery (ISIC, Rev.2 : 382) and transport equipment (ISIC, Rev.2 : 384). Electrical machinery (ISIC, Rev.2 : 383) includes precision instruments (ISIC, Rev.2 : 385) and manufacture of office, computing and accounting machinery.
d/ At 1980 prices.
e/ Metal products (ISIC, Rev.2 : 381) include railroad equipment and structural steel erection. Transport equipment (ISIC, Rev.2 : 384) excludes railroad equipment.
f/ In billions of national currency.
g/ Transport equipment (ISIC, Rev.2 : 384) includes precision instruments (ISIC, Rev.2 : 385).
h/ At 1979 prices.

Table II.48 Share of gross output of the five engineering subsectors within total engineering industries in ECE member countires and Japan in 1982

SUBSECTORS / COUNTRY	Percentage distribution a/					TOTAL ISIC 38 (in millions of national currency at current prices)
	ISIC 381	ISIC 382	ISIC 383	ISIC 384	ISIC 385	
Austria	21.3	33.8	30.2	13.0	1.8	221420
Belgium	...	...	...	...	...	771427
Bulgaria	...	...	...	...	...	8462
Cyprus	55.4	21.9	15.4	7.3	0.0	60
Czechoslovakia b/	13.4	44.7	14.5	26.4	1.1	218740
Denmark	20.6	34.3	16.3	22.9	5.9	53772
Finland	19.3	36.7	15.8	25.8	2.4	40024
France	10.4	32.2	19.2	35.9	2.3	729200
German Dem.Rep. c/	70.0	.	30.0	.	.	123500
Germany,F.R.of d/	11.9	28.3	23.8	32.2	3.7	552950
Greece	...	...	...	...	...	...
Hungary	9.3	24.0	29.4	28.2	9.0	246097
Ireland	16.4	36.8	23.5	13.1	10.3	2481
Italy	15.0	25.5	22.9	31.2	5.5	85772 e/
Luxembourg f/	13.2	66.9	9.2	10.7	.	13168
Malta	23.6	5.5	42.4	8.2	20.3	55
Netherlands	19.2	22.2	35.5	21.6	1.6	53680
Norway	15.3	36.0	15.6	32.3	0.7	53029
Poland	15.3	35.2	18.7	28.8	2.0	1524200 g/
Portugal	22.7	16.2	27.3	32.6	1.2	236813
Romania	...	...	...	...	...	...
Spain	24.2	16.1	21.1	37.7	0.9	3117000
Sweden	18.1	27.0	18.6	33.9	2.3	137956
Switzerland	...	...	...	...	...	...
Turkey	17.3	29.3	20.7	32.0	0.7	803291
USSR	...	...	...	...	...	182400
United Kingdom	15.7	28.4	22.7	29.9	3.3	64230
Yugoslavia	27.5	22.0	23.5	26.0	1.1	940900
Canada	17.9	17.6	15.2	47.1	2.2	63760
United States	16.9	26.5	20.9	28.4	7.3	708428
Japan	12.9	24.2	29.1	30.4	3.4	94115 e/

Source : ECE/ITD - Data bank of the Engineering Industries and Technology Section. Government replies to ECE questionnaires for the Annual Reviews of Engineering Industries and Automation.

a/ Based on values expressed in national currency at current prices.
b/ State national industry only.
c/ Metal products (ISIC, Rev.2 : 381) include non-electrical machinery (ISIC, Rev.2 : 382) and transport equipment (ISIC, Rev.2 : 384). Electrical machinery (ISIC, Rev.2 : 383) includes precision instruments (ISIC, Rev.2 : 385) and manufacture of office, computing and accounting machinery.
d/ Metal products (ISIC, Rev.2 : 381) include railroad equipment and structural steel erection. Transport equipment (ISIC, Rev.2 : 384) excludes railroad equipment.
e/ In billions of national currency.
f/ Transport equipment (ISIC, Rev.2 : 384) includes precision instruments (ISIC, Rev.2 : 385).
g/ At 1982 prices.

Table II.49 Share of gross output of the five engineering subsectors within total engineering industries in ECE member countries and Japan in 1983

SUBSECTORS / COUNTRY	Percentage distribution a/					TOTAL ISIC 38 (in millions of national currency at current prices)
	ISIC 381	ISIC 382	ISIC 383	ISIC 384	ISIC 385	
Austria	28.6	24.6	30.6	14.6	1.6	228485
Belgium	...	...	...	...	...	826025
Bulgaria b/	10.0	41.0	24.0	23.9	.	9302
Cyprus	55.2	22.0	15.9	6.9	0.0	61
Czechoslovakia c/	13.2	45.2	13.9	26.5	1.2	228140
Denmark	20.1	34.9	16.0	22.7	6.2	59902
Finland	20.3	33.9	16.6	26.7	2.5	41763
France	10.0	31.5	19.4	36.7	2.4	785800
German Dem.Rep. d/	69.4	.	30.6	.	.	127800
Germany,F.R.of e/	11.3	27.9	24.1	33.2	3.6	568020
Greece	...	...	...	...	...	...
Hungary	9.8	24.1	29.5	27.6	9.1	260961
Ireland	13.0	42.7	23.0	10.8	10.5	3172
Italy	13.6	32.0	23.1	26.5	4.9	105469 f/
Luxembourg g/	23.4	55.7	10.6	10.3	.	14976
Malta	19.4	5.9	44.9	8.1	21.8	58
Netherlands	18.4	22.3	36.8	20.9	1.5	53860
Norway	17.3	36.4	17.5	28.0	0.9	50366
Poland	15.0	35.4	19.1	28.4	2.1	1629200 h/
Portugal	23.0	14.3	31.1	30.3	1.3	280472
Romania	...	...	...	...	...	...
Spain	19.7	18.2	20.9	40.3	0.9	3591301
Sweden	17.4	27.6	18.7	34.0	2.4	162441
Switzerland	...	...	...	...	...	...
Turkey	16.3	27.4	22.9	32.8	0.5	1127493
USSR	...	...	...	...	...	193500
United Kingdom	15.6	27.6	23.6	30.3	2.9	68150
Yugoslavia	26.4	22.0	24.0	26.4	1.2	1318900
Canada	15.8	15.3	13.8	53.1	2.1	70070
United States	16.1	23.6	21.0	32.2	7.1	757959
Japan	12.4	23.9	30.7	29.6	3.4	98596 f/

Source : ECE/ITD - Data bank of the Engineering Industries and Technology Section. Government replies to ECE questionnaires for the Annual Reviews of Engineering Industries and Automation.

a/ Based on values expressed in national currency at current prices.
b/ Non-electrical machinery (ISIC, Rev.2 : 382) includes precision instruments (ISIC, Rev.2 : 385).
c/ State national industry only.
d/ Metal products (ISIC, Rev.2 : 381) include non-electrical machinery (ISIC, Rev.2 : 382) and transport equipment (ISIC, Rev.2 : 384). Electrical machinery (ISIC, Rev.2 : 383) includes precision instruments (ISIC, Rev.2 : 385) and manufacture of office, computing and accounting machinery.
e/ Metal products (ISIC, Rev.2 : 381) include railroad equipment and structural steel erection. Transport equipment (ISIC, Rev.2 : 384) excludes railroad equipment.
f/ In billions of national currency.
g/ Transport equipment (ISIC, Rev.2 : 384) includes precision instruments (ISIC, Rev.2 : 385).
h/ At 1982 prices.

Table II.50 Share of gross output of the five engineering subsectors within total engineering industries in ECE member countries and Japan in 1984

SUBSECTORS / COUNTRY	Percentage distribution a/					TOTAL ISIC 38 (in millions of national currency at current prices)
	ISIC 381	ISIC 382	ISIC 383	ISIC 384	ISIC 385	
Austria	18.8	32.0	31.9	15.5	1.9	236594
Belgium	...	...	...	...	...	856578
Bulgaria b/	9.9	39.7	26.5	23.8	.	10147
Cyprus	55.3	23.8	14.5	6.3	0.0	67
Czechoslovakia c/	11.8	45.4	14.7	26.9	1.2	236320
Denmark	22.6	35.3	16.7	18.9	6.6	66788
Finland	19.3	35.9	16.8	25.0	3.0	45046
France	10.1	32.0	20.3	35.1	2.5	842462
German Dem.Rep. d/	69.3	.	30.7	.	.	130100
Germany,F.R.of e/	11.1	28.5	24.3	32.6	3.5	601210
Greece	...	...	...	...	...	...
Hungary	9.7	23.5	30.4	27.6	8.8	280760
Ireland	...	...	...	...	...	...
Italy	13.8	34.7	21.0	25.8	4.7	121115 f/
Luxembourg	...	...	...	...	...	...
Malta	...	...	...	...	...	...
Netherlands	17.4	23.6	38.1	19.4	1.4	60772
Norway	15.7	41.2	16.8	25.2	1.0	55912
Poland	14.7	35.5	19.8	28.0	2.1	1744300 g/
Portugal	22.1	14.8	34.6	27.1	1.5	327297
Romania	...	...	...	...	...	...
Spain	19.5	19.8	20.5	39.3	0.9	3863481
Sweden	18.2	27.6	18.1	33.9	2.3	185269
Switzerland	...	...	...	...	...	...
Turkey	15.1	30.0	24.8	29.7	0.4	1832000
USSR	...	...	...	...	...	205000
United Kingdom	14.9	28.4	23.8	30.0	2.9	75951
Yugoslavia	25.8	22.0	25.2	25.8	1.2	2158700
Canada	13.7	14.7	13.1	56.7	1.8	88010
United States	15.5	24.1	21.5	32.1	6.8	873960
Japan	11.2	23.9	33.2	28.6	3.1	111681 f/

Source : ECE/ITD - Data bank of the Engineering Industries and Technology Section. Government replies to ECE questionnaires for the Annual Reviews of Engineering Industries and Automation.

a/ Based on values expressed in national currency at current prices.
b/ Non-electrical machinery (ISIC, Rev.2 : 382) includes precision instruments (ISIC, Rev.2 : 385).
c/ State national industry only.
d/ Metal products (ISIC, Rev.2 : 381) include non-electrical machinery (ISIC, Rev.2 : 382) and transport equipment (ISIC, Rev.2 : 384). Electrical machinery (ISIC, Rev.2 : 383) includes precision instruments (ISIC, Rev.2 : 385) and manufacture of office, computing and accounting machinery.
e/ Metal products (ISIC, Rev.2 : 381) include railroad equipment and structural steel erection. Transport equipment (ISIC, Rev.2 : 384) excludes railroad equipment.
f/ In billions of national currency.
g/ At 1982 prices.

Table II.51 Share of gross output of the five engineering subsectors within total engineering industries in ECE member countries and Japan in 1985

SUBSECTORS / COUNTRY	Percentage distribution a/					TOTAL ISIC 38 (in millions of national currency at current prices)
	ISIC 381	ISIC 382	ISIC 383	ISIC 384	ISIC 385	
Austria	...	...	...	...	...	...
Belgium	...	...	...	...	...	901300
Bulgaria b/	9.5	41.0	26.7	22.8	...	11226
Cyprus	56.0	21.1	16.9	5.9	0.0	81
Czechoslovakia c/	11.8	45.8	14.8	26.6	1.0	245550
Denmark	23.2	35.8	16.4	18.0	6.5	77340
Finland	17.9	36.9	18.0	24.1	3.1	51056
France	9.8	32.2	20.3	35.1	2.6	890167
German Dem.Rep. d/	69.1	.	30.9	.	.	143600
Germany,F.R.of	...	...	...	...	...	...
Greece	...	...	...	...	...	...
Hungary	9.7	22.8	30.6	27.8	9.2	312550
Ireland	...	...	...	...	...	...
Italy	...	...	...	...	...	...
Luxembourg	...	...	...	...	...	...
Malta	...	...	...	...	...	...
Netherlands	17.7	22.9	39.3	18.8	1.4	66176
Norway	14.1	50.3	14.3	20.3	1.0	72066
Poland	14.5	35.9	20.2	27.2	2.2	1867800 e/
Portugal	23.5	14.9	31.2	29.0	1.5	390000
Romania	...	...	...	...	...	...
Spain	...	...	...	...	...	...
Sweden	17.6	28.3	17.2	34.1	2.9	206207
Switzerland	...	...	...	...	...	...
Turkey	...	...	...	...	...	...
USSR	...	...	...	...	...	216900
United Kingdom	14.0	29.0	23.0	31.2	2.9	84402
Yugoslavia	25.3	21.8	26.7	25.0	1.2	4279200
Canada	...	...	...	...	...	...
United States	15.3	23.6	21.2	33.1	6.7	908786
Japan	11.6	24.4	31.5	29.3	3.2	121353 f/

Source : ECE/ITD - Data bank of the Engineering Industries and Technology Section. Government replies to ECE questionnaires for the Annual Reviews of Engineering Industries and Automation.

a/ Based on values expressed in national currency at current prices.
b/ Non-electrical machinery (ISIC, Rev.2 : 382) includes precision instruments (ISIC, Rev.2 : 385).
c/ State national industry only.
d/ Metal products (ISIC, Rev.2 : 381) include non-electrical machinery (ISIC, Rev.2 : 382) and transport equipment (ISIC, rev.2 : 384). Electrical machinery (ISIC, Rev.2 : 383) includes precision instruments (ISIC, Rev.2 : 385) and manufacture of office, computing and accounting machinery.
e/ At 1982 prices.
f/ In billions of national currency.

Table II.52 Share of gross output of the five engineering subsectors within total engineering industries in ECE member countries and Japan in 1986

SUBSECTORS / COUNTRY	Percentage distribution a/					TOTAL ISIC 38 (in millions of national currency at current prices)
	ISIC 381	ISIC 382	ISIC 383	ISIC 384	ISIC 385	
Austria	...	...	...	...	...	...
Belgium	...	...	...	...	...	...
Bulgaria b/	9.3	44.5	24.0	22.2	.	11640
Cyprus	58.4	20.8	14.9	6.0	0.0	85
Czechoslovakia c/	11.4	44.4	15.2	27.9	1.0	258835
Denmark	...	...	...	...	...	...
Finland	18.1	33.6	20.9	24.3	3.1	52690
France	...	...	...	...	...	...
German Dem.Rep.	...	...	...	...	...	...
Germany,F.R.of	...	...	...	...	...	...
Greece	...	...	...	...	...	...
Hungary	9.4	22.5	30.5	27.9	9.6	342470
Ireland	...	...	...	...	...	...
Italy	...	...	...	...	...	...
Luxembourg	...	...	...	...	...	...
Malta	...	...	...	...	...	...
Netherlands	17.5	22.4	40.7	18.0	1.4	67490
Norway	...	...	...	...	...	...
Poland	13.7	37.3	20.3	26.4	2.3	2561000 d/
Portugal	...	...	...	...	...	...
Romania	...	...	...	...	...	...
Spain	...	...	...	...	...	...
Sweden	17.0	27.8	16.2	36.4	2.6	221827
Switzerland	...	...	...	...	...	...
Turkey	...	...	...	...	...	...
USSR	...	...	...	...	...	...
United Kingdom	13.9	28.9	23.3	30.9	3.0	88080
Yugoslavia	...	...	...	...	...	...
Canada	...	...	...	...	...	...
United States	...	...	...	...	...	...
Japan	...	...	...	...	...	...

Source : ECE/ITD - Data bank of the Engineering Industries and Technology Section. Government replies to ECE questionnaires for the Annual Reviews of Engineering Industries and Automation.

a/ Based on values expressed in national currency at current prices.
b/ Non-electrical machinery (ISIC, Rev.2 : 382) includes precision instruments (ISIC, Rev.2 : 385).
c/ State national industry only.
d/ At 1984 prices.

REFERENCES

II.1 *Capital goods industry in developing countries: A second study.* UNIDO Publication IS.530, Vienna 1985.

II.2 According to a rough secretariat estimate, the consumer durables for all types of non-industrial use (including military) accounted for some 20-30% of total engineering output in terms of value. See also figure II.1.

II.3 *Survey of current business, No. 5, 1984.* U.S. Department of Commerce, Washington, D.C.

II.4 *Problems of economics and organization in engineering industries* (Problemy ekonomiki i organisatsii mashinostroenia). Nauka publishers, Moscow 1985. (In Russian)

II.5 In the absence of a precise definition of "high-technology industries", the figures presented should be considered a rough approximation mainly indicating the growth rates in the field of information technologies and automation in general. See also table II.23.

II.6 Priority areas of technological development in CMEA member countries have been included in the "Comprehensive Programme for the Scientific and Technical Progress of the CMEA Member Countries up to the Year 2000". See subchapter IV.4 of the present study.

II.7 An increasing share of total investment and production costs in engineering industries has been recorded in the field of software development, maintenance and related services. This concerns the basic software, systems software and applications software. See *Software for Industrial Automation.* UN/ECE publication, Sales No. E.87.II.E.19.

II.8 See *Annual Review of Engineering Industries and Automation, 1983-1984,* vol.I, chapter I. UN/ECE publication, Sales No. E.85.II.E.43.

II.9 Production statistics in chapters II and III of the present study are mainly based on Government replies to regular questionnaires for the *Annual Review.* The first issue of the *Annual Review,* covering 1979, was published in 1981. UN/ECE publication, Sales No. E.II.81.E.16.

II.10 Weights calculated on the basis of 1980 data. 1980 has been set as the base year throughout the study wherever possible. This is also the case for other indicators (indices).

II.11 In tables II.3 and II.4, figures for Japan are included in total data for developed market economies. As from table II.5, in order to illustrate significant growth rates recorded by Japanese engineering industries, Japan is listed (where reasonable) separately.

II.12 If expressed in 1985 prices, the share of engineering industries in total industrial production in the German Democratic Republic increased continuously: 25.6% in 1980, 26.1%(1981), 26.6%(1982), 26.9%(1983), 27.2%(1984), 27.9%(1985) and 28.3% in 1986. In this respect it should be noted that the figures in table II.12 are based on current prices.

II.13 The selection of four "standard" countries (i.e. the United States, the Federal Republic of Germany, Czechoslovakia and Japan) as representatives of different regions is retained in figures presented in subchapter II.4 and in the whole of chapter III.

II.14 The data on investments (capital formation) presented in subchapter II.3 are based on several different sources and should not be used for any detailed comparisons. They only illustrate general trends, e.g. high oscillation in individual years resulting from the overall economic climate, relatively lower investments in engineering industries compared with other industrial sectors, etc.

II.15 NMP produced is used in the ECE *Economic Surveys of Europe* as the main measure of output of centrally planned economies. It is in fact the net value added in the material sphere.

II.16 For detailed comments see *Economic Survey of Europe in 1986-1987,* subchapter 3.4. UN/ECE publication, Sales No. E.87.II.E.1.

II.17 See *Economic Survey of Europe in 1985-1986,* chapters 3 and 4. UN/ECE publication, Sales No. E.86.II.E.1.

II.18 ECE Engineering Industries and Technology Section data bank; and *Science Indicators* 1985 (see also *II.24*).

II.19 ECE Engineering Industries and Technology Section data bank - calculations based on current prices.

II.20 *The national economy of the USSR in 1985* (Narodnoe Khoziaistvo SSSR v 1985 godu), Moscow 1986. (In Russian)

II.21 For comparative statistics on investments in individual engineering subsectors (ISIC 381-385), see subchapters III.2-III.6.

II.22 *Bulletin on Foreign Trade Information* (Byulletin inostrannoi kommercheskoi informatsii), No. 50, Moscow 1985. (In Russian)

II.23 *Bulletin on Foreign Trade Information* (Byulletin inostrannoi kommercheskoi informatsii), No. 8, Moscow 1986. (In Russian)

II.24 *Science Indicators, The 1985 Report*. National Science Board, Washington, D.C.. *National Patterns of Science and Technology Resources*. National Science Foundation, Washington, D.C. 1986. *Annual Abstract of Statistics*. National Statistical Office, London, 1986. *Statistisches Jahrbuch fur die Bundesrepublic Deutschland*. Statistisches Bundesamt, Berlin 1986. *Japan Statistical Yearbook*. Statistics Bureau, Tokyo, 1985.

II.25 *Science and Technology Indicators,* OECD 1984. *IFO - Schnelldienst,* No. 17-18, 1984.

II.26 *The Telecommunication Industry: Growth and Structural Change*. UN/ECE publication, Sales No. E.87.II.E.35.

II.27 Sources as for table II.29.

II.28 *Statistisches Jahrbuch,* 1986 (data for 1983). *Annual Abstract of Statistics*. National Statistical Office, London, 1986.

II.29 *The Engineer,* 10 January 1985.

II.30 *Current Developments in Science and Technology Policies: Review of Changes in Overall National Science and Technology Policies 1980-1985.* UN/ECE document SC.TECH/R.175/Rev.1.

II.31 For a more detailed description of international programmes in the field of Science and Technology (both European Economic Community and CMEA), see subchapter IV.4.

II.32 See also the report of ECE/ILO Symposium on Management Training Programmes and Methods: Implications of New Technologies, held at Geneva in November 1987. UN/ECE document ECE/SEM.7/3.

II.33 Between 1950 and 1980 the proportion of production workers in the total work-force of the engineering industries decreased from 81 to 70% in the United States, and from 74 to 66% in the Federal Republic of Germany.

II.34 These factors include, *inter alia*, advances in science and technology, the quality of the work-force, the level of demand and the scale of production, structural changes, state control, the quality of raw and partly processed materials, international competition, labour organization and management.

II.35 *Kommunist,* No. 24, 1986, p.60. (In Russian)

CHAPTER III STRUCTURAL TRENDS AND DEVELOPMENTS IN ENGINEERING INDUSTRIES AND IN THEIR OUTPUT

III.1. TECHNOLOGICAL CHANGES AND THEIR IMPACT ON ENGINEERING INDUSTRIES

The period under review saw substantial improvement in many basic engineering technologies. Labour-saving and capital-intensive process innovations completely changed the picture of traditional engineering industries. The rapid advances in engineering technologies were the result, basically, of the application of microelectronics and, in particular, of an increasing number of ever more powerful, sophisticated and cheaper microprocessors. These latter have become the key elements in a whole range of new products and machine systems, such as numerically controlled machine tools, industrial robots, flexible manufacturing systems, computer-aided design and computer-aided manufacturing systems, telecommunication equipment and energy and power systems, which in their turn have transformed the industrial environment. Being exceptionally reliable, they have facilitated the rationalization and automation of manufacturing and the corresponding growth in productivity to a hitherto unknown degree. [III.1]

The digitalization of control processes at all manufacturing stages was the prerequisite for the advent of computerized manufacturing technologies throughout the engineering industries. The introduction of these technologies has been accompanied by rapidly growing information processing requirements. Material processing systems have been supplemented by powerful information processing networks. Computerized manufacturing technologies have made it possible to achieve integrated information and material processing systems.

Some of the major implications of computerized manufacturing technologies were reviewed in a recent ECE study on flexible manufacturing systems: [III.2]

- In mass-production environments, computer-controlled machines will make it possible to add flexibility to the production system in the sense that the system can be used to manufacture several different product variants with minimal set-up times. This opens up important potentials for dividing large-scale production into many smaller batches with the obvious purpose of reducing in-process inventory and achieving a faster adaptation to consumer preference; and

- The greatest potential for computerized manufacturing technologies is, however, found in traditional job-shops with small- and medium-sized batch production. Until recently, these manually operated but very flexible job-shops were characterized by long lead times, large in-process inventories, low machine utilization and very little automation.

This is of particular importance, as most of the manufacturing in engineering industries is carried out in smaller batches.

Although the fully automated factory with no human intervention is still a thing of the future, there already exist numerous examples of integrated manufacturing processes controlled by computerized systems. While mechanization at earlier stages of technological development was mainly concerned with the decrease of labour cost per produced unit, computer-controlled automation is aimed at reducing all items that make up total production costs:

- Reduction of labour cost per unit of production;

- Reduction of capital cost through reduction of both in-process and finished goods inventory;

- Higher capital utilization through a higher degree of machine as well as overall plant utilization;

- Saving in energy and raw material;

- Faster product development; and

- Higher and more even quality of products.

But new technologies do more than merely change the face of the engineering industries themselves. As the engineering industries are the suppliers of capital goods, needed to maintain the productive capacity of virtually all sectors of the national economy, any improvement of the engineering industries' products in respect of quality, effectiveness etc. will have a direct impact on, amongst other things, the technological level, quality of products, productivity and employment in those sectors. This applies not only to manufacturing sectors but equally to services. The current revolution in information technology affecting all levels of national economies is based on the mass introduction of computers and the availability of powerful telecommunication networks - both key products supplied by the engineering industries.

While steel has remained the dominant input material for traditional manufacturing technologies, [III.3] a significant shift towards the use of new materials characterized the period under review. These include not only a wide variety of other metals and metal composites, but also various plastics, ceramics etc. [III.4] The introduction of new materials influenced practically all engineering subsectors; material innovation was however most marked in the development of the electrical and electronics industries and in all types of transport equipment.

Engineering products are increasingly complex as concerns both their design and their functions. Built-in computer power significantly improves the adaptability of machines to changing requirements and the accuracy of tasks to be carried out by them. It helps to accelerate the work process and, at the same time, to prevent breakdown. The possibility of linking single machines to form technological systems improves considerably. Consumer goods are increasingly equipped with electronics, which enables them to carry out more complex and sophisticated functions in shorter times. As a consequence, the intrinsic value of the goods is growing and operating conditions greatly improving.

Because of the increasing number of different products integrated into one final product, it is becoming difficult clearly to allocate certain complex engineering products to the predefined product groups for statistical purposes. Furthermore, the merger of mechanical engineering and electronics has led to new requirements in the qualifications of operating personnel responsible for supervision and maintenance of technological equipment.

III.2. MANUFACTURE OF METAL PRODUCTS

This major group (ISIC 381) covers a wide range of traditional metal semi-products, as well as final products for industrial and other uses. It includes, in particular:

- Hand tools, cutlery and general hardware, components and parts for buildings, etc.;
- Metal furniture and fixtures for professional, public and household use;
- Structural metal components and products for the construction and assembly of bridges, buildings, ships, including tanks, boilers, air conditioning etc.; and
- Other metal products for joining and fitting (screws, bolts, nuts, springs etc.), cans, barrels, sanitary and other small metal ware.

It represents a well-established industry with many specialized enterprises, providing opportunities for private and co-operative manufacturers in responding to a flexible market demand.

In spite of increased orientation towards the use of non-metallic materials in engineering industries (plastics, ceramics, fibres, composites etc.), as well as of measures adopted in many countries in order to reduce the metal input into engineering products, the metal products industry continues to be the major supplier of components and parts to all other industrial sectors.

Tables III.1-III.6 illustrate the development of selected production-related indicators characterizing the metal products industry (ISIC 381) between 1979 and 1986. In the light of the discussion in chapter II, the value-added indicator has been selected, as it reflects more accurately than that of gross output the input of engineering in the final product. In the ECE region, during the period under review, several relatively smaller countries showed the largest growth, e.g. the 1986 index of industrial production (1980 = 100) was 286 in Luxembourg, 153 in Denmark, 148 in Turkey, 140 in Finland, 134 in Bulgaria and 132 in Yugoslavia. Amongst developing economies, three countries, Argentina, Brazil and Mexico, accounted for nearly 50% of the total production of all developing countries.

As was done in chapter II, the functional relationship between indices of production, employment and labour productivity by individual major groups of ISIC 38 is presented throughout chapter III in graphical form. Four "standard" engineering-oriented countries for which the relevant data were available were selected, i.e. the Federal Republic of Germany (as representative of the west European market economies), Czechoslovakia (east European centrally planned economies), the United States (North American market economy) and Japan (non-ECE region market economy), in order to indicate the development of ISIC major groups 381-385 in traditionally structured engineering industries (for ISIC 381 see figures III.1 (a-d)). For each major group, four "additional" countries were selected, in order to illustrate either the significant production or productivity increase or the employment decrease (or the interesting overall performance in the subsector) (for ISIC 381 see figures III.2 (a-d) - Bulgaria, Denmark, Finland and Yugoslavia). With the exception of Denmark, in all other selected countries the growth of labour productivity in the metal products industry was below the total productivity in engineering industries.

Tables III.7-III.9 show the share of the metal products industry in total engineering industries, in terms of value added, employment and gross fixed capital formation. Among metal-products-oriented countries, one may find Cyprus with a 49.4% share of value added in 1985 (for some 3,000 employees only), Austria with a 27.9% share in 1983 (65,000 employees), Yugoslavia with a 27.2% share in 1984 (232,000 employees) and Portugal with a 25.3% share in 1983 (40,000 employees).

In absolute terms, the largest producer of metal products remained the United States (1.72 million employees in 1979 and 1.47 million in 1985). The structure of United States shipments is shown in table III.10.

Tables III.11 and III.12 show the geographic distribution of metal product production (ISIC 381). In contrast to total engineering production (ISIC 38 - see chapter II), the period 1980-1985 saw further growth of the metal products industry (highest increase in developing countries of Asia). Western Europe's share in total world production fell from 26.4% in 1975 to 19.4 in 1985. However, owing to the increased share of North America and European centrally planned economies, the total share of the ECE region remained high: in 1985 it amounted to 81.2%, i.e. 4.5 percentage points above its share in total engineering production. This fact, amongst others, points to the continuing importance of the metal products industry in supplying other industrial sectors with basic metal-based components and parts. In other words, it will probably not be before the last decade of the century that the introduction of new non-metallic materials into manufacturing technologies will contribute more significantly to structural changes.

TABLE III.1 Index of industrial production in metal products industries (ISIC, Rev.2:381) in ECE member countries and Japan, 1979-1986

(1980=100)

COUNTRY	1979	1980	1981	1982	1983	1984	1985	1986
Austria	97	100	97	96	103	113	120	...
Belgium	100	100	95	95	96	97	99	96
Bulgaria	114	100	109	115	116	125	133	134
Cyprus	95	100	103	125	100	103	104	96
Czechoslovakia	97	100	103	104	107	111	116	119
Denmark	104	100	94	96	105	132	150	153
Finland	88	100	106	121	129	128	131	140
France	100	100	94	94	94	92	93	...
German Dem.Rep.	96	100	104	108	108	113	116	...
Germany,Fed.Rep. of	98	100	95	91	90	91	94	97
Greece	94	100	101	91	88	98	90	96
Hungary	115	100	98	91	96	96	98	100
Ireland	103	100	91	83	76	77	79	72
Italy	90	100	102	89	79	75	75	72
Luxembourg	102	100	101	101	145	244	254	286
Malta	46	100	113	120	102	105	...	...
Netherlands a/	98	100	97	93	88	94	99	100
Norway	93	100	95	94	84	87	91	95
Poland	101	100	85	81	84	89	94	98
Portugal	88	100	94	100	92	87	99	118
Romania	91	100	104	107	112	125	138	...
Spain	99	100	98	93	94	87	89	90
Sweden	93	100	98	94	96	105	109	106
Switzerland	...	100	98	91	89	98	...	...
Turkey b/	...	...	100	93	97	121	128	148
USSR	...	100	105	108	115	121	...	...
United Kingdom	114	100	95	101	103	105	105	105
Yugoslavia	97	100	104	113	109	107	109	132
Canada c/	102	100	100	92	87	92	99	102
United States	107	100	100	85	88	101	105	105
Japan	102	100	96	99	94	98	99	97

Source : ECE/ITD - Data bank of the Engineering Industries and Technology Section.
Government replies to ECE questionnaires for the Annual Reviews of Engineering Industries and Automation.

a/ Including foundries.
b/ 1981=100.
c/ Excluding manufacture of metal fixtures.

TABLE III.2 Value added in metal products industries (ISIC, Rev.2:381) in ECE member countries and Japan, 1979-1986

(Percentage change) a/

COUNTRY	1980/1979	1981/1980	1982/1981	1983/1982	1984/1983	1985/1984	1986/1985	1986/1979
Austria	-12.6	5.6	8.8	21.4	-22.8	...	...	...
Belgium	...	...	...	...	...	...	...	...
Bulgaria	...	...	...	...	...	...	...	...
Cyprus	19.1	22.2	25.9	3.3	11.8	7.7	...	14.7 b/
Czechoslovakia	5.0	-19.1	1.3	9.3	3.4	-3.4	-0.1	-0.9
Denmark	17.4	-3.8	23.2	8.5	22.1	15.8	...	13.5 b/
Finland	21.7	14.3	24.3	12.6	-0.1	5.6	5.6	11.7
France	16.3	5.4	15.9	4.0	7.7	2.8	...	8.5 b/
German Dem.Rep.	...	...	...	...	...	...	...	...
Germany,Fed.Rep. of	32.2	-0.3	1.6	-1.6	3.0	5.3	...	6.1 b/
Greece	...	25.7	...	...	...	...	...	...
Hungary	-26.3	4.3	2.7	12.1	10.2	17.3	8.7	3.2
Ireland	21.3	6.9	3.5	2.0	...	...	...	...
Italy	24.5	17.7	6.9	-8.9	12.6	...	...	...
Luxembourg	18.2	3.1	0.4	39.0	...	...	...	...
Malta	-0.0	28.2	-11.2	-12.1	...	...	...	...
Netherlands	5.5	-1.8	2.6	-3.9	7.6	6.7	...	2.7 b/
Norway	17.4	8.6	3.1	0.4	8.4	12.9	...	8.3 b/
Poland	-0.8	-2.9	...	17.5	27.6	18.8	...	...
Portugal	30.7	17.2	16.4	21.2	4.5	...	...	...
Romania	...	...	...	...	...	...	...	...
Spain	20.0	10.1	5.0	-9.5	5.7	...	...	...
Sweden	17.4	7.4	-1.0	8.9	25.7	8.7	6.4	10.2
Switzerland	...	...	...	...	...	...	...	...
Turkey	52.7	42.2	39.3	22.7	46.0	...	...	...
USSR	...	...	...	...	...	...	...	...
United Kingdom	-15.2	3.0	9.6	6.5	9.0	6.0	4.0	2.9
Yugoslavia c/	38.7	47.5	35.4	32.7	55.6	99.2	...	50.0 b/
Canada	10.0	10.3	-13.4	-2.2	8.5	...	...	...
United States	1.8	6.3	-4.3	4.1	10.3	2.2	...	3.3 b/
Japan	2.6	7.7	2.4	-12.9	5.9	28.5	...	5.0 b/

Source : ECE/ITD - Data bank of the Engineering Industries and Technology Section.
Year to year percentage changes and annual average increase 1986/1979 are calculated from Government replies to ECE questionnaires for the Annual Reviews of Engineering Industries and Automation.

a/ Based on values expressed in national currency at current prices.
b/ 1985/1979.
c/ Social product including turnover taxes.

Table III.3 Index of total number of persons engaged in metal products industries (ISIC, Rev.2:381) in ECE member countries and Japan, 1979-1986

(1980 = 100)

COUNTRY	1979	1980	1981	1982	1983	1984	1985	1986
Austria a/	116	100	99	94	103	87	...	...
Belgium	101	100	93	90	88	85	84	...
Bulgaria	109	100	102	97	97	100	101	108
Cyprus	93	100	107	116	116	119	129	125
Czechoslovakia a/ b/	101	100	101	101	102	96	97	97
Denmark	101	100	92	94	99	108	120	119
Finland	97	100	100	107	112	109	105	104
France	102	100	96	93	90	86	83	...
German Dem.Rep. a/ c/	100	100	101	102	103	103	105	...
Germany, Fed.Rep. of	99	100	97	93	88	88	89	92
Greece	...	100	101	...	...	...	...	...
Hungary	104	100	97	89	89	88	89	89
Ireland	98	100	97	92	81	80	76	...
Italy	103	100	95	85	94	91	...	...
Luxembourg	94	100	95	94	105	...	...	...
Malta	104	100	106	80	82	...	...	...
Netherlands	100	100	94	87	81	81	83	...
Norway	98	100	102	99	91	91	93	...
Poland	105	100	93	86	83	81	79	78
Portugal	94	100	102	101	99	93	...	...
Romania	...	...	...	...	...	...	...	...
Spain	103	100	95	89	71	66	...	...
Sweden	97	100	97	93	93	94	93	90
Switzerland	97	100	101	101	98	97	96	...
Turkey	100	100	108	114	101	103	...	...
USSR	...	...	...	...	...	...	...	...
United Kingdom d/	122	100	89	83	79	76	76	75
Yugoslavia	94	100	107	111	112	112	114	...
Canada	101	100	98	82	76	75	...	...
United States	107	100	99	88	85	91	91	89
Japan	111	100	103	101	103	100	106	...

Source : ECE/ITD - Data bank of the Engineering Industries and Technology Section.
Government replies to ECE questionnaires for the Annual Reviews of Engineering Industries and Automation.

a/ Including homeworkers.
b/ State national industry only.
c/ Including non-electrical machinery (ISIC, Rev.2 : 382) and transport equipment (ISIC, Rev.2 : 384).
d/ At June of each year.

Table III.4 Total number of persons engaged in metal products industries (ISIC, Rev.2:381) in ECE member countries and Japan, 1979-1986

(Percentage change)

COUNTRY	1980/1979	1981/1980	1982/1981	1983/1982	1984/1983	1985/1984	1986/1985	1986/1979
Austria a/	-13.5	-0.9	-5.3	9.6	-15.5	...	...	...
Belgium	-1.3	-6.7	-4.0	-2.2	-3.0	-1.4	...	-3.1 b/
Bulgaria	-7.8	2.4	-4.9	0.0	2.9	1.0	6.8	-0.1
Cyprus	8.0	6.7	8.4	0.1	3.0	8.0	-2.9	4.4
Czechoslovakia a/ c/	-0.6	0.6	0.6	0.6	-5.8	1.2	0.0	-0.5
Denmark	-1.0	-8.4	2.5	5.1	9.4	10.7	-0.5	2.4
Finland	3.0	0.3	7.0	4.2	-2.3	-3.8	-1.5	0.9
France	-2.1	-4.2	-3.3	-3.1	-4.7	-3.3	...	-3.5 b/
German Dem.Rep. a/ d/	0.3	1.0	0.6	1.2	0.5	1.6	...	0.9 b/
Germany,Fed.Rep. of	0.9	-3.1	-4.0	-4.9	-0.7	1.6	3.0	-1.1
Greece	...	1.2	...	...	...	...	...	...
Hungary	-3.6	-2.9	-7.9	0.0	-1.9	1.2	0.0	-2.2
Ireland	1.8	-2.9	-4.8	-12.6	-0.7	-5.8	...	-4.3 b/
Italy	-3.3	-4.8	-10.7	10.8	-3.6	...	...	...
Luxembourg	6.9	-4.7	-1.1	11.5	...	...	...	...
Malta	-3.6	5.8	-24.2	2.1	...	...	...	...
Netherlands	-0.1	-6.2	-7.4	-7.2	1.0	1.6	...	-3.1 b/
Norway	2.5	1.6	-2.4	-8.5	0.4	1.8	...	-0.8 b/
Poland	-4.7	-6.5	-8.4	-3.0	-2.0	-2.8	-2.1	-4.2
Portugal	6.3	2.5	-1.4	-1.9	-6.2	...	...	...
Romania	...	...	...	...	...	...	...	...
Spain	-2.9	-5.5	-6.3	-20.2	-6.4	...	...	...
Sweden	2.6	-3.1	-3.7	-0.6	1.4	-0.9	-3.3	-1.1
Switzerland	2.8	1.2	0.2	-3.3	-1.0	-0.9	...	-0.2 b/
Turkey	-0.3	7.7	5.8	-11.3	1.9	...	...	...
USSR	...	...	...	...	...	...	...	...
United Kingdom e/	-18.1	-11.5	-6.0	-4.8	-3.5	-0.7	-1.6	-6.8
Yugoslavia	6.7	6.8	3.6	0.9	0.4	2.2	...	3.4 b/
Canada	-1.3	-2.0	-15.9	-7.4	-1.8	...	...	...
United States	-6.1	-1.4	-10.3	-4.0	6.9	0.2	-1.9	-2.5
Japan	-9.7	2.7	-1.2	1.1	-2.1	5.2	...	-0.8 b/

Source : ECE/ITD - Data bank of the Engineering Industries and Technology Section.
Year to year percentage changes and annual average increase 1986/1979 are calculated from Government replies to ECE questionnaires for the Annual Reviews of Engineering Industries and Automation.

a/ Including homeworkers.
b/ 1985/1979.
c/ State national industry only.
d/ Including non-electrical machinery (ISIC, Rev.2 : 382) and transport equipment (ISIC, Rev.2 : 384).
e/ At June of each year.

Table III.5 Index of productivity in metal products industries (ISIC, Rev.2:381) in ECE member countries and Japan, 1979-1986 a/

(1980 = 100)

COUNTRY	1979	1980	1981	1982	1983	1984	1985	1986
Austria	84	100	98	102	100	130	...	...
Belgium	99	100	101	106	109	114	118	...
Bulgaria	105	100	106	118	119	125	131	124
Cyprus	103	100	97	108	86	86	81	76
Czechoslovakia	96	100	102	103	105	115	119	122
Denmark	103	100	103	102	106	122	125	129
Finland	91	100	106	113	115	117	125	136
France	98	100	98	102	105	108	112	...
German Dem.Rep.	...	...	...	...	...	...	...	...
Germany,Fed.Rep. of	99	100	98	97	101	103	105	105
Greece	...	100	100	...	...	...	...	...
Hungary	111	100	101	102	108	109	110	113
Ireland	105	100	94	90	94	96	105	...
Italy	87	100	108	105	84	82	...	...
Luxembourg	109	100	106	107	138	...	...	...
Malta	44	100	107	150	125	...	...	...
Netherlands b/	98	100	103	107	109	115	120	...
Norway	95	100	94	95	93	95	98	...
Poland	96	100	91	95	101	109	119	126
Portugal	94	100	92	99	93	94	...	...
Romania	...	...	...	...	...	...	...	...
Spain	96	100	104	105	132	131	...	...
Sweden	95	100	101	100	104	112	117	117
Switzerland	...	100	97	90	91	101	...	...
Turkey c/	...	...	100	88	103	126	...	...
USSR	...	...	...	...	...	...	...	...
United Kingdom	93	100	107	121	130	137	138	140
Yugoslavia	104	100	97	102	98	95	95	...
Canada d/	101	100	102	112	114	123	...	...
United States	100	100	101	96	104	111	115	118
Japan	92	100	94	97	92	98	93	...

Source : ECE/ITD - Data bank of the Engineering Industries and Technology Section. Calculated on the basis of indices of industrial production and indices of total number of persons engaged. See tables III.1 and III.3.

a/ Productivity expressed in terms of labour productivity.
b/ Indexes of industrial production include foundries.
c/ 1981=100.
d/ Indexes of industrial production exclude manufacture of metal fixtures.

Table III.6 Gross fixed capital formation in metal products industries (ISIC, Rev.2:381) in ECE member countries and Japan, 1979-1986

(Percentage change) a/

COUNTRY	1980/1979	1981/1980	1982/1981	1983/1982	1984/1983	1985/1984	1986/1985	1986/1979
Austria	-6.2	23.6	-18.7	31.5	-22.2	...	...	...
Belgium	...	...	...	...	...	...	...	...
Bulgaria b/	224.6	-2.7	-67.8	126.0	-42.5	15.5	...	7.3 c/
Cyprus	-11.8	47.2	-23.9	-16.7	78.6	64.0	-24.4	9.0
Czechoslovakia	...	...	...	...	...	...	...	...
Denmark	21.8	-9.6	17.5	-4.4	60.4	43.2	...	19.0 c/
Finland	39.9	27.5	34.1	8.8	2.9	-20.0	10.5	13.1
France	11.6	22.9	6.8	3.2	7.7	20.8	...	11.9 c/
German Dem.Rep. d/	7.5	11.0	-5.9	-3.1	0.4	4.8	...	2.3 c/
Germany,Fed.Rep. of	11.4	-6.6	21.8	5.3	...	...	...	9.3 c/
Greece	...	9.6	...	...	...	...	...	...
Hungary	27.0	-7.3	-3.2	0.3	8.8	-13.3	-34.9	-4.9
Ireland	-6.1	7.4	-28.1	-47.6	...	...	...	...
Italy	17.6	13.3	14.5	57.8	-9.9	...	...	...
Luxembourg	-11.5	877.6	-5.4	-84.8	92.4	-34.8	...	7.8 c/
Malta	-33.5	-38.0	72.3	502.4	...	...	...	...
Netherlands	6.5	-6.1	-4.8	0.8	17.1	20.2	16.3	6.6
Norway	-3.1	-11.5	23.3	-24.6	14.5	37.9	...	3.9 c/
Poland	-26.3	-43.5	...	3.7	4.1	-1.2	...	...
Portugal	69.5	25.5	19.2	-10.7	30.2	...	...	...
Romania	...	...	...	...	...	...	...	...
Spain	-21.1	30.4	-22.6	1.0	-12.9	...	...	...
Sweden	47.7	-6.1	-18.0	17.0	52.8	-2.7	42.7	16.0
Switzerland	...	...	...	...	...	...	...	...
Turkey	6.9	120.1	59.1	2.5	62.7	...	...	...
USSR	...	...	...	...	...	...	...	...
United Kingdom	-8.1	-16.7	-6.9	-3.1	25.5	6.5	-7.9	-2.3
Yugoslavia	24.3	13.5	42.0	12.2	54.3	45.3	...	30.9 c/
Canada	33.3	-12.3	-34.8	-13.5	80.8	...	...	...
United States	-3.0	0.0	-12.5	8.1	22.9	3.8	13.4	4.1
Japan	7.6	3.5	15.5	-21.3	29.6	23.9	...	8.4 c/

Source : ECE/ITD - Data bank of the Engineering Industries and Technology Section. Year to year percentage changes and annual average increase 1986/1979 are calculated from Government replies to ECE questionnaires for the Annual Reviews of Engineering Industries and Automation.

a/ Based on values expressed in national currency at current prices.
b/ Excluding land.
c/ 1985/1979.
d/ Including non-electrical machinery (ISIC, Rev.2 : 382) and transport equipment (ISIC, Rev.2 : 384).

Figure III.1 (a-d)

Development of production, employment and productivity indices in the metal products industries (ISIC 381) of four (standard) countries, 1979-1986 [a]

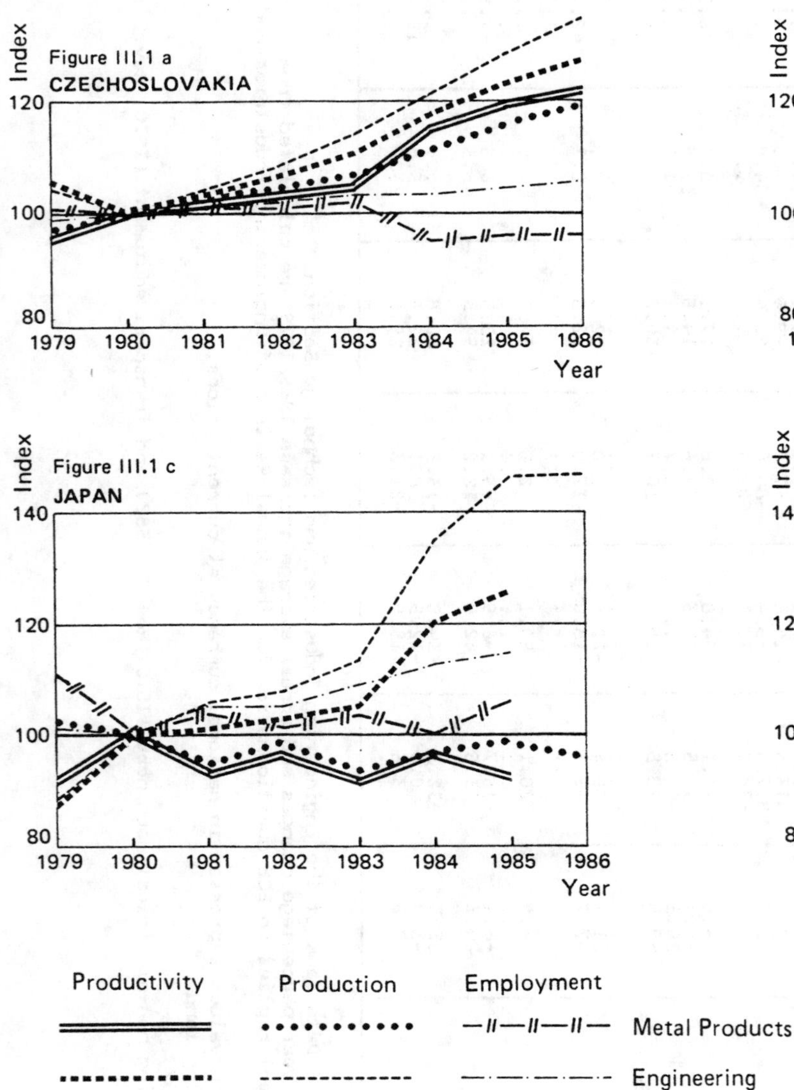

Index
Figure III.1 b
FEDERAL REPUBLIC OF GERMANY
120
100
80
1979 1980 1981 1982 1983 1984 1985 1986
Year

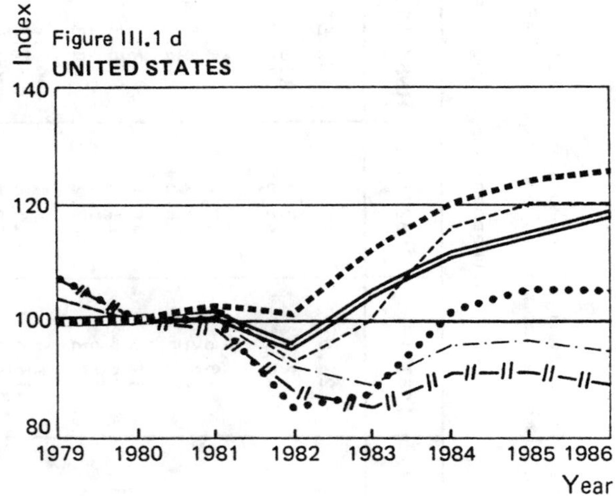

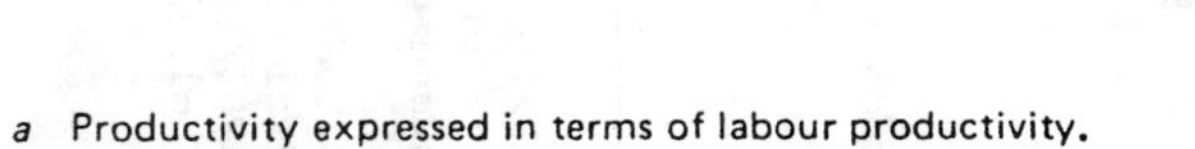

a Productivity expressed in terms of labour productivity. 1980 = 100.

Figure III.2 (a-d)

Development of production, employment and productivity indices in the metal products industries (ISIC 381) of four (additional) countries, 1979-1986 [a]

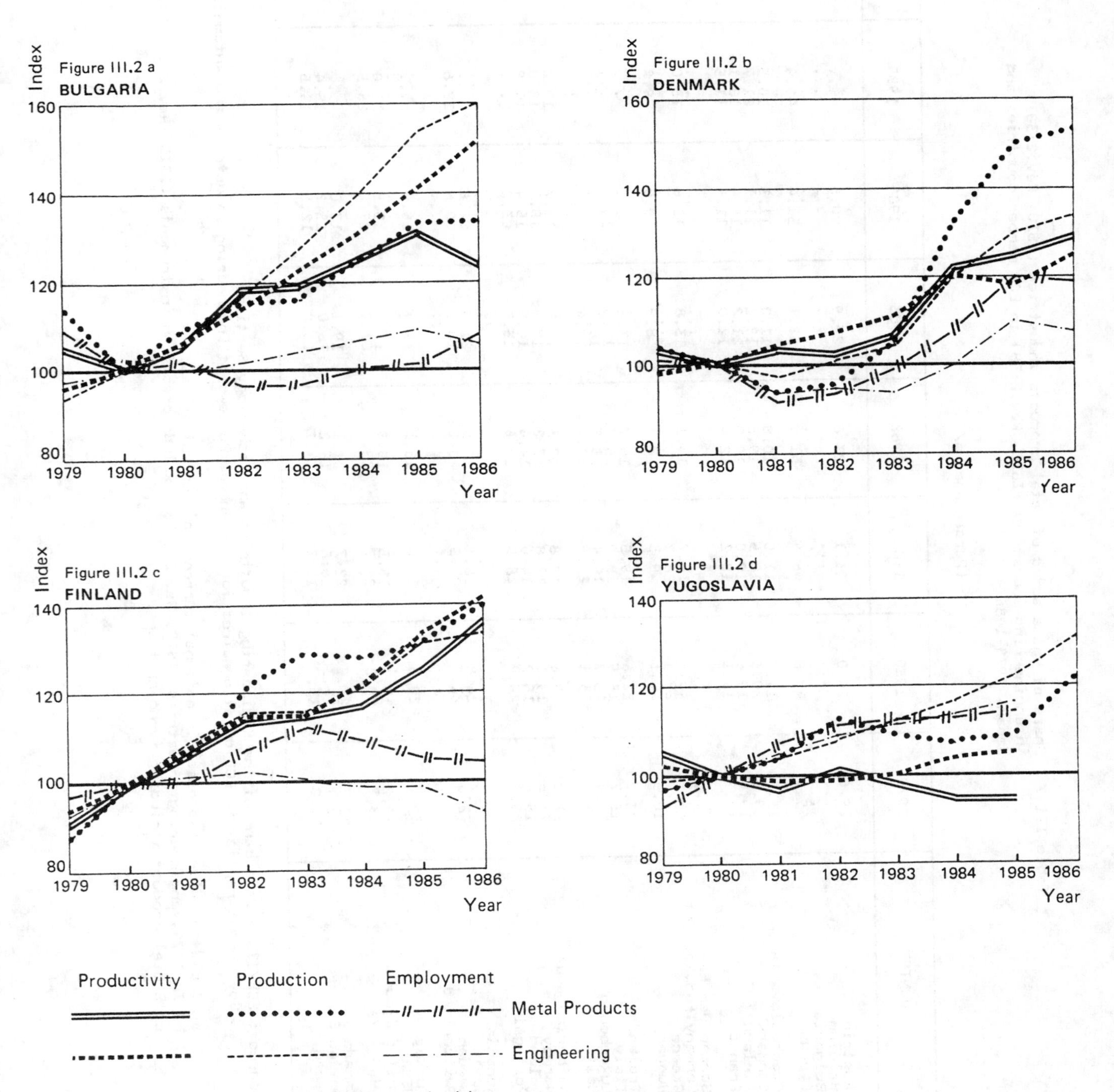

a Productivity expressed in terms of labour productivity.
1980 = 100.

Table III.7 Share of value added of metal products industries (ISIC, Rev.2:381) in engineering industries (ISIC, Rev.2:38) in ECE member countries and Japan, 1979-1986

(Percentage) a/

COUNTRY	1979	1980	1981	1982	1983	1984	1985	1986
Austria	29.6	24.0	23.1	23.8	27.9	20.5	...	...
Belgium	...	...	...	...	...	...	...	...
Bulgaria	...	...	...	...	...	...	...	...
Cyprus	49.6	47.8	46.9	51.0	52.1	51.4	49.4	...
Czechoslovakia	11.5	11.5	11.0	10.8	11.0	11.3	10.5	10.7
Denmark	19.6	20.3	18.5	19.8	19.2	21.1	21.7	...
Finland	19.0	19.6	19.0	19.8	21.2	19.4	18.1	19.0
France	12.7	13.1	12.2	12.4	12.0	11.9	11.5	...
German Dem.Rep.	...	...	...	...	...	...	...	...
Germany,Fed.Rep. of	12.8	15.4	14.7	14.3	13.8	13.6	13.0	...
Greece	...	29.4	28.7	...	...	...	...	...
Hungary	10.5	10.1	9.3	8.5	8.9	9.0	9.2	9.1
Ireland	23.7	22.6	19.5	16.7	13.2	...	...	...
Italy	15.7	15.9	15.9	15.3	14.6	14.2	...	...
Luxembourg	17.4	15.7	14.4	13.5	18.7	...	...	...
Malta	23.1	23.1	25.9	19.7	17.4	...	...	...
Netherlands b/	21.0	20.6	19.8	19.3	18.0	18.7	19.2	...
Norway	17.5	19.1	17.8	17.2	17.9	18.1	18.0	...
Poland	15.6	15.2	14.6	14.6	14.9	15.4	14.9	14.7
Portugal	26.3	25.9	25.5	25.6	25.3	23.2	...	...
Romania	...	...	...	...	...	...	...	...
Spain	25.7	26.8	28.1	26.5	21.8	22.6	...	...
Sweden	19.9	21.7	21.3	20.2	18.8	20.6	18.6	19.5
Switzerland	...	...	...	...	...	...	...	...
Turkey	23.7	20.6	19.0	18.1	17.4	16.4	...	...
USSR	...	...	...	...	...	...	...	...
United Kingdom	19.2	15.3	15.5	15.9	16.0	15.8	15.0	14.9
Yugoslavia c/	28.7	29.8	29.7	29.2	27.9	27.2	26.5	...
Canada	22.7	23.5	22.7	21.1	19.1	17.0	...	...
United States	17.7	17.3	16.8	16.2	16.0	15.2	15.4	...
Japan	17.0	15.7	15.2	14.7	13.2	12.5	13.5	...

Source : ECE/ITD - Data bank of the Engineering Industries and Technology Section. Government replies to ECE questionnaires for the Annual Reviews of Engineering Industries and Automation.

a/ Based on values expressed in national currency at current prices.
b/ In 1979, engineering industries (ISIC, Rev.2 : 38) exclude precision instruments (ISIC, Rev.2 : 385).
c/ Social product including turnover taxes.

Table III.8 Share of total number of persons engaged in metal products industries (ISIC, Rev.2:381) in total number engaged in engineering industries (ISIC, Rev.2:38) in ECE member countries and Japan, 1979-1986

(Percentage)

COUNTRY	1979	1980	1981	1982	1983	1984	1985	1986
Austria a/	29.1	24.8	25.2	24.7	27.9	23.3	...	...
Belgium	27.0	26.9	26.7	26.9	26.9	26.6	26.5	...
Bulgaria	15.1	13.7	13.9	13.1	12.8	12.8	12.7	14.0
Cyprus	52.0	52.1	52.4	55.6	56.6	56.5	56.2	56.4
Czechoslovakia a/ b/	15.7	15.5	15.4	15.4	15.3	14.4	14.4	14.2
Denmark	20.9	20.6	20.0	20.4	21.6	22.3	22.4	22.7
Finland	19.0	18.2	18.1	19.1	20.3	20.2	19.4	20.1
France	13.1	13.0	12.8	12.5	12.4	12.2	12.2	...
German Dem.Rep. a/ c/	68.3	68.2	68.2	68.0	67.9	67.4	67.6	...
Germany, Fed.Rep. of	17.1	17.1	16.8	16.6	16.4	16.4	16.1	16.0
Greece	...	30.7	30.4	...	...	...	...	...
Hungary	11.5	11.4	11.3	10.6	11.1	11.1	10.9	10.9
Ireland	27.8	27.1	26.2	25.0	23.0	23.5	22.8	...
Italy	16.0	15.6	15.5	14.7	16.0	15.6	...	...
Luxembourg	16.6	16.7	15.9	16.2	18.2	...	...	...
Malta	26.6	26.6	27.7	22.1	23.3	...	...	...
Netherlands	21.6	21.6	21.0	20.1	19.9	20.0	19.7	...
Norway	19.6	20.1	20.4	20.3	20.2	20.5	20.3	...
Poland	19.8	19.0	18.1	17.8	17.6	17.5	17.3	17.0
Portugal	27.1	28.4	28.1	27.6	28.3	28.5	...	...
Romania	...	...	...	...	...	...	...	...
Spain	30.4	30.2	30.5	29.8	24.3	24.3	...	...
Sweden	19.8	20.5	20.2	20.1	20.7	20.6	20.2	19.4
Switzerland	17.4	17.4	17.6	17.8	18.0	18.6	17.9	...
Turkey	22.4	22.4	22.9	23.2	21.9	21.6	...	...
USSR	...	...	...	...	...	...	...	...
United Kingdom d/	18.0	15.7	15.6	15.9	16.1	15.2	15.3	15.6
Yugoslavia	28.8	29.9	30.3	30.3	29.9	29.6	29.4	...
Canada	25.0	24.7	24.0	22.6	21.9	20.4	...	...
United States	18.9	18.3	18.0	17.6	17.4	17.2	17.2	17.1
Japan	19.9	18.1	17.7	17.5	17.1	16.1	16.5	...

Source : ECE/ITD - Data bank of the Engineering Industries and Technology Section.
Government replies to ECE questionnaires for the Annual Reviews of Engineering Industries and Automation.

a/ Including homeworkers.
b/ State national industry only.
c/ Metal products (ISIC, Rev.2 : 381) include non-electrical machinery (ISIC, Rev.2 : 382) and transport equipment (ISIC, Rev.2 : 384).
d/ At June of each year.

Table III.9 Share of gross fixed capital formation of metal products industries (ISIC, Rev.2:381) in engineering industries (ISIC, Rev.2:38) in ECE member countries and Japan, 1979-1986

(Percentage) a/

COUNTRY	1979	1980	1981	1982	1983	1984	1985	1986
Austria	26.0	19.5	15.3	15.4	27.3	21.3	...	...
Belgium	...	...	...	...	...	...	...	...
Bulgaria b/	17.9	24.2	41.1	10.1	14.0	9.3	9.1	...
Cyprus	48.6	53.6	57.6	56.5	56.0	64.1	64.1	54.4
Czechoslovakia	...	...	...	...	...	...	...	...
Denmark	21.5	20.8	20.2	20.8	18.6	23.0	22.5	...
Finland	17.3	15.5	18.5	19.9	21.0	19.8	13.3	16.5
France	13.2	12.1	15.0	15.1	15.1	13.1	13.2	...
German Dem.Rep. c/	72.2	71.0	68.0	69.9	67.3	68.5	71.3	...
Germany,Fed.Rep. of	10.3	9.8	9.2	10.6	10.8	...	11.2	...
Greece	...	21.4	19.4	...	...	...	...	...
Hungary	7.2	10.9	11.6	10.6	10.4	11.8	11.9	7.5
Ireland	29.9	20.3	19.7	11.4	7.9	...	...	...
Italy	16.0	15.8	15.8	14.8	19.3	15.6	...	...
Luxembourg	20.7	15.0	70.6	73.9	22.1	31.2	21.4	...
Malta	17.9	38.7	12.8	13.8	45.7	...	...	...
Netherlands	21.5	19.3	20.4	19.3	20.5	17.4	15.5	16.1
Norway	23.2	22.5	17.2	21.1	19.3	17.8	18.3	...
Poland	17.4	16.0	11.4	19.1	17.0	15.7	15.1	...
Portugal	22.7	27.2	24.3	18.9	21.5	26.8	...	...
Romania	...	...	...	...	...	...	...	...
Spain	33.6	28.9	33.5	27.1	22.4	12.0	...	...
Sweden	20.6	23.6	19.8	17.0	17.5	19.6	15.7	19.5
Switzerland	...	...	...	...	...	...	...	...
Turkey	10.5	14.8	13.9	15.5	10.8	9.5	...	...
USSR	...	...	...	...	...	...	...	...
United Kingdom	14.7	14.0	13.1	12.5	11.3	11.5	11.6	10.0
Yugoslavia	31.3	30.1	28.3	32.8	29.7	32.3	25.9	...
Canada	22.0	17.6	15.3	14.1	11.7	19.4	...	...
United States	8.4	7.0	6.6	6.3	7.2	6.9	6.6	7.9
Japan	11.4	8.6	7.2	7.7	6.3	6.6	6.9	...

Source : ECE/ITD - Data bank of the Engineering Industries and Technology Section.
Government replies to ECE questionnaires for the Annual Reviews of Engineering Industries and Automation.

a/ Based on values expressed in national currency at current prices.
b/ Excluding land.
c/ Metal products (ISIC, Rev.2 : 381) include non-electrical machinery (ISIC, Rev.2 : 382) and transport equipment (ISIC, Rev.2 : 384).

Table III.10

Value of shipments by selected component industries in the United States, 1979-1986

Product type	*Shipments in billions of 1982 US dollars*				*Index*
	1979	*1982*	*1984*	*1986* [a]	*1986-1979*
Screw machine products	2.44	2.16	2.84	3.01	123.4
Industrial fasteners	4.13	3.40	4.29	4.31	104.4
Valves and pipe fittings	7.09	8.71	9.03	8.83	124.5
Ball and roller bearings [b]	3.32	2.97	3.70	3.44	103.6
Total	16.98	17.24	19.86	19.59	115.4

Source: *U.S Industrial Outlook 1988*. Department of Commerce, Washington, D.C.

a Estimated by International Trade Administration (ITA).

b In ISIC included in major group 382.

Table III.11

Production of metal products (ISIC 381) by geographic region, 1975-1985

Region	*Value in billions of constant 1975 US dollars* [a]			*Index*	
	1975	*1980*	*1985*	*1980/1975*	*1985/1980*
Total world	258.0	311.5	409.7	120.7	131.5
of which:					
Developed market economies	145.9	165.9	212.9	113.7	128.3
of which:					
North America	51.5	62.3	87.2	121.0	140.0
Western Europe	68.1	70.9	79.4	104.1	112.0
Japan and others	26.3	32.7	46.3	124.3	141.6
Centrally planned economies	95.7	125.2	165.9	130.8	132.6
of which:					
European	82.6	119.6	158.5	144.8	132.6
Asian	13.1	5.6	7.4	42.8	132.1
Developing countries	16.4	20.4	30.9	124.4	151.5
of which:					
African	1.6	1.7	1.4	106.3	82.4
Asian	5.4	5.8	12.1	107.4	208.6
Latin America	9.4	12.9	17.4	137.2	134.9

Source: Sectoral Studies Series, No.15, vol.II, UNIDO, Vienna 1986 (based on United Nations Industrial Statistics).

a See table II.1.

Table III.12

Share of geographic regions in world production of metal products, 1975-1985

(Percentage)

Region	*1975*	*1980*	*1985*
Total world	100.0	100.0	100.0
of which:			
Developed market economies	56.5	53.3	52.0
of which:			
North America	19.9	20.0	21.3
Western Europe	26.4	22.8	19.4
Japan and others	10.2	10.5	11.3
Centrally planned economies	37.1	40.2	40.5
of which:			
European	32.1	38.4	38.7
Asian	5.0	1.8	1.8
Developing countries	6.4	6.5	7.5
of which:			
African	0.6	0.6	0.3
Asian	2.1	1.8	2.9
Latin America	3.7	4.1	4.3
ECE region	78.4	83.0	81.2

Source: As for table III.11.

III.3. MANUFACTURE OF NON-ELECTRICAL MACHINERY

The major group ISIC 382 still constitutes the core of the engineering industries. Its share in total engineering production - in terms of value added - accounted in 1986 for 50.2% in Czechoslovakia, 40.3% in Norway (1985), 38.3% in Finland (1985), 38.1% in Poland (1986) and 37.8% in Denmark (1985). The traditional west European suppliers of capital goods also recorded shares over 30%: France 34.8% in 1985, Italy 34.5% in 1984, Sweden 34.1% in 1985, United Kingdom 32.1% in 1985 and Federal Republic of Germany 31.1% in 1985.

The manufacture of non-electrical machinery includes, in particular:

- Various types of engines (except electrical) and turbines (steam, gas, hydraulic);
- Agricultural machinery and equipment;
- Metal and wood-working machinery (including machine tools, rolling mills, forges and presses etc.);
- Special machinery for chemical, food, textile, paper industries etc.;
- Office, computing and accounting equipment; and
- Other engineering equipment and machines (industrial furnaces, lifting and hoisting machinery, pumps and compressors, service industry machines etc.).

With the rapid development of automated manufacturing technologies, many countries have already included, in their statistical nomenclatures related to non-electrical machinery, numerically controlled machine tools, industrial robots and manipulators, automated machining centres and manufacturing lines, and even entire flexible manufacturing systems.

Tables III.13-III.18 illustrate the development of selected indicators characterizing the production of non-electrical machinery from 1979 to 1986, as was done for the metal products industry (see subchapter III.2). With only a few exceptions (e.g. Canada, Portugal, Luxembourg), the large majority of ECE member countries showed continuing growth of production in this traditional field of engineering, the largest in such countries as Ireland (with an index of industrial production of 258 in 1985, 1980 = 100), Bulgaria (187 in 1986), Cyprus (173 in 1985), Japan (135 in 1985), Czechoslovakia (135 in 1985), Finland (133 in 1985), and Norway (132 in 1986).

Amongst developing countries, Brazil and India and, to a lesser degree, Argentina and Mexico have significantly increased their self-sufficiency in non-electrical machinery. Most of the other developing countries, however, still rely strongly on imports from the ECE region and Japan.

Figures III.3 (a-d) illustrate, as was done in subchapter III.2 above for major group 381, the relationship between production, employment and productivity indices for the period 1979-1986 in non-electrical machinery industries of four "standard" countries, i.e. Czechoslovakia, the Federal Republic of Germany, Japan and the United States. In the United States, the development since 1983 seems to follow the "optimum model", i.e. the modest increase of the total production volume (influenced by highly specialized market demand) is accompanied by a significant productivity increase. Figures III.4 (a-d) show the evolution in four selected "additional" countries of the ECE region, i.e. Finland, Poland, Sweden and the United Kingdom.

Tables III.19-III.21 show the share of non-electrical machinery in total engineering production - in terms of value added, employment and capital formation. The non-electrical machinery branches have remained highly labour and capital intensive. With the introduction of high technologies (numerical controls, robotics, flexible manufacturing systems), the share of software costs in total installation costs of machining centres and integrated production systems grew significantly (research, design, computer programming, user training, etc.). [III.5]

The *Annual Review for 1986* [III.6] analyses the overall performance (including historical trends) of non-electrical machinery (ISIC 382) with emphasis on machine tools, FMS and robots. It shows, in particular, that world machine-tool production has remained at the 1980 level, indicating that markets for traditional machinery are more or less satisfied and that future growth will be oriented towards the production of machines with new functional capabilities and improved controls (including interfaces with other equipment). In the major producer countries, the machine-tool industry recorded a significant recovery in 1986, as may be seen from figure III.5. [III.7]

The ECE Working Party on Engineering Industries and Automation, as from its creation in 1980/1981, has recognized the basic role of automated manufacturing in the restructuring of the engineering industries. This is also the reason why the initial general assessment of the level of automation by individual branches was followed up by in-depth studies in two major, rapidly developing areas - industrial robotics and flexible manufacturing systems (FMS). During the period 1983-1986, two most successful seminars (FMS, Bulgaria 1984, and robotics, Czechoslovakia 1986) were organized by the Working Party and two major studies were issued in the corresponding fields. [III.8]

The Working Party compiles, on a regular basis, within the scope of its annual reviews, statistics on the diffusion of industrial robots; since 1986, this work has been carried out in close co-operation with the International Federation of Robotics (IFR). The world population of programmable (by means of software) robots overstepped the 130,000 unit level in 1986. However, while many countries of the ECE region, and Japan, accounted during the past years for annual growth rates of some 30-40% (e.g. in the Federal Republic of Germany, the programmable-robot population increased from 2,300 units in 1981 to 12,400 in 1986), some of the original forecasts (e.g. 1 million robots in use by 1990) have proved unrealistic (see table III.22).

The diffusion of machine tools (including those numerically controlled) and of industrial robots, and their integration with computing and controlling devices and automated handling equipment into automated manufacturing systems continued to dominate the general development in this subsector. However, the rate of application was slightly overestimated, regarding in particular the process of introduction of integrated systems (FMS, CIM) - the impact of high software costs, complicated machine-machine and man-machine interfaces, associated investments and demanding personnel retraining. The effective use of automated technologies remained the major challenge throughout the period under review.

Tables III.23 and III.24 illustrate the geographic distribution of the production of non-electrical machinery (ISIC 382). With the exception of developing countries (accounting for only 5.4% of 1985 total world production) and centrally planned economies of Asia (6.7%), all other regions recorded a decline in 1980-1985 compared with the period 1975-1980. The ECE region's share decreased from 78.4% in 1975 to 76.9 in 1980 and 74.7 in 1985.

Table III.13 Index of industrial production in non-electrical machinery industries (ISIC, Rev.2:382) in ECE member countries and Japan, 1979-1986

(1980=100)

COUNTRY	1979	1980	1981	1982	1983	1984	1985	1986
Austria	98	100	89	95	87	86	96	...
Belgium	105	100	97	92	91	101	109	105
Bulgaria a/	101	100	110	124	138	146	166	187
Cyprus	98	100	103	121	126	140	173	169
Czechoslovakia	96	100	104	109	115	122	129	135
Denmark	100	100	99	100	102	112	116	120
Finland	88	100	114	117	109	120	133	128
France	98	100	106	102	97	95	96	...
German Dem.Rep.	93	100	106	111	115	119	126	...
Germany,Fed.Rep. of	96	100	100	98	97	101	111	117
Greece	119	100	103	94	80	93	101	91
Hungary	107	100	107	114	113	113	115	117
Ireland	88	100	134	139	176	243	258	...
Italy	91	100	97	89	86	90	98	103
Luxembourg	100	100	90	87	73	85	100	93
Malta	69	100	83	82	100	105	...	...
Netherlands	100	100	100	95	93	99	103	104
Norway	95	100	104	105	102	118	128	132
Poland	100	100	91	91	98	105	114	124
Portugal	96	100	114	140	127	95	93	95
Romania a/ b/	91	100	101	105	112	121	129	...
Spain	90	100	108	98	88	91	98	104
Sweden	103	100	98	94	98	108	116	115
Switzerland c/	...	100	97	92	88	88	94	101
Turkey d/	...	...	100	102	112	119	108	109
USSR	...	100	...	...	...	...	...	...
United Kingdom	107	100	87	87	88	97	107	107
Yugoslavia	97	100	107	108	114	116	120	119
Canada e/	95	100	102	83	78	89	95	98
United States	99	100	105	94	96	115	118	115
Japan	91	100	104	105	108	124	135	133

Source : ECE/ITD - Data bank of the Engineering Industries and Technology Section.
Government replies to ECE questionnaires for the Annual Reviews of Engineering Industries and Automation.

a/ Including precision instruments (ISIC, Rev.2 : 385).
b/ Including transport equipment (ISIC, Rev.2 : 384).
c/ Including electrical machinery (ISIC, Rev.2 : 383).
d/ 1981=100.
e/ Excluding electronic computers and their accessories.

Table III.14 Value added in non-electrical machinery industries (ISIC, Rev.2:382) in ECE member countries and Japan, 1979-1986

(Percentage change) a/

COUNTRY	1980/1979	1981/1980	1982/1981	1983/1982	1984/1983	1985/1984	1986/1985	1986/1979
Austria	37.6	11.6	9.1	-22.2	21.9	...	...	...
Belgium	...	...	...	...	...	...	...	...
Bulgaria	...	...	...	...	...	...	...	...
Cyprus	18.5	40.3	16.0	-1.5	20.3	14.3	...	17.3 b/
Czechoslovakia	5.5	-15.1	1.7	7.4	-1.3	7.6	-2.4	0.2
Denmark	15.8	4.4	5.8	14.0	9.0	17.6	...	11.0 b/
Finland	17.9	21.0	15.4	-2.3	17.3	14.4	-5.8	10.7
France	11.9	22.6	15.1	6.2	8.4	6.6	...	11.7 b/
German Dem.Rep.	...	...	...	...	...	...	...	...
Germany,Fed.Rep. of	6.7	4.8	3.1	0.4	6.1	11.5	...	5.4 b/
Greece	...	19.6	...	...	...	...	...	...
Hungary	-26.0	14.5	18.1	7.0	8.8	12.1	8.0	5.0
Ireland	42.5	29.0	22.1	54.3	...	...	...	...
Italy	24.3	22.4	8.0	21.5	19.2	...	...	...
Luxembourg	40.8	18.3	6.7	-10.0	...	...	...	...
Malta	2.7	12.5	16.1	17.3	...	...	...	...
Netherlands	3.6	0.7	8.9	-0.4	7.3	-0.2	...	3.3 b/
Norway	5.6	20.8	10.0	-0.3	16.1	31.0	...	13.4 b/
Poland	1.8	5.9	...	15.7	22.5	26.6	...	...
Portugal	42.6	25.3	39.4	7.4	15.8	...	...	...
Romania	...	...	...	...	...	...	...	...
Spain	17.2	8.0	6.3	34.2	9.3	...	...	...
Sweden	9.6	6.8	0.7	18.7	14.2	41.0	-13.3	10.0
Switzerland	...	...	...	...	...	...	...	...
Turkey	92.3	61.9	44.2	13.2	59.3	...	...	...
USSR	...	...	...	...	...	...	...	...
United Kingdom	18.3	-2.8	4.7	2.5	14.7	17.9	3.9	8.2
Yugoslavia c/	30.0	42.8	40.1	41.8	59.9	93.4	...	50.1 b/
Canada	18.2	11.3	3.9	-4.7	21.0	...	...	...
United States	7.5	12.0	-8.2	-7.4	18.6	-1.9	...	3.0 b/
Japan	12.1	9.0	7.8	-4.1	11.2	17.9	...	8.8 b/

Source : ECE/ITD - Data bank of the Engineering Industries and Technology Section.
Year to year percentage changes and annual average increase 1986/1979 are calculated from Government replies to ECE questionnaires for the Annual Reviews of Engineering Industries and Automation.

a/ Based on values expressed in national currency at current prices.
b/ 1985/1979.
c/ Social product including turnover taxes.

Table III.15 Index of total number of persons engaged in non-electrical machinery industries (ISIC, Rev.2:382) in ECE member countries and Japan, 1979-1986

(1980 = 100)

COUNTRY	1979	1980	1981	1982	1983	1984	1985	1986
Austria a/	82	100	98	94	80	93	...	...
Belgium	102	100	93	89	84	85	84	...
Bulgaria b/	97	100	82	85	86	87	90	90
Cyprus	94	100	117	123	116	120	128	125
Czechoslovakia a/ c/	98	100	102	102	104	105	106	107
Denmark	99	100	93	89	90	97	108	106
Finland	90	100	101	100	93	92	90	83
France	99	100	99	99	96	92	89	...
German Dem.Rep. d/	.	.	.	.	.	.	.	...
Germany, Fed.Rep. of	99	100	100	98	94	92	94	98
Greece	...	100	98	...	...	...	...	...
Hungary	102	100	99	102	99	96	97	96
Ireland	89	100	99	98	101	101	104	...
Italy	100	100	98	93	103	124	...	...
Luxembourg	92	100	102	98	93	...	...	...
Malta	96	100	103	96	83	...	...	...
Netherlands	100	100	97	96	89	89	93	...
Norway	104	100	105	105	105	110	125	...
Poland	99	100	99	94	92	90	90	90
Portugal	90	100	103	112	99	87	...	...
Romania	...	...	...	...	...	...	...	...
Spain	103	100	91	84	108	103	...	...
Sweden	104	100	98	93	90	91	92	92
Switzerland e/	97	100	101	103	99	96	98	...
Turkey	92	100	106	113	101	101	...	...
USSR	...	...	...	...	...	...	...	...
United Kingdom f/	99	100	88	80	74	90	90	88
Yugoslavia	97	100	104	108	111	113	117	...
Canada	94	100	101	106	98	102	...	...
United States	100	100	100	90	82	88	87	83
Japan	103	100	102	103	107	110	116	...

Source : ECE/ITD - Data bank of the Engineering Industries and Technology Section.
Government replies to ECE questionnaires for the Annual Reviews of Engineering Industries and Automation.

a/ Including homeworkers.
b/ Including precision instruments (ISIC, Rev.2 : 385).
c/ State national industry only.
d/ Included with metal products (ISIC, Rev.2 : 381).
e/ Including electrical machinery (ISIC. Rev.2 : 383).
f/ At June of each year.

Table III.16 Total number of persons engaged in non-electrical machinery industries (ISIC, Rev.2:382) in ECE member countries and Japan, 1979-1986

(Percentage change)

COUNTRY	1980/1979	1981/1980	1982/1981	1983/1982	1984/1983	1985/1984	1986/1985	1986/1979
Austria a/	22.1	-2.5	-3.8	-14.4	15.3	...	...	...
Belgium	-1.9	-6.8	-4.0	-5.8	1.3	-1.1	...	-3.1 b/
Bulgaria c/	3.6	-17.8	3.6	1.2	1.4	3.3	-0.7	-1.1
Cyprus	6.3	16.9	5.4	-6.0	3.9	5.9	-2.0	4.1
Czechoslovakia a/ d/	1.6	1.6	0.8	1.9	0.7	0.9	1.3	1.3
Denmark	1.4	-7.2	-3.7	0.2	8.6	10.6	-1.1	1.1
Finland	10.7	0.9	-0.9	-6.9	-1.6	-2.0	-8.0	-1.3
France	0.6	-0.8	-0.3	-3.1	-4.0	-2.8	...	-1.7 b/
German Dem.Rep. e/	...	...	...	...	...	...	...	...
Germany,Fed.Rep. of	1.4	0.3	-2.5	-4.1	-1.9	2.7	4.1	-0.1
Greece	...	-2.1	...	...	...	...	...	...
Hungary	-2.0	-0.6	2.7	-3.1	-2.8	1.1	-1.6	-0.9
Ireland	12.6	-1.5	-0.8	3.1	0.0	3.0	...	2.6 b/
Italy	-0.3	-2.0	-5.4	11.3	19.7	...	...	...
Luxembourg	8.5	1.9	-3.6	-4.9	...	...	...	...
Malta	3.8	3.3	-7.1	-13.8	...	...	...	...
Netherlands	-0.2	-3.4	-0.7	-7.2	0.1	4.7	...	-1.2 b/
Norway	-3.6	5.3	-0.3	-0.3	4.8	13.7	...	3.1 b/
Poland	0.7	-0.5	-5.9	-2.3	-1.2	-0.8	0.2	-1.4
Portugal	11.7	2.9	8.9	-11.6	-11.8	...	...	...
Romania	...	...	...	...	...	...	...	...
Spain	-2.7	-9.2	-7.8	28.6	-4.9	...	...	...
Sweden	-4.0	-2.2	-5.3	-3.4	1.7	0.8	-0.2	-1.8
Switzerland f/	3.4	0.6	1.9	-3.3	-3.5	2.2	...	0.2 b/
Turkey	8.5	5.5	6.6	-10.2	0.2	...	...	...
USSR	...	...	...	...	...	...	...	...
United Kingdom g/	0.8	-11.7	-9.9	-6.9	20.9	1.1	-2.6	-1.7
Yugoslavia	2.9	3.9	3.8	3.1	2.0	3.0	...	3.1 b/
Canada	6.1	0.8	4.9	-7.8	5.0	...	...	...
United States	0.4	0.2	-10.2	-9.4	8.1	-0.7	-4.5	-2.5
Japan	-2.6	2.5	0.8	3.9	2.8	4.7	...	2.0 b/

Source : ECE/ITD - Data bank of the Engineering Industries and Technology Section.
Year to year percentage changes and annual average increase 1986/1979 are calculated from Government replies to ECE questionnaires for the Annual Reviews of Engineering Industries and Automation.

a/ Including homeworkers.
b/ 1985/1979.
c/ Including precision instruments (ISIC, Rev.2 : 385).
d/ State national industry only.
e/ Included with metal products (ISIC, Rev.2 : 381).
f/ Including electrical machinery (ISIC, Rev.2 : 383).
g/ At June of each year.

Table III.17 Index of productivity in non-electrical machinery industries (ISIC, Rev.2:382) in ECE member countries and Japan, 1979-1986 a/

(1980 = 100)

COUNTRY	1979	1980	1981	1982	1983	1984	1985	1986
Austria	120	100	91	101	108	93	...	...
Belgium	103	100	104	103	108	119	129	...
Bulgaria b/	105	100	134	146	160	167	184	209
Cyprus	104	100	88	98	109	117	136	135
Czechoslovakia	98	100	102	107	110	116	122	126
Denmark	101	100	107	112	114	115	108	113
Finland	97	100	113	117	117	131	148	155
France	99	100	107	103	101	103	107	...
German Dem.Rep.	...	...	...	...	...	...	...	...
Germany,Fed.Rep. of	97	100	100	100	103	110	118	119
Greece	...	100	105	...	...	...	...	...
Hungary	105	100	107	112	115	118	118	122
Ireland	99	100	136	142	175	241	249	...
Italy	91	100	99	96	84	73	...	...
Luxembourg	109	100	89	88	78	...	...	...
Malta	72	100	80	86	121	...	...	...
Netherlands	100	100	104	99	105	111	110	...
Norway	92	100	99	100	97	108	103	...
Poland	101	100	91	97	107	116	127	138
Portugal	107	100	111	125	128	109	...	...
Romania	...	...	...	...	...	...	...	...
Spain	88	100	119	117	82	88	...	...
Sweden	99	100	100	101	110	119	127	125
Switzerland c/	...	100	96	90	89	92	96	...
Turkey d/	...	...	100	96	117	124	...	...
USSR	...	...	...	...	...	...	...	...
United Kingdom	108	100	98	110	119	108	119	121
Yugoslavia	100	100	103	100	102	102	103	...
Canada e/	101	100	101	78	80	87	...	...
United States	99	100	105	104	118	131	135	138
Japan	89	100	101	102	101	113	117	...

Source : ECE/ITD - Data bank of the Engineering Industries and Technology Section. Calculated on the basis of indices of industrial production and indices of total number of persons engaged. See tables III.13 and III.15.

a/ Productivity expressed in terms of labour productivity.
b/ Including precision instruments (ISIC, Rev.2 : 385).
c/ Including electrical machinery (ISIC, Rev.2 : 383).
d/ 1981=100.
e/ Indices of industrial production exclude electronic computers and their accessories.

Table III.18 Gross fixed capital formation in non-electrical machinery industries (ISIC, Rev.2:382) in ECE member countries and Japan, 1979-1986

(Percentage change) a/

COUNTRY	1980/1979	1981/1980	1982/1981	1983/1982	1984/1983	1985/1984	1986/1985	1986/1979
Austria	28.6	12.5	-0.3	-17.6	9.9	...	...	...
Belgium	...	...	...	...	...	...	...	...
Bulgaria b/ c/	94.6	-59.3	127.6	66.5	1.6	18.5	...	23.9 d/
Cyprus	-13.5	27.2	2.3	-21.3	28.6	44.4	84.6	17.2
Czechoslovakia	...	...	...	...	...	...	...	...
Denmark	21.5	-5.1	-0.5	9.7	23.6	60.9	...	16.5 d/
Finland	76.8	13.4	23.9	-20.4	9.7	19.9	-11.2	12.7
France c/	28.6	2.8	-2.7	5.6	8.7	20.4	...	10.1 d/
German Dem.Rep. e/	...	...	...	...	...	...	...	...
Germany,Fed.Rep. of	15.5	-2.0	-4.1	5.2	1.0	16.2	...	5.0 d/
Greece	...	-21.2	...	...	...	...	...	...
Hungary	-21.9	-16.0	-17.9	-13.2	0.9	-20.5	5.6	-12.4
Ireland	20.7	34.9	37.6	-17.4	...	...	...	...
Italy	19.7	21.3	-4.2	56.7	27.6	...	...	...
Luxembourg	...	...	...	...	...	...	...	...
Malta	-71.0	20.3	-40.5	22.7	...	...	...	...
Netherlands	9.5	-18.8	7.1	-5.9	44.4	30.4	8.3	9.0
Norway	-24.2	25.4	7.3	-41.7	134.2	54.9	...	13.7 d/
Poland	-22.0	-20.6	...	13.2	12.4	-4.3	...	...
Portugal	161.6	61.1	30.6	-31.1	46.4	...	...	...
Romania	...	...	...	...	...	...	...	...
Spain	-11.1	46.9	-9.2	54.6	-4.6	...	...	...
Sweden	31.8	19.0	-16.5	6.9	30.3	37.6	7.4	15.2
Switzerland	...	...	...	...	...	...	...	...
Turkey	-39.3	307.9	-30.6	64.7	37.6	...	...	...
USSR	...	...	...	...	...	...	...	...
United Kingdom	-4.0	-13.0	2.6	3.7	28.2	21.4	17.3	7.1
Yugoslavia	9.5	45.9	18.3	30.1	29.6	80.3	...	33.8 d/
Canada	38.5	15.5	-2.4	-23.1	34.6	...	...	...
United States	10.2	14.1	-2.5	4.7	14.1	3.7	-13.8	3.9
Japan	37.2	16.1	13.9	-0.8	20.0	26.5	...	18.2 d/

Source : ECE/ITD - Data bank of the Engineering Industries and Technology Section.
Year to year percentage changes and annual average increase 1986/1979 are calculated from Government replies to ECE questionnaires for the Annual Reviews of Engineering Industries and Automation.

a/ Based on values expressed in national currency at current prices.
b/ Excluding land.
c/ Including precision instruments (ISIC, Rev.2 : 385).
d/ 1985/1979.
e/ Included with metal products (ISIC, Rev.2 : 381).

Figure III.3 (a-d)

Development of production, employment and productivity indices in the non-electrical machinery industries (ISIC 382) of four (standard) countries, 1979-1986 [a]

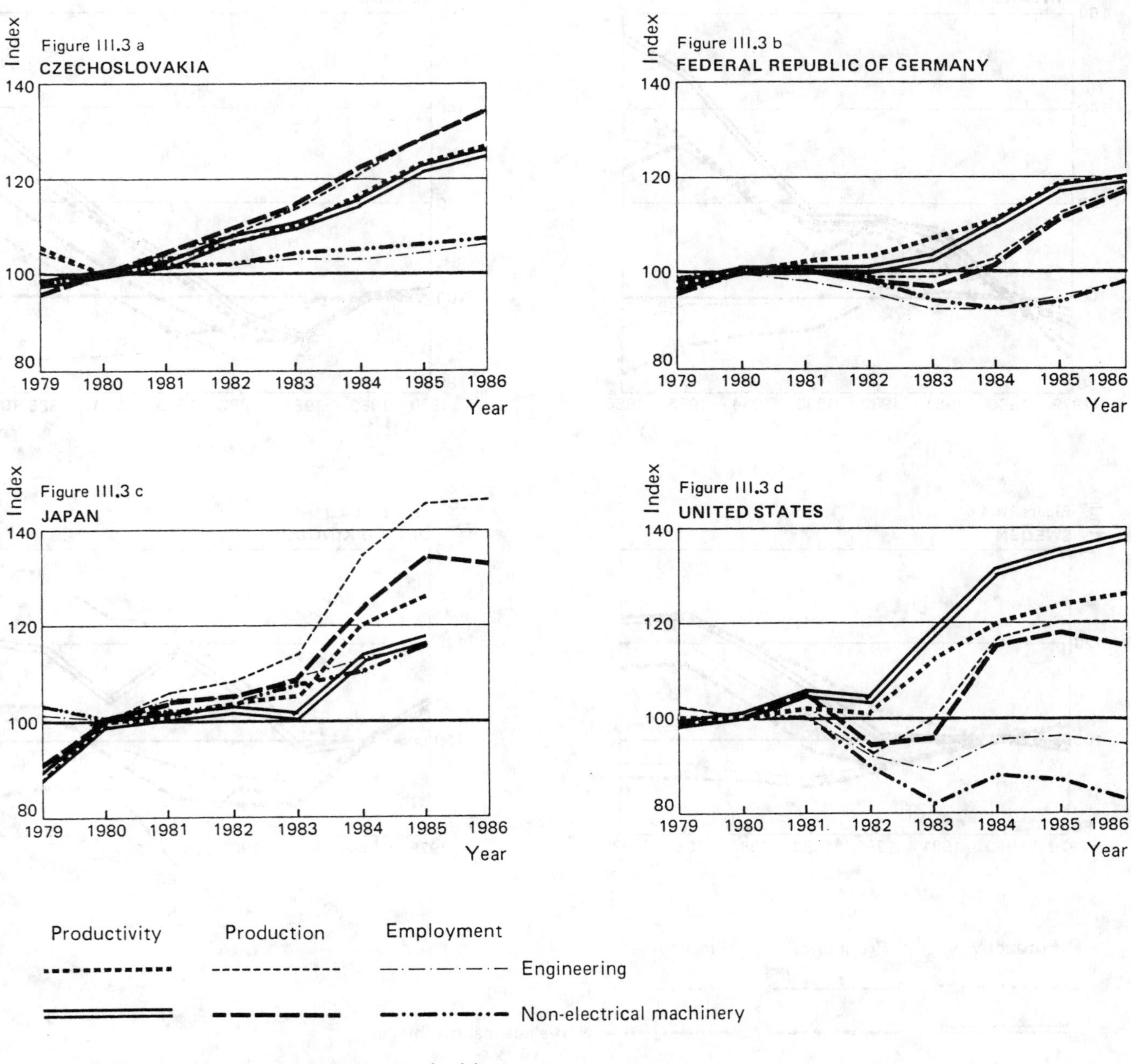

a Productivity expressed in terms of labour productivity. 1980 = 100.

Figure III.4 (a-d)

Development of production, employment and productivity indices in the non-electrical machinery industries (ISIC 382) of four (additional) countries, 1979-1986 [a]

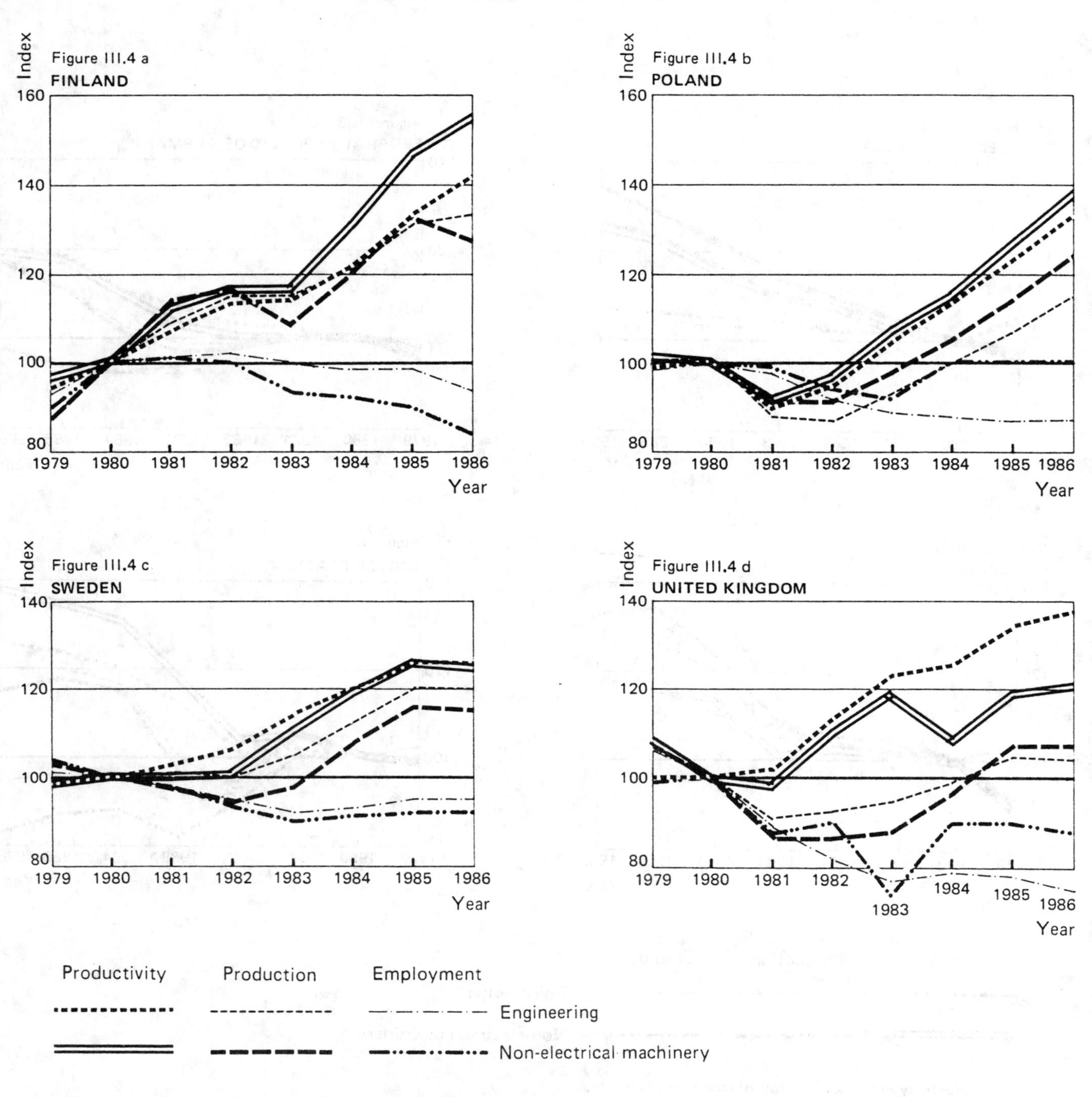

a Productivity expressed in terms of labour productivity. 1980 = 100.

Table III.19 Share of value added of non-electrical machinery industries (ISIC, Rev.2:382) in engineering industries (ISIC, Rev.2:38) in ECE member countries and Japan, 1979-1986

(Percentage) a/

COUNTRY	1979	1980	1981	1982	1983	1984	1985	1986
Austria	24.2	30.9	31.6	32.5	24.4	28.4	...	...
Belgium	...	...	...	...	...	...	...	...
Bulgaria	...	...	...	...	...	...	...	...
Cyprus	24.9	23.9	26.9	27.0	26.2	27.9	28.4	...
Czechoslovakia	50.0	50.3	50.3	49.7	49.5	48.5	50.2	50.2
Denmark	38.9	39.8	39.3	36.1	36.7	36.1	37.8	...
Finland	38.2	38.1	39.2	37.8	35.2	37.7	38.3	35.7
France	32.3	32.2	34.8	35.2	34.7	34.8	34.8	...
German Dem.Rep.	...	...	...	...	...	...	...	...
Germany,Fed.Rep. of	31.9	31.0	31.1	30.7	30.2	30.6	31.1	...
Greece	...	10.0	9.3	...	...	...	...	...
Hungary	24.2	23.4	23.7	24.9	25.0	25.0	24.4	23.8
Ireland	28.7	32.2	33.4	33.8	40.3	...	...	...
Italy	25.7	26.1	27.0	26.3	33.5	34.5	...	...
Luxembourg	57.3	61.4	64.6	64.6	57.7	...	...	...
Malta	8.5	8.7	8.5	8.5	10.0	...	...	...
Netherlands b/	25.3	24.3	24.0	24.8	23.9	24.9	23.8	...
Norway	29.8	29.1	30.4	31.3	32.4	35.0	40.3	...
Poland	37.0	36.9	38.7	36.2	36.2	36.1	37.3	38.1
Portugal	12.7	13.6	14.4	17.3	15.1	15.3	...	...
Romania	...	...	...	...	...	...	...	...
Spain	18.2	18.5	19.1	18.2	22.1	23.7	...	...
Sweden	30.2	30.6	29.9	28.9	29.3	29.2	34.1	29.2
Switzerland	...	...	...	...	...	...	...	...
Turkey	24.1	26.5	27.7	27.4	24.2	25.0	...	...
USSR	...	...	...	...	...	...	...	...
United Kingdom	28.9	32.1	30.8	30.1	29.2	30.4	32.1	31.9
Yugoslavia c/	26.6	25.8	25.0	25.4	25.9	26.0	24.6	...
Canada	18.9	21.0	20.5	22.8	20.1	20.1	...	...
United States	28.8	29.7	30.3	28.1	24.7	25.3	24.5	...
Japan	28.0	28.3	27.6	28.2	27.8	27.7	27.5	...

Source : ECE/ITD - Data bank of the Engineering Industries and Technology Section. Government replies to ECE questionnaires for the Annual Reviews of Engineering Industries and Automation.

a/ Based on values expressed in national currency at current prices.
b/ In 1979, engineering industries (ISIC, Rev.2 : 38) exclude precision instruments (ISIC, Rev.2 : 385).
c/ Social product including turnover taxes.

Table III.20 Share of total number of persons engaged in non-electrical machinery industries (ISIC, Rev.2:382) in total number engaged in engineering industries (ISIC, Rev.2:38) in ECE member countries and Japan, 1979-1986

(Percentage)

COUNTRY	1979	1980	1981	1982	1983	1984	1985	1986
Austria a/	23.7	28.5	28.5	28.4	25.0	28.5	...	...
Belgium	18.4	18.2	18.0	18.1	17.5	18.1	18.1	...
Bulgaria b/	48.0	48.7	39.9	40.7	40.3	39.8	40.3	41.3
Cyprus	23.9	23.5	25.9	26.7	25.6	25.8	25.1	25.5
Czechoslovakia a/ c/	46.5	46.9	47.1	47.0	47.4	47.7	47.6	47.6
Denmark	38.2	38.6	38.0	36.5	36.7	37.7	37.7	38.0
Finland	37.0	38.2	38.1	37.4	35.4	35.4	34.7	33.6
France	28.4	28.8	29.4	29.6	29.3	29.2	29.4	...
German Dem.Rep. d/	...	...	...	...	...	...	...	...
Germany,Fed.Rep. of	29.5	29.5	30.1	30.1	30.0	29.6	29.4	29.5
Greece	...	11.8	11.2	...	...	...	...	...
Hungary	24.8	25.0	25.5	26.5	27.0	26.6	26.1	25.8
Ireland	19.6	21.1	20.7	20.6	22.3	23.0	24.3	...
Italy	22.8	23.0	23.5	23.6	25.9	31.2	...	...
Luxembourg	57.1	58.3	59.5	58.8	56.4	...	...	...
Malta	11.5	12.4	12.6	12.3	11.0	...	...	...
Netherlands	23.4	23.4	23.4	24.0	23.7	23.7	24.1	...
Norway	26.6	25.7	27.1	27.5	29.9	31.5	34.9	...
Poland	34.7	35.0	35.5	35.9	35.9	35.9	36.2	36.4
Portugal	15.1	16.6	16.6	18.0	16.7	15.7	...	...
Romania	...	...	...	...	...	...	...	...
Spain	18.2	18.1	17.6	16.9	22.2	22.6	...	...
Sweden	30.3	29.3	29.2	28.6	28.6	28.5	28.4	28.3
Switzerland e/	59.6	60.1	60.1	61.9	62.7	63.0	62.8	...
Turkey	26.5	28.8	28.8	29.5	28.2	27.3	...	...
USSR	...	...	...	...	...	...	...	...
United Kingdom f/	27.0	28.9	28.7	28.1	27.8	32.8	33.6	33.8
Yugoslavia	25.7	25.7	25.4	25.4	25.6	25.8	25.8	...
Canada	19.1	20.4	20.3	23.8	23.1	22.9	...	...
United States	27.3	28.3	28.4	27.6	25.9	25.9	25.5	24.8
Japan	26.5	26.0	25.3	25.6	25.7	25.5	25.9	...

Source : ECE/ITD - Data bank of the Engineering Industries and Technology Section.
Government replies to ECE questionnaires for the Annual Reviews of Engineering Industries and Automation.

a/ Including homeworkers.
b/ Non-electrical machinery (ISIC, Rev.2 : 382) includes precision instruments (ISIC, Rev.2 : 385).
c/ State national industry only.
d/ Non-electrical machinery (ISIC, Rev.2 : 382) is included with metal products (ISIC, Rev.2 : 381).
e/ Non-electrical machinery (ISIC, Rev.2 : 382) includes electrical machinery (ISIC, Rev.2 : 383).
f/ At June of each year.

Table III.21 Share of gross fixed capital formation of non-electrical machinery industries (ISIC, Rev.2:382) in engineering industries (ISIC, Rev.2:38) in ECE member countries and Japan, 1979-1986

(Percentage) a/

COUNTRY	1979	1980	1981	1982	1983	1984	1985	1986
Austria	22.9	23.5	16.8	20.7	23.1	25.4	...	...
Belgium	...	...	...	...	...	...	...	...
Bulgaria b/ c/	54.7	44.4	31.6	54.7	55.8	65.7	65.7	...
Cyprus	22.8	25.0	22.7	29.9	28.0	23.1	20.3	42.1
Czechoslovakia	...	...	...	...	...	...	...	...
Denmark	44.0	42.5	43.3	37.8	38.7	36.9	40.6	...
Finland	31.6	36.0	38.1	37.8	29.2	29.4	29.7	29.5
France c/	17.2	18.1	18.8	17.3	17.7	15.5	15.5	...
German Dem.Rep. d/	...	...	...	...	...	...	...	...
Germany,Fed.Rep. of	29.9	29.5	29.0	26.3	26.7	27.3	25.5	...
Greece	...	6.0	3.9	...	...	...	...	...
Hungary	28.6	26.4	25.4	19.8	16.8	17.6	16.4	16.7
Ireland	34.1	29.7	36.3	40.2	43.6	...	...	...
Italy	22.7	22.8	24.4	19.1	24.8	28.4	...	...
Luxembourg	...	...	...	...	...	...	...	...
Malta	11.5	10.9	7.0	2.6	1.8	...	...	...
Netherlands	21.4	19.7	18.0	19.1	19.0	19.9	19.2	18.6
Norway	34.4	26.1	28.2	30.2	21.3	40.3	46.4	...
Poland	52.1	50.8	51.0	46.9	45.5	45.5	42.3	...
Portugal	7.1	13.2	15.1	12.9	11.3	15.9	...	...
Romania	...	...	...	...	...	...	...	...
Spain	13.2	12.8	16.7	15.8	20.1	11.8	...	...
Sweden	27.4	28.0	29.8	26.1	24.6	23.4	26.6	24.8
Switzerland	...	...	...	...	...	...	...	...
Turkey	35.5	28.5	49.4	24.1	26.9	20.0	...	...
USSR	...	...	...	...	...	...	...	...
United Kingdom	24.9	24.7	24.2	25.5	24.8	25.6	29.4	32.4
Yugoslavia	24.1	20.4	24.7	23.9	25.0	22.9	22.7	...
Canada	15.1	12.5	14.3	19.7	14.6	17.9	...	...
United States	29.1	27.4	29.5	31.2	34.6	31.0	29.3	26.8
Japan	22.2	21.3	20.0	21.0	21.9	21.3	22.6	...

Source : ECE/ITD - Data bank of the Engineering Industries and Technology Section.
Government replies to ECE questionnaires for the Annual Reviews of Engineering Industries and Automation.

a/ Based on values expressed in national currency at current prices.
b/ Excluding land.
c/ Non-electrical machinery (ISIC, Rev.2 : 382) includes precision instruments (ISIC, Rev.2 : 385).
d/ Non-electrical machinery (ISIC, Rev.2 : 382) is included with metal products (ISIC, Rev.2 : 381).

Figure III.5

Machine-tool shipments from the nine leading producer countries, 1973-1986

(Millions of US dollars at current prices)

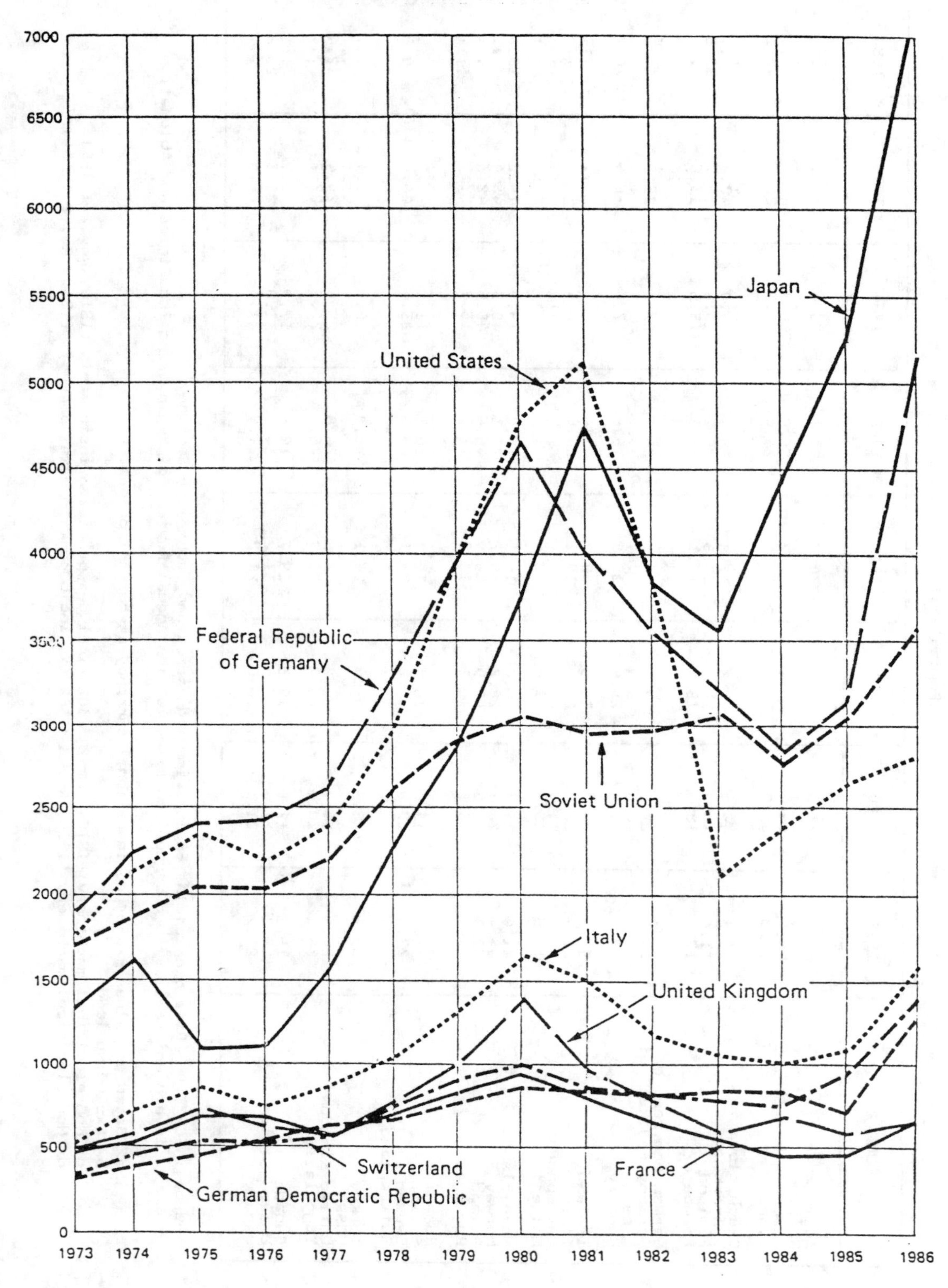

Source : Annual Review of Engineering Industries and Automation, 1986, vol.1 (UN/ECE publication, Sales No. E.88.II.E.16). Based on *American Machinist*, February 1987.

Table III.22

ECE estimate of the growth of the world-wide programmable-robot population, 1981-1986

Region	Total number of units						
	1981	*1982*	*1983*	*1984*	*1985*	*1986*	*1986 share of total population (%)*
Asia (including Japan) [a]	11 500	13 500	16 500	23 500	30 000	42 000	32.3
North America [b]	5 500	7 000	8 500	15 500	21 500	26 500	20.4
Western Europe [b]	6 500	9 000	13 000	20 000	25 500	33 000	25.4
Eastern Europe (including USSR) [c]	6 000	8 500	10 500	16 000	20 000	24 500	18.8
Other	500	1 000	1 500	2 500	3 000	4 000	3.1
Total	30 000	39 000	50 000	77 500	100 000	130 000	100.0

Source: *Annual Review of Engineering Industries and Automation 1983-1984, 1985 and 1986. Production and Use of Industrial Robots.* UN/ECE publication, Sales No. E.84.II.E.33.

a ECE estimate in accordance with the ECE/ISO definition of programmable robot. According to the JIRA definition, the robot population in Japan (1986) attained some 90,000 units.

b Based on data supplied by national robotics associations as collected by IFR.

c ECE rough estimate, in some countries it might include simple robots and manipulators (e.g. pick and place).

Table III.23

Production of non-electrical machinery (ISIC 382) by geographic region, 1975-1985

Region	Value in billions of constant 1975 US dollars [a]			Index	
	1975	*1980*	*1985*	*1980/1975*	*1985/1980*
Total world	400.0	528.0	660.0	132.0	125.0
of which:					
Developed market economies	243.7	307.2	365.5	126.1	119.0
of which:					
North America	83.3	108.7	130.8	130.5	120.3
Western Europe	117.9	133.1	147.7	112.9	111.0
Japan and others	42.5	65.4	87.0	153.9	133.0
Centrally planned economies	138.7	197.9	259.0	142.7	130.9
of which:					
European	112.5	164.3	214.5	146.0	130.6
Asian	26.2	33.6	44.5	128.2	132.4
Developing countries	17.2	23.0	35.5	133.7	154.3
of which:					
African	0.5	0.6	0.5	120.0	83.3
Asian	5.7	7.7	12.2	135.1	158.4
Latin America	11.0	14.7	22.8	133.6	155.1

Source: As for table III.11.

a See table II.1.

Table III.24

Share of geographic regions in world production of non-electrical machinery, 1975-1985

(Percentage)

Region	*1975*	*1980*	*1985*
Total world	100.0	100.0	100.0
of which:			
Developed market economies	60.9	58.2	55.4
of which:			
North America	20.8	20.6	19.8
Western Europe	29.5	25.2	22.4
Japan and others	10.6	12.4	13.2
Centrally planned economies	34.7	37.5	39.2
of which:			
European	28.1	31.1	32.5
Asian	6.6	6.4	6.7
Developing countries	4.3	4.4	5.4
of which:			
African	0.1	0.1	0.1
Asian	1.4	1.5	1.8
Latin America	2.8	2.8	3.5
ECE region	78.4	76.9	74.7

Source: As for table III.11.

III.4. MANUFACTURE OF ELECTRICAL MACHINERY

Manufacture of electrical machinery (ISIC major group 383), despite the continuously decreasing prices of electronics-based products, had, by the end of 1985, already reached nearly 90% of the world production in the traditionally dominant non-electrical machinery (compared with less than 75% in 1975 and some 82% in 1980). Its share - in terms of value added - in the total engineering production of several countries has exceeded 30%, e.g. in Malta it accounted for 44.4% (1982), in the Netherlands 39.4% (1985) in Portugal 36.9% (1984), in Japan 34.5% (1984), in Hungary 32.9% (1986), and in Austria 32.3% (1984).

The major group ISIC 383 includes the following groups of electrical and electronics products:

- Electrical machinery and apparatus for use in industry, e.g. electrical motors, generators, transformers, electrical transmission and distribution equipment, welding apparatus etc.;
- Radio, television and communication equipment, radar equipment, X-ray apparatus (including electronic components and parts used in products classified in this group) etc.;
- Electrical appliances and housewares; and
- Other apparatus and supplies (wires and cables, batteries, lighting equipment etc.).

Tables III.25-III.30 show the development of selected indicators characterizing the growth of the electrical machinery industry from 1979 to 1986, as was done in the preceding subchapters.

All countries of the region, with the exception only of Belgium, recorded significant growth of production during the period 1979-1986, the largest in Turkey, with an index of industrial production of 328 in 1986 (1981 = 100), Japan (222 in 1986, 1980 = 100 as also for all the following countries), Ireland (184 in 1986), Cyprus (181 in 1985), Bulgaria (164 in 1985), Czechoslovakia (163 in 1986) and Finland (152 in 1986).

The ongoing structural development within the electrical and electronics engineering subsector is analysed in a recently published ECE study. [III.9] The study demonstrated, among other things, the priority growth in the fields of professional and consumer electronics, including the wide range of components and parts used in practically all types of machinery and equipment. The dominant producers of electronic components remained the United States, followed by Japan. The rate of development in the United States electronic components industry (1979-1986) is shown in table III.31.

A further rapidly growing area is telecommunications, the subject of another recent ECE study. [III.10] Entirely new concepts were introduced during the period under review, both in switching technologies (network digitalization) and in transmission systems (optical fibres, telecommunication satellites). The traditional indicator, the number of telephone sets per capita, recorded significant growth between 1975 and 1984 (see table III.32).

The priority development in the fields of communication equipment and the electronic components industry in the United States (1975-1985) may also be seen from table III.33; in this respect one should note in particular the dramatic increase in the production of semiconductors.

Since 1983, when an ECE seminar on innovation in biomedical engineering was held in Hungary, the Working Party has included in its programme of work the continuous study of developments in electromedical equipment. The recent study on digital imaging in health care drew attention to the enormous research and application efforts in this branch, characterized by important economic and social implications. [III.11] In this respect, the ECE seminar held in Czechoslovakia in 1987 demonstrated further the decisive role of automation in the improvement of health-care services. [III.12]

Figures III.6 (a-d) show the relationship of production, employment and productivity indices in electrical machinery (ISIC 383) in four "standard" countries in 1979-1986. Figures III.7 (a-d) illustrate, similarly, the development in four "additional" countries (Hungary, Netherlands, Poland and the United Kingdom). It seems that the high annual production growth rates (e.g. Czechoslovakia and Japan) are accompanied by increased labour input (the process of automation is slower here than in some other engineering branches). In other words, the productivity level achieved so far is still relatively low (except in several countries such as Hungary and the United Kingdom).

Tables III.34-III.36 reveal the share of electrical machinery (ISIC 383) in total engineering - in terms of value added, employment and gross fixed capital formation - in the years 1979-1985. The largest share of capital investments was registered in electronics-oriented countries: Netherlands with a 47.4% share of total engineering investments in 1986, Japan 45.1% in 1984, France 37.7% in 1984, Hungary 36.1% in 1984 and the United States 29.4% in 1984.

Tables III.37-III.38 present the geographic distribution of electrical machinery production (ISIC 383). In comparison with the period 1975-1980 (characterized e.g. by a strong increase of production in Japan), the period 1980-1985 was marked by a relative slow-down. Only the regions of North America (owing to the expansion of telecommunication and electronic components industries in the United States) and of Asia (both centrally planned and developing countries) showed important growth. The ECE region's share stabilized at the level of 76.2% of total world production.

Table III.25 Index of industrial production in electrical machinery industries (ISIC, Rev.2:383) in ECE member countries and Japan, 1979-1986

(1980=100)

COUNTRY	1979	1980	1981	1982	1983	1984	1985	1986
Austria	96	100	106	109	112	121	132	...
Belgium	102	100	93	93	96	93	94	98
Bulgaria	82	100	107	114	127	147	164	152
Cyprus	81	100	159	190	198	170	181	149
Czechoslovakia	95	100	107	112	121	136	149	163
Denmark	97	100	94	97	105	120	133	145
Finland	85	100	102	104	108	114	134	152
France	96	100	101	105	106	112	114	...
German Dem.Rep. a/ b/	92	100	111	118	129	140	155	...
Germany,Fed.Rep. of	96	100	98	98	99	107	121	127
Greece	86	100	103	92	93	91	105	105
Hungary	106	100	108	113	115	122	128	133
Ireland	84	100	111	120	135	177	183	184
Italy	90	100	95	103	104	108	109	113
Luxembourg b/	...	100	102	129	140	171	177	191
Malta	75	100	80	89	115	149	...	...
Netherlands	95	100	105	104	107	116	121	124
Norway	108	100	110	113	113	119	128	136
Poland	99	100	88	87	95	105	115	125
Portugal	86	100	101	116	119	108	155	131
Romania	89	100	102	99	100	107	116	...
Spain	101	100	102	92	89	92	92	109
Sweden	93	100	108	108	115	124	133	130
Switzerland c/	.	100	.	.	.	.	.	.
Turkey d/	...	...	100	99	136	205	268	328
USSR	...	100	...	...	...	...	...	...
United Kingdom	103	100	92	98	105	113	116	118
Yugoslavia	101	100	103	105	110	117	124	133
Canada e/	100	100	107	99	95	102	114	115
United States	96	100	103	99	110	131	129	128
Japan	86	100	114	123	147	193	213	222

Source : ECE/ITD - Data bank of the Engineering Industries and Technology Section.
Government replies to ECE questionnaires for the Annual Reviews of Engineering Industries and Automation.

a/ Including transport equipment (ISIC, Rev.2 : 384).
b/ Including precision instruments (ISIC, Rev.2 : 385).
c/ Included with non-electrical machinery (ISIC, Rev.2 : 382).
d/ 1981=100.
e/ Including electronic computers and their accessories.

Table III.26 Value added in electrical machinery industries (ISIC, Rev.2:383) in ECE member countries and Japan, 1979-1986

(Percentage change) a/

COUNTRY	1980/1979	1981/1980	1982/1981	1983/1982	1984/1983	1985/1984	1986/1985	1986/1979
Austria	8.7	10.8	5.7	5.2	11.7	...	...	...
Belgium	...	...	...	...	...	...	...	...
Bulgaria	...	...	...	...	...	...	...	...
Cyprus	42.2	48.8	27.9	-3.0	12.5	27.8	...	24.8 b/
Czechoslovakia	6.1	-7.9	5.2	6.6	8.9	3.7	0.6	3.2
Denmark	6.1	4.2	18.2	5.8	15.4	10.6	...	9.9 b/
Finland	22.4	15.6	16.6	8.4	6.0	17.8	28.6	16.3
France	13.3	9.2	12.2	8.3	11.4	7.4	...	10.3 b/
German Dem.Rep.	...	...	...	...	...	...	...	...
Germany,Fed.Rep. of	11.2	2.9	6.5	3.4	5.0	9.5	...	6.4 b/
Greece	...	28.6	...	...	...	...	...	...
Hungary	-21.7	15.9	10.5	10.1	8.6	16.6	12.1	6.6
Ireland	31.0	36.6	35.1	25.2	...	...	...	...
Italy	19.7	14.2	13.9	0.5	9.8	...	...	...
Luxembourg	15.1	-0.2	4.0	6.9	...	...	...	...
Malta	33.5	-1.9	63.8	-3.6	...	...	...	...
Netherlands	1.3	7.9	1.8	13.4	8.2	4.9	...	6.2 b/
Norway	1.4	28.4	5.6	4.8	4.5	6.6	...	8.2 b/
Poland	3.1	1.2	...	18.5	25.6	24.1	...	...
Portugal	45.9	18.0	30.2	37.4	32.3	...	...	...
Romania	...	...	...	...	...	...	...	...
Spain	10.7	4.4	10.3	15.2	0.3	...	...	...
Sweden	7.3	21.0	5.7	19.4	10.2	-1.7	1.9	8.8
Switzerland	...	...	...	...	...	...	...	...
Turkey	77.8	42.9	39.7	41.5	78.7	...	...	...
USSR	...	...	...	...	...	...	...	...
United Kingdom	24.8	3.2	10.7	7.5	13.9	7.5	5.9	10.3
Yugoslavia c/	36.0	46.3	37.1	39.4	65.0	129.2	...	56.0 b/
Canada	11.9	17.3	-6.4	-2.4	20.1	...	...	...
United States	10.0	9.0	6.1	9.4	19.2	-0.4	...	8.7 b/
Japan	14.4	14.7	7.9	5.2	19.9	11.4	...	12.2 b/

Source : ECE/ITD - Data bank of the Engineering Industries and Technology Section.
Year to year percentage changes and annual average increase 1986/1979 are calculated from Government replies to ECE questionnaires for the Annual Reviews of Engineering Industries and Automation.

a/ Based on values expressed in national currency at current prices.
b/ 1985/1979.
c/ Social product including turnover taxes.

Table III.27 Index of total number of persons engaged in electrical machinery industries (ISIC, Rev.2:383) in ECE member countries and Japan, 1979-1986

(1980 = 100)

COUNTRY	1979	1980	1981	1982	1983	1984	1985	1986
Austria a/	98	100	100	97	95	97	...	...
Belgium	102	100	95	90	88	87	85	...
Bulgaria	92	100	116	119	125	131	133	118
Cyprus	67	100	105	105	110	115	137	127
Czechoslovakia a/ b/	99	100	102	103	102	106	109	113
Denmark	104	100	95	94	93	98	110	108
Finland	93	100	100	97	95	96	103	103
France	99	100	99	99	98	97	95	...
German Dem.Rep. a/ c/	99	100	101	103	104	107	108	...
Germany,Fed.Rep. of	99	100	97	93	90	90	95	99
Greece	...	100	101	...	...	...	...	...
Hungary	103	100	97	95	92	91	94	93
Ireland	94	100	111	116	113	117	120	...
Italy	102	100	94	89	95	89	...	...
Luxembourg	93	100	94	86	86	...	...	...
Malta	112	100	92	107	98	...	...	...
Netherlands	99	100	96	96	94	94	100	...
Norway	99	100	99	100	98	94	95	...
Poland	98	100	99	94	92	91	90	90
Portugal	98	100	106	106	106	104	...	...
Romania	...	...	...	...	...	...	...	...
Spain	110	100	93	90	87	79	...	...
Sweden	101	100	101	98	93	96	97	99
Switzerland d/	.	.	.	.	.	.	.	...
Turkey	106	100	103	109	107	115	...	...
USSR	...	...	...	...	...	...	...	...
United Kingdom e/	97	100	89	83	82	78	76	72
Yugoslavia	95	100	105	109	111	113	119	...
Canada	98	100	105	89	84	90	...	...
United States	101	100	100	96	96	106	106	104
Japan	95	100	110	111	120	131	131	...

Source : ECE/ITD - Data bank of the Engineering Industries and Technology Section.
Government replies to ECE questionnaires for the Annual Reviews of Engineering Industries and Automation.

a/ Including homeworkers.
b/ State national industry only.
c/ Including precision instruments (ISIC, Rev.2 : 385) and manufacture of office, computing and accounting machinery.
d/ Included with non-electrical machinery (ISIC, Rev.2 : 382).
e/ At June of each year.

Table III.28 Total number of persons engaged in electrical machinery industries (ISIC, Rev.2:383) in ECE member countries and Japan, 1979-1986

(Percentage change)

COUNTRY	1980/1979	1981/1980	1982/1981	1983/1982	1984/1983	1985/1984	1986/1985	1986/1979
Austria a/	1.7	0.3	-3.1	-2.4	2.5	...	...	...
Belgium	-1.6	-4.9	-5.0	-2.8	-1.0	-2.4	...	-2.9 b/
Bulgaria	9.0	16.3	2.2	5.1	5.0	1.3	-11.3	3.6
Cyprus	49.7	5.5	-0.6	4.6	4.8	19.3	-7.6	9.6
Czechoslovakia a/ c/	0.7	2.0	0.7	-0.6	3.9	3.1	3.0	1.8
Denmark	-3.5	-5.2	-0.4	-1.7	5.6	11.9	-1.1	0.7
Finland	7.3	0.3	-3.5	-2.0	1.0	7.1	0.6	1.5
France	0.6	-1.4	0.0	-0.8	-0.6	-2.4	...	-0.8 b/
German Dem.Rep. a/ d/	0.9	0.9	1.8	1.6	2.7	0.9	...	1.5 b/
Germany, Fed.Rep. of	0.5	-2.8	-4.1	-3.9	0.5	5.1	4.2	-0.1
Greece	...	1.1	...	...	...	...	...	...
Hungary	-3.0	-3.4	-1.7	-3.3	-0.5	2.7	-1.3	-1.5
Ireland	5.9	11.2	4.4	-2.4	3.7	1.8	...	4.0 b/
Italy	-1.5	-6.3	-4.8	7.1	-6.3	...	...	...
Luxembourg	7.6	-6.2	-8.5	0.0	...	...	...	...
Malta	-10.7	-7.9	16.5	-8.3	...	...	...	...
Netherlands	0.7	-3.7	-0.1	-2.7	0.8	5.4	...	0.0 b/
Norway	1.0	-0.9	1.4	-2.3	-4.3	1.5	...	-0.6 b/
Poland	2.0	-0.7	-5.6	-2.1	-1.1	-0.4	0.0	-1.1
Portugal	1.6	5.8	0.6	-0.3	-1.8	...	...	...
Romania	...	...	...	...	...	...	...	...
Spain	-8.8	-7.1	-2.8	-3.7	-9.7	...	...	...
Sweden	-0.7	0.7	-2.5	-4.9	2.3	1.7	1.8	-0.3
Switzerland e/	.	.	.	.	.	.	...	.
Turkey	-5.8	2.6	6.4	-2.1	7.3	...	...	...
USSR	...	...	...	...	...	...	...	...
United Kingdom f/	3.2	-10.8	-6.5	-2.0	-4.7	-2.5	-4.7	-4.1
Yugoslavia	4.9	5.3	3.1	2.4	1.8	5.3	...	3.8 b/
Canada	2.5	4.8	-15.3	-5.4	7.6	...	...	...
United States	-1.2	0.1	-4.1	0.2	9.7	-0.0	-1.7	0.3
Japan	4.8	10.2	1.0	7.4	9.3	0.6	...	5.5 b/

Source : ECE/ITD - Data bank of the Engineering Industries and Technology Section.
Year to year percentage changes and annual average increase 1986/1979 are calculated from Government replies to ECE questionnaires for the Annual Reviews of Engineering Industries and Automation.

a/ Including homeworkers.
b/ 1985/1979.
c/ State national industry only.
d/ Including precision instruments (ISIC, Rev.2 : 385) and manufacture of office and accounting machinery.
e/ Included with non-electrical machinery (ISIC, Rev.2 :382).
f/ At June of each year.

Table III.29 Index of productivity in electrical machinery industries (ISIC, Rev.2:383) in ECE member countries and Japan, 1979-1986 a/

(1980 = 100)

COUNTRY	1979	1980	1981	1982	1983	1984	1985	1986
Austria	98	100	106	112	118	125	...	...
Belgium	100	100	97	103	110	107	111	...
Bulgaria	89	100	92	96	102	112	123	129
Cyprus	121	100	151	181	180	148	132	118
Czechoslovakia	96	100	105	109	118	129	136	144
Denmark	94	100	99	103	113	122	121	134
Finland	91	100	102	107	114	119	131	147
France	97	100	102	106	108	115	120	...
German Dem.Rep.	...	...	...	...	...	...	...	...
Germany,Fed.Rep. of	96	100	101	105	110	119	127	128
Greece	...	100	102	...	...	...	...	...
Hungary	103	100	112	119	125	133	137	144
Ireland	89	100	100	103	119	151	153	...
Italy	89	100	101	115	109	121	...	...
Luxembourg	...	...	...	...	...	...	...	...
Malta	67	100	87	83	117	...	...	...
Netherlands	96	100	109	108	114	123	122	...
Norway	109	100	111	112	115	127	134	...
Poland	101	100	89	93	104	116	127	138
Portugal	87	100	95	109	112	104	...	...
Romania	...	...	...	...	...	...	...	...
Spain	92	100	110	102	103	117	...	...
Sweden	92	100	107	110	123	130	137	132
Switzerland b/	.	.	.	.	.	.	.	...
Turkey c/	...	...	100	93	131	183	...	...
USSR	...	...	...	...	...	...	...	...
United Kingdom	106	100	103	118	128	145	153	163
Yugoslavia	106	100	98	97	99	103	104	...
Canada d/	102	100	102	111	113	113	...	...
United States	95	100	103	103	114	124	122	123
Japan	90	100	103	111	123	147	162	...

Source : ECE/ITD - Data bank of the Engineering Industries and Technology Section. Calculated on the basis of indices of industrial production and indices of total number of persons engaged. See tables III.25 and III.27.

a/ Productivity expressed in terms of labour productivity.
b/ Included with non-electrical machinery (ISIC, Rev.2 : 382).
c/ 1981=100.
d/ Indices of industrial production include electronic computers and their accessories.

Table III.30 Gross fixed capital formation in electrical machinery industries (ISIC, Rev.2:383) in ECE member countries and Japan, 1979-1986

(Percentage change) a/

COUNTRY	1980/1979	1981/1980	1982/1981	1983/1982	1984/1983	1985/1984	1986/1985	1986/1979
Austria	27.4	11.0	-23.9	13.4	18.3	...	...	...
Belgium	...	...	...	...	...	...	...	...
Bulgaria b/	267.4	-70.7	82.2	52.9	-37.7	31.3	...	16.1 c/
Cyprus	0.0	-18.7	22.5	-33.1	50.0	100.0	-83.3	-14.5
Czechoslovakia	...	...	...	...	...	...	...	...
Denmark	6.7	-15.5	37.3	7.6	37.9	36.6	...	16.6 c/
Finland	100.4	4.2	-6.0	19.0	35.3	8.9	-14.8	26.6
France	18.2	3.8	9.6	6.8	27.7	15.8	...	13.4 c/
German Dem.Rep. d/	14.3	27.7	-13.8	9.4	-5.2	-8.1	...	3.1 c/
Germany,Fed.Rep. of	17.4	-4.1	-0.4	12.4	16.4	...	...	...
Greece	...	-0.3	...	...	...	...	...	...
Hungary	-23.1	-27.5	21.8	12.2	16.1	-19.1	-4.3	-5.3
Ireland	129.2	10.8	32.4	-15.0	...	...	...	...
Italy	20.3	6.0	5.8	37.9	28.6	...	...	...
Luxembourg	...	...	...	...	...	...	...	...
Malta	-80.7	273.0	87.9	28.7	...	...	...	...
Netherlands	14.6	-8.3	-7.6	10.9	38.8	51.2	15.1	14.6
Norway	1.4	31.2	2.2	20.9	-2.0	27.6	...	12.8 c/
Poland	-16.0	-4.9	...	32.3	10.2	23.5	...	...
Portugal	-6.6	37.0	33.7	20.7	3.3	...	...	...
Romania	...	...	...	...	...	...	...	...
Spain	11.3	-7.0	20.4	35.1	2.5	...	...	...
Sweden	-1.3	31.4	9.0	20.3	14.4	22.4	-6.9	12.0
Switzerland	...	...	...	...	...	...	...	...
Turkey	-54.7	33.1	99.6	254.1	24.8	...	...	...
USSR	...	...	...	...	...	...	...	...
United Kingdom e/	11.0	0.8	4.8	14.9	43.0	-7.5	5.5	9.4
Yugoslavia	44.9	22.7	7.7	14.2	50.5	134.2	...	40.5 c/
Canada	60.2	22.3	-1.4	21.5	5.6	...	...	...
United States	31.7	7.5	3.0	7.5	27.9	5.9	-8.5	10.0
Japan	50.9	33.3	10.7	5.9	53.7	8.5	...	25.7 c/

Source : ECE/ITD - Data bank of the Engineering Industries and Technology Section.
Year to year percentage changes and annual average increase 1986/1979 are calculated from Government replies to ECE questionnaires for the Annual Reviews of Engineering Industries and Automation.

a/ Based on values expressed in national currency at current prices.
b/ Excluding land.
c/ 1985/1979.
d/ Including precision instruments (ISIC, Rev.2 : 385) and manufacture of office, computing and accounting machinery.
e/ From 1984 onwards, including precision instruments (ISIC, Rev.2 : 385).

Table III.31

Growth of the electronic components industry (SIC 367) - production, employment and international trade - in the United States, 1979-1986

Indicator	*1979*	*1980*	*1981*	*1982*	*1983*	*1984*	*1985*	*1986*
Value of shipments product data [a], [b]	22.2	24.8	28.1	32.2	35.1	45.8	50.2	52.6
Value of imports [b]	3.6	4.5	4.9	5.8	6.9	10.4	8.5	9.3
Value of exports [b]	3.9	5.0	5.2	5.5	6.1	7.4	6.2	7.1
Total employment [c]	468	499	504	516	515	582	558	531

Source: *U.S. Industrial Outlook 1988*. Department of Commerce, Washington, D.C.

a Estimate for 1987: 59.1 and forecast for 1988: 66.5 (billions of 1982 United States dollars).
b In billions of 1982 United States dollars.
c In thousands.

Table III.32

Number of telephone sets per 100 inhabitants in 1975 and 1984, number of main lines per 100 inhabitants in 1984 and total number of telephone sets and main lines in 1984, in ECE member countries, Australia, Japan and New Zealand

	Number of			*Total number of*	
	telephone sets per 100 inhabitants		*main lines (100)*	*telephone sets (1 000)*	*main lines (1 000)*
Country	*1984*	*1975*	*1984*	*1984*	*1984*
ECE countries:					
Austria	47.59	28.13	34.97	3 594	2 641
Belgium	43.04	28.51	29.94	4 243	2 951
Bulgaria	20.00 [a]	8.90	15.12 [a]	1 790 [a]	1 353 [a]
Canada	66.41 [a]	57.53	41.84 [a]	16 618 [a]	10 468 [a]
Cyprus	29.58	11.59	19.39	196	128
Czechoslovakia	22.60	17.61	11.78	3 499	1 824
Denmark	74.90	45.16	48.25	3 828	2 466
Finland	59.19	38.89	42.94	2 899	2 103
France	60.80 [e]	26.37	41.20 [e]	34 000 [e]	23 100 [e]
German Democratic Republic	21.12	15.18	9.11	3 527	1 521
Germany, Federal Republic of	59.88	31.70	40.27	36 582	24 603
Greece	35.54	22.13	29.48	3 529	2 927
Hungary	13.45	9.91	6.61	1 433	705
Iceland	52.50 [a]	41.74	40.43	125 [a]	98
Ireland	23.49 [a]	13.83	17.49 [a]	824 [a]	613 [a]
Italy	42.61	25.88	28.93	24 331	16 521
Luxembourg	54.82 [b]	41.18	39.22	199 [b]	147
Malta	34.66	16.25 [c]	24.08	115	80
Netherlands	59.05	36.75	39.04	8 535	5 643
Norway	62.20	35.25	39.03	2 579	1 618
Poland	10.87	7.54	6.34	4 028	2 349
Portugal	17.32	11.29	13.01	1 764	1 325
Romania	...	...	6.71 [b]	...	1 480 [b]
Spain	36.02	21.97	23.14	13 825	8 882
Sweden	88.95 [a]	66.07	61.46	7 410 [a]	5 128
Switzerland	81.01	61.09	48.95	5 270	3 184
Turkey	6.41	2.52	4.02	3 091	1 941
USSR	9.83 [d]	6.63 [d]	9.05 [d]	26 667 [d]	24 540 [d]
United Kingdom	52.36	37.63 [c]	35.82	29 518	20 193
United States	76.03 [d]	68.64	40.91 [d]	176 391	94 906
Yugoslavia	13.15	6.09	9.86	3 031	2 273
Other countries:					
Australia	53.58	38.25	38.81	8 329	6 033
Japan	53.54	38.44 [c]	36.67	63 976	43 811
New Zealand	61.57	52.13 [c]	37.53	2 011	1 226

Source: *The Telecommunication Industry - Growth and Structural Change*, UN/ECE publication, Sales No. E.87.II.E.35 (data in table based on *ITU Yearbook of Common Carrier Telecommunication Statistics, 1986*).

a 1983.
b 1979.
c 1976.
d 1982.
e 1985.

Table III.33

Production of selected electrical machinery and equipment in the United States, 1975-1985

Type of product	*Value of shipments in billions of 1982 US dollars* [a]			*Index*	
	1975	*1980*	*1985**	*1980/1975*	*1985/1980*
Transformers	2.9	3.6	3.2	124.1	88.[illegible]
Switchgear, apparatus	4.5	5.7	5.6	126.7	98.2
Motors and generators	5.5	6.7	5.8	121.8	86.6
Industrial controls	3.7	4.4	5.1	118.9	115.9
Radio and TV communication equipment	16.5	28.4	40.6	172.1	143.0
Telephone and telegraph apparatus	7.6	13.3	17.2	175.0	129.3
Electronic components and parts	12.6	26.2	46.2	207.9	176.3
of which:					
Semiconductors and related devices	1.3	6.9	24.5	530.8	355.1
X-ray and electromedical apparatus	1.6	3.0	4.8	187.5	160.0

Source: U.S. Industrial Outlook 1987. Department of Commerce, Washington, D.C.

a Industry data.

Figure III.6 (a-d)

Development of production, employment and productivity indices in the electrical machinery industries (ISIC 383) of four (standard) countries, 1979-1986 [a]

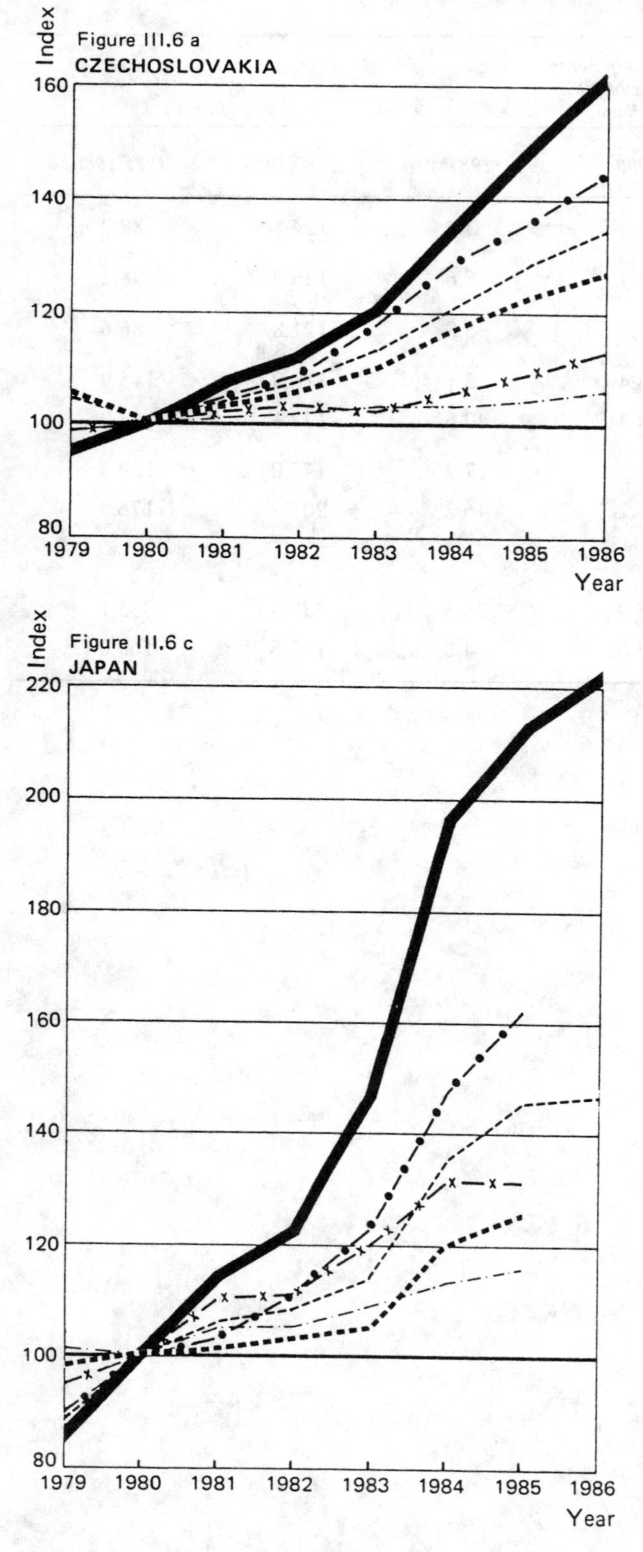

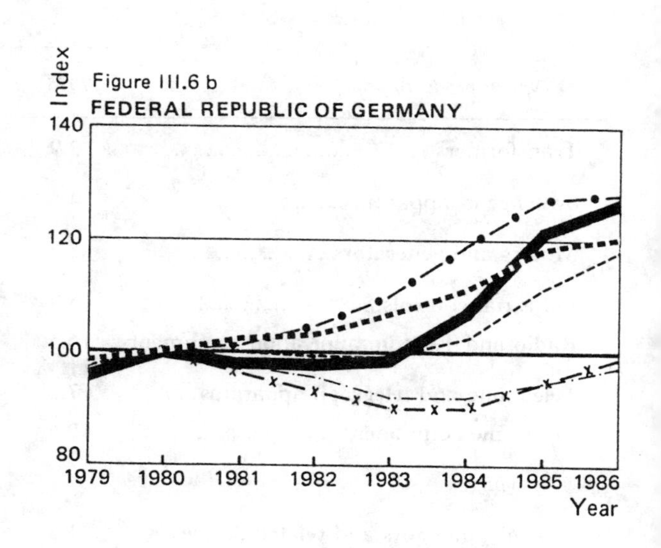

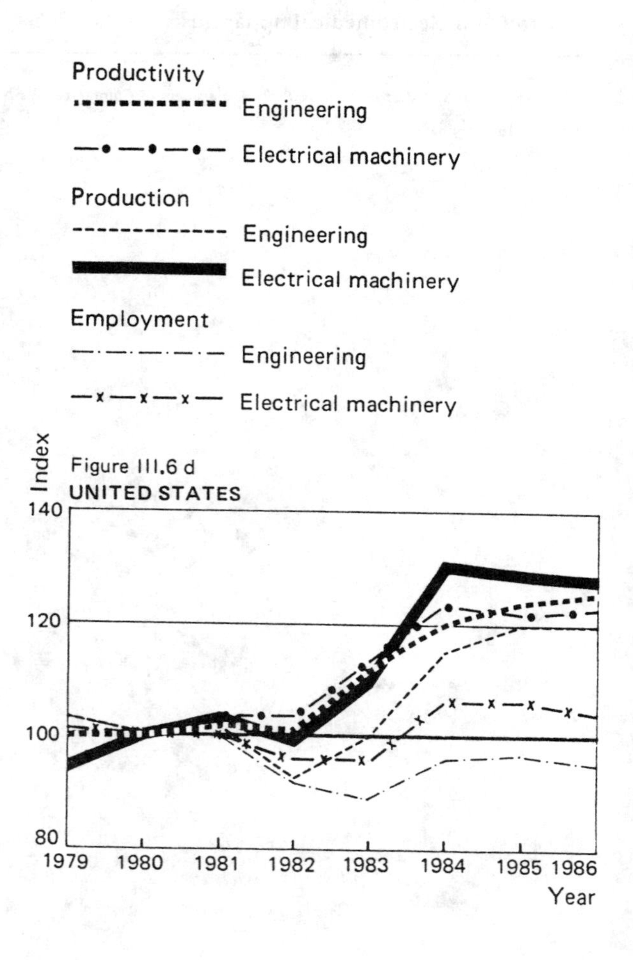

a Productivity expressed in terms of labour productivity. 1980 = 100.

Figure III.7 (a-d)

Development of production, employment and productivity indices in the electrical machinery industries (ISIC 383) of four (additional) countries, 1979-1986 [a]

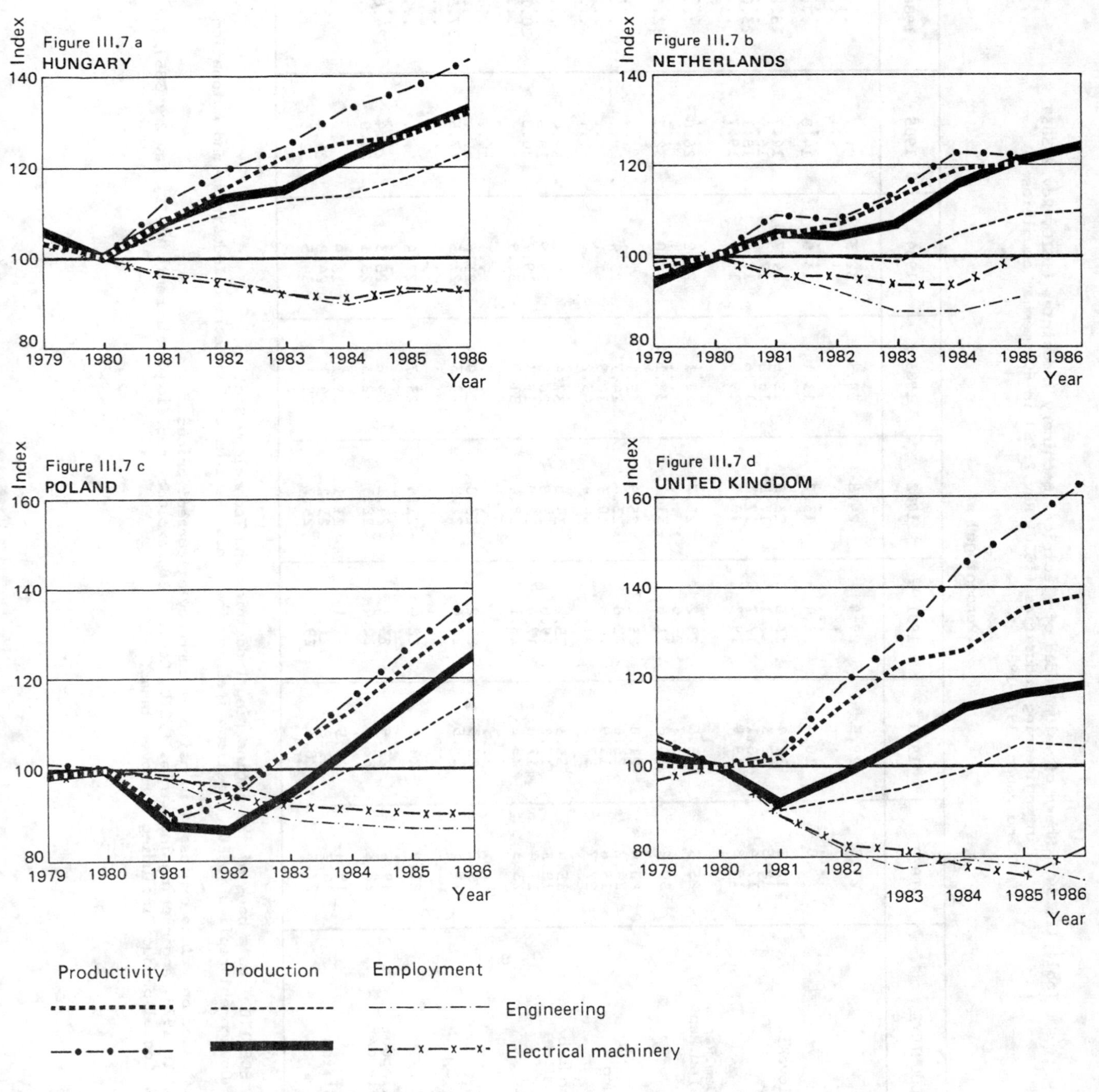

a Productivity expressed in terms of labour productivity.
1980 = 100.

Table III.34 Share of value added of electrical machinery industries (ISIC, Rev.2:383) in engineering industries (ISIC, Rev.2:38) in ECE member countries and Japan, 1979-1986

(Percentage) a/

COUNTRY	1979	1980	1981	1982	1983	1984	1985	1986
Austria	29.3	29.5	29.9	29.8	30.3	32.3	...	...
Belgium	...	...	...	...	...	...	...	...
Bulgaria	...	...	...	...	...	...	...	...
Cyprus	9.0	10.4	12.4	13.7	13.1	13.0	14.8	...
Czechoslovakia	12.3	12.4	13.5	13.8	13.6	14.7	14.7	15.2
Denmark	18.5	17.3	17.1	17.5	16.5	17.2	16.9	...
Finland	17.4	18.1	17.7	17.3	17.8	17.3	18.1	23.0
France	19.8	19.9	19.2	18.9	19.0	19.6	19.7	...
German Dem.Rep.	...	...	...	...	...	...	...	...
Germany,Fed.Rep. of	25.3	25.6	25.2	25.7	26.1	26.2	26.0	...
Greece	...	23.7	23.6	...	...	...	...	...
Hungary	30.2	30.8	31.6	31.1	32.1	32.0	32.5	32.9
Ireland	20.4	21.0	23.1	25.9	25.1	...	...	...
Italy	24.2	23.6	22.8	23.4	24.7	23.4	...	...
Luxembourg	13.6	11.9	10.6	10.3	10.9	...	...	...
Malta	27.7	37.0	31.7	44.4	43.2	...	...	...
Netherlands b/	35.4	33.3	35.2	34.0	37.4	39.1	39.4	...
Norway	18.3	17.2	19.0	18.8	20.5	19.9	18.6	...
Poland	17.4	17.6	17.6	18.9	19.3	19.7	20.0	20.1
Portugal	23.3	25.5	25.3	28.4	31.9	36.9	...	...
Romania	...	...	...	...	...	...	...	...
Spain	24.2	23.3	23.2	23.0	24.0	23.6	...	...
Sweden	19.6	19.5	21.6	21.8	22.3	21.5	17.5	17.5
Switzerland	...	...	...	...	...	...	...	...
Turkey	23.9	24.2	22.4	21.5	23.7	27.4	...	...
USSR	...	...	...	...	...	...	...	...
United Kingdom	19.5	22.9	23.3	24.1	24.6	25.3	24.4	24.8
Yugoslavia c/	22.2	22.6	22.4	22.3	22.3	23.1	26.0	...
Canada	19.4	20.5	21.1	21.1	19.1	18.9	...	...
United States	20.7	21.8	21.7	23.2	24.1	24.8	24.4	...
Japan	27.5	28.3	29.1	29.7	32.1	34.5	32.3	...

Source : ECE/ITD - Data bank of the Engineering Industries and Technology Section.
Government replies to ECE questionnaires for the Annual Reviews of Engineering Industries and Automation.

a/ Based on values expressed in national currency at current prices.
b/ In 1979, engineering industries (ISIC, Rev.2 : 38) exclude precision instruments (ISIC, Rev.2 : 385).
c/ Social product including turnover taxes.

Table III.35 Share of total number of persons engaged in electrical machinery industries (ISIC, Rev.2:383) in total number engaged in engineering industries (ISIC, Rev.2:38) in ECE member countries and Japan, 1979-1986

(Percentage)

COUNTRY	1979	1980	1981	1982	1983	1984	1985	1986
Austria a/	29.8	29.9	30.8	30.8	31.0	31.4	...	...
Belgium	27.1	26.9	27.2	27.1	27.0	27.3	26.9	...
Bulgaria	21.0	22.4	26.0	26.2	26.9	27.5	27.4	25.1
Cyprus	7.2	9.9	9.9	9.6	10.2	10.4	11.4	10.9
Czechoslovakia a/ b/	13.7	13.7	13.8	13.8	13.5	14.0	14.3	14.6
Denmark	17.2	16.5	16.6	16.4	16.3	16.2	16.4	16.6
Finland	18.1	18.1	17.9	17.1	17.1	17.6	18.8	19.9
France	22.8	23.2	23.5	23.7	24.1	24.8	25.0	...
German Dem.Rep. a/ c/	31.7	31.8	31.8	32.0	32.1	32.6	32.4	...
Germany,Fed.Rep. of	26.1	26.0	25.6	25.2	25.2	25.5	25.9	26.0
Greece	...	21.8	21.5	...	...	...	...	...
Hungary	31.9	31.9	31.6	31.5	31.9	32.3	32.2	31.9
Ireland	22.2	22.6	25.0	26.1	26.8	28.7	29.9	...
Italy	25.1	25.0	24.4	24.6	25.9	24.5	...	...
Luxembourg	10.7	10.9	10.2	9.6	9.7	...	...	...
Malta	34.1	31.5	28.6	35.1	33.2	...	...	...
Netherlands	32.2	32.5	32.4	33.5	34.6	34.9	35.6	...
Norway	16.9	17.0	16.9	17.4	18.5	17.9	17.7	...
Poland	18.3	18.8	19.0	19.2	19.3	19.3	19.5	19.6
Portugal	21.6	21.6	22.1	22.2	23.1	24.3	...	...
Romania	...	...	...	...	...	...	...	...
Spain	21.2	19.7	19.6	19.9	19.6	18.9	...	...
Sweden	19.8	19.8	20.3	20.5	20.2	20.3	20.4	20.7
Switzerland d/	.	.	.	.	.	.	.	...
Turkey	19.9	18.7	18.2	18.6	19.4	20.1	...	...
USSR	...	...	...	...	...	...	...	...
United Kingdom e/	21.0	23.1	23.1	23.5	24.4	22.8	22.5	22.2
Yugoslavia	21.4	21.8	21.8	21.7	21.8	21.8	22.4	...
Canada	20.3	20.9	21.7	20.5	20.4	20.7	...	...
United States	23.3	23.7	23.8	24.7	25.6	26.0	25.8	25.8
Japan	28.7	30.3	31.8	32.1	33.4	35.2	34.3	...

Source : ECE/ITD - Data bank of the Engineering Industries and Technology Section.
Government replies to ECE questionnaires for the Annual Reviews of Engineering Industries and Automation.

a/ Including homeworkers.
b/ State national industry only.
c/ Electrical machinery (ISIC, Rev.2 : 383) includes precision instruments (ISIC. Rev.2 : 385) and manufacture of office, computing and accounting machinery.
d/ Electrical machinery (ISIC, Rev.2 : 383) is included with non-electrical machinery (ISIC, Rev.2 :382).
e/ At June of each year.

Table III.36 Share of gross fixed capital formation of electrical machinery industries (ISIC, Rev.2:383) in engineering industries (ISIC, Rev.2:38) in ECE member countries and Japan, 1979-1986

(Percentage) a/

COUNTRY	1979	1980	1981	1982	1983	1984	1985	1986
Austria	30.2	30.8	21.7	20.4	31.3	37.0	...	...
Belgium	...	...	...	...	...	...	...	...
Bulgaria b/	15.7	24.0	12.3	17.1	16.0	11.6	12.8	...
Cyprus	8.6	10.7	6.4	10.1	8.0	7.7	9.4	1.8
Czechoslovakia	...	...	...	...	...	...	...	...
Denmark	20.2	17.2	15.5	18.8	18.8	20.0	18.7	...
Finland	17.4	22.5	21.9	16.5	19.0	23.6	21.6	20.6
France	33.8	32.7	34.4	35.5	36.7	37.7	36.5	...
German Dem.Rep. c/	27.8	29.0	32.0	30.1	32.7	31.5	28.7	...
Germany,Fed.Rep. of	21.7	21.7	20.9	19.7	21.4	25.2	...	...
Greece	...	12.0	9.9	...	...	...	...	...
Hungary	31.3	28.4	23.6	27.3	30.0	36.1	34.1	31.7
Ireland	14.8	24.5	24.5	26.1	29.2	...	...	...
Italy	21.3	21.5	20.1	17.4	19.8	22.9	...	...
Luxembourg	...	...	...	...	...	...	...	...
Malta	45.3	28.5	56.6	66.7	47.1	...	...	...
Netherlands	38.3	36.9	38.2	35.0	41.0	41.2	46.3	47.4
Norway	17.4	17.7	20.0	20.4	29.8	23.6	22.4	...
Poland	14.0	14.7	17.7	13.1	14.9	14.6	17.5	...
Portugal	36.5	24.1	23.5	20.6	31.6	31.3	...	...
Romania	...	...	...	...	...	...	...	...
Spain	17.4	21.1	17.4	21.9	24.3	15.3	...	...
Sweden	20.0	15.4	18.1	20.6	21.8	18.3	18.5	14.9
Switzerland	...	...	...	...	...	...	...	...
Turkey	34.6	20.7	11.7	16.4	39.5	26.6	...	...
USSR	...	...	...	...	...	...	...	...
United Kingdom d/	17.6	20.2	22.9	24.6	26.5	30.5	26.7	26.5
Yugoslavia	21.2	23.7	24.1	21.2	19.5	20.8	26.8	...
Canada	12.9	12.4	15.1	20.9	24.4	23.6	...	...
United States	20.1	22.7	23.0	25.7	29.2	29.4	28.4	27.5
Japan	28.1	29.6	31.9	32.6	36.3	45.1	41.2	...

Source : ECE/ITD - Data bank of the Engineering Industries and Technology Section.
Government replies to ECE questionnaires for the Annual Reviews of Engineering Industries and Automation.

a/ Based on values expressed in national currency at current prices.
b/ Excluding land.
c/ Electrical machinery (ISIC, Rev.2 : 383) includes precision instruments (ISIC. Rev.2 : 385) and manufacture of office, computing and accounting machinery.
d/ From 1984 onwards, electrical machinery (ISIC, Rev.2 : 383) includes precision instruments (ISIC, Rev.2 : 385).

Table III.37

Production of electrical machinery (ISIC 383) by geographic region, 1975-1985

Region	*Value in billions of constant 1975 US dollars* [a] 1975	1980	1985	*Index* 1980/1975	1985/1980
Total world	297.2	434.0	587.9	146.0	135.5
of which:					
Developed market economies	182.5	262.9	351.3	144.1	134.[illegible]
of which:					
North America	58.5	84.8	133.1	145.0	157.0
Western Europe	88.6	107.2	120.8	121.0	112.7
Japan and others	35.4	70.9	97.4	200.3	137.4
Centrally planned economies	97.4	144.9	194.3	148.8	134.1
of which:					
European	92.6	138.7	186.2	149.8	134.2
Asian	4.8	6.2	8.1	129.2	130.6
Developing countries	17.3	26.2	42.3	151.4	161.5
of which:					
African	0.8	1.3	1.8	162.5	138.5
Asian	7.7	12.8	25.8	166.2	201.6
Latin America	8.8	12.1	14.7	137.5	121.5

Source: As for table III.11.

a See table II.1.

Table III.38

Share of geographic regions in world production of electrical machinery, 1975-1985

(Percentage)

Region	1975	1980	1985
Total world	100.0	100.0	100.0
of which:			
Developed market economies	61.4	60.6	59.8
of which:			
North America	19.7	19.5	22.6
Western Europe	29.8	24.7	20.5
Japan and others	11.9	16.4	16.6
Centrally planned economies	32.8	33.4	33.0
of which:			
European	31.2	32.0	31.6
Asian	1.6	1.4	1.4
Developing countries	5.8	6.0	7.2
of which:			
African	0.3	0.3	0.3
Asian	2.6	2.9	4.4
Latin America	2.9	2.8	2.5
ECE region	80.7	76.2	76.2

Source: As for table III.11.

III.5 MANUFACTURE OF TRANSPORT EQUIPMENT

ISIC major group 384 includes the following groups of transport equipment:

- Shipbuilding;
- Railroad equipment (locomotives, wagons, tramways etc.);
- Motor vehicles (passenger cars, trucks, buses, motor vehicle parts and accessories);
- Motorcycles and bicycles; and
- Aircraft.

Transport equipment represents a traditional branch of engineering, the third largest in terms of production volume after the non-electrical and electrical machinery industry.

Tables III.39-III.44 show the development of selected production-oriented indicators characterizing the developments in the transport equipment industry (ISIC 384) in the period under review. Most ECE member countries recorded continuing production growth, the largest in Turkey with an index of industrial production of 177 in 1986 (1981 = 100), Cyprus (159 in 1984, 1980 = 100 as also for the following countries), Yugoslavia (142 in 1986), Canada (42 in 1986), Bulgaria (138 in 1986); United States (130 in 1986) etc. Transport equipment also contributed significantly to total engineering output in Asia and Latin America, e.g. in shipbuilding (Republic of Korea, Singapore, Brazil), railroad equipment (Brazil, Argentina, Mexico, India), aircraft (Argentina, Brazil, India, Indonesia), motor vehicles (Brazil, Mexico, Republic of Korea and others), and motorcycles and bicycles (mainly China and India).

Figures III.8 (a-d) illustrate, as was done for ISIC major groups 381-383, the relationship between production, employment and productivity indices for the period 1979-1986 in the transport equipment industries of four "standard" countries (Czechoslovakia, Federal Republic of Germany, Japan and United States). One may see that, despite the increased application of robotics and other automated technologies, the employment level remained relatively high. Figures III.9 (a-d) show the development in four selected "additional" countries (Belgium, Canada, Sweden and Yugoslavia). It will be noted that, contrary to the "standard" countries, Belgium and Sweden have reached high productivity levels.

Tables III.45-III.47 show the share of transport equipment in total engineering industries production - in terms of value added, employment and capital formation. The largest share of value added in total engineering was registered in 1985-1986: in France (31.2% in 1985), Sweden (30.7% in 1986), United States (26.8% in 1985), Federal Republic of Germany (26.7% in 1985), Poland and the United Kingdom (both 25.0% in 1985). In the absence of relevant 1985 and 1986 data for further "transport equipment countries", one should not forget other important producers, such as Canada (41.3% share in 1984), Turkey (34.0% in 1983), Spain (31.1% in 1982), Italy (29.1% in 1982) etc.

Table III.48 presents the production growth of motor vehicles in selected countries. It shows that, in passenger-car manufacture, three groups of main producers have been created over the past years: United States and Japan (with a production volume of between 5 and 9 million units per year), Federal Republic of Germany and France (2.5-4.5 million units) and five other ECE member countries with a production level of between 1 and 1.5 million units.

The largest world truck manufacturer remained the Daimler-Benz Company (with subsidiaries in North and South America and South Africa), followed by Isuzu (Japan) and Ford. [III.13] While the automotive industry recorded significant growth during the period under review, including an important increase in the functional capabilities of motor road vehicles - this applies to a lesser extent also to the aerospace industries - the use of traditional means of transport, such as railroad equipment (in spite of the growing popularity of high-speed trains) and passenger and merchant vessels, declined. [III.14]

Tables III.49 and III.50 present the geographic distribution of transport equipment production. They disclose in particular that, in comparison with the years 1975-1980, the period 1980-1985 showed increased dynamics in Asia (including Japan) and Latin America. The North American recovery as from 1982 (see also table III.48) was not high enough to compensate the decline of 1980-1981. The ECE region share in total world production decreased from 80.1% in 1975 to 73.8% in 1985, mainly owing to the decline in some non-automotive industry branches such as shipbuilding, motorcylces and bicycles.

Table III.39 Index of industrial production in transport equipment industries (ISIC, Rev.2:384) in ECE member countries and Japan, 1979-1986

(1980=100)

COUNTRY	1979	1980	1981	1982	1983	1984	1985	1986
Austria	81	100	98	92	103	110	134	..
Belgium	106	100	103	114	122	113	120	130
Bulgaria	85	100	103	108	119	129	136	138
Cyprus	90	100	160	148	149	159	142	124
Czechoslovakia	95	100	104	107	111	118	124	130
Denmark a/	101	100	114	126	110	115	124	128
Finland	89	100	105	119	119	114	117	110
France	104	100	95	96	99	94	90	...
German Dem.Rep. b/	.	.	.	.	.	.	.	...
Germany,Fed.Rep. of c/	101	100	108	110	110	108	121	127
Greece	90	100	103	96	81	80	63	71
Hungary	104	100	103	109	109	111	114	119
Ireland	118	100	102	95	91	93	82	79
Italy	88	100	103	99	103	102	100	109
Luxembourg	...	100	91	88	95	81	76	86
Malta	99	100	110	100	96	...	...	...
Netherlands	96	100	97	109	106	105	102	105
Norway	99	100	94	95	82	77	75	73
Poland	100	100	86	86	91	96	100	105
Portugal	77	100	110	106	79	63	71	75
Romania d/	.	.	.	.	.	.	.	.
Spain e/	99	100	93	96	104	100	105	116
Sweden	109	100	102	108	110	113	120	128
Switzerland	...	100	...	...	...	...	...	...
Turkey f/	...	...	100	123	156	166	174	177
USSR	...	100	...	...	...	...	...	...
United Kingdom	105	100	92	90	90	87	91	88
Yugoslavia	100	100	102	103	115	134	138	142
Canada	122	100	101	91	104	124	133	134
United States	112	100	98	90	102	116	125	130
Japan	83	100	106	100	95	104	108	105

Source : ECE/ITD - Data bank of the Engineering Industries and Technology Section.
Government replies to ECE questionnaires for the Annual Reviews of Engineering Industries and Automation.

a/ Excluding shipbuilding and repairing.
b/ Included with electrical machinery (ISIC, Rev.2 : 383).
c/ Excluding manufacture of aircraft.
d/ Included with non-electrical machinery (ISIC, Rev.2 : 382).
e/ From 1983 onwards, data are not comparable with previous years.
f/ 1981=100.

Table III.40 Value added in transport equipment industries (ISIC, Rev.2:384) in ECE member countries and Japan, 1979-1986

(Percentage change) a/

COUNTRY	1980/1979	1981/1980	1982/1981	1983/1982	1984/1983	1985/1984	1986/1985	1986/1979
Austria	4.3	9.4	-5.8	34.7	12.4	...	...	...
Belgium	...	...	...	...	...	...	...	...
Bulgaria	...	...	...	...	...	...	...	...
Cyprus	33.3	-3.3	-30.8	5.0	0.0	9.5	...	0.4 b/
Czechoslovakia	4.1	-17.4	4.6	9.4	-0.8	0.6	-4.6	-0.9
Denmark	11.6	18.2	25.6	15.2	-2.8	0.5	...	11.0 b/
Finland	12.0	14.2	27.6	5.9	5.2	14.0	-13.8	8.6
France	11.9	10.0	12.8	10.9	5.2	7.2	...	9.6 b/
German Dem.Rep.	...	...	...	...	...	...	...	...
Germany,Fed.Rep. of	1.8	8.7	7.2	4.8	4.2	10.6	...	6.2 b/
Greece	...	34.4	...	...	...	...	...	...
Hungary	-26.3	11.8	13.4	0.8	10.6	13.7	9.2	3.7
Ireland	13.4	14.2	8.2	2.0	...	...	...	...
Italy	23.3	18.0	12.5	-18.7	15.5	...	...	...
Luxembourg c/	23.4	7.0	17.0	11.8	...	...	...	...
Malta	-31.2	54.5	-41.8	7.1	...	...	...	...
Netherlands	13.2	-3.3	9.8	1.6	-15.6	5.0	...	1.3 b/
Norway	9.5	9.8	6.8	-14.8	-1.5	-4.7	...	0.4 b/
Poland	3.0	-6.4	...	12.7	19.0	17.5	...	...
Portugal	22.9	18.7	-7.7	17.9	-2.3	...	...	...
Romania	...	...	...	...	...	...	...	...
Spain	14.5	-1.4	22.3	9.1	-4.7	...	...	...
Sweden	-1.8	5.4	12.0	18.0	11.8	23.8	17.0	12.0
Switzerland	...	...	...	...	...	...	...	...
Turkey	77.3	66.1	54.8	35.1	39.2	...	...	...
USSR	...	...	...	...	...	...	...	...
United Kingdom	1.6	4.4	3.6	7.5	3.0	12.0	3.4	5.0
Yugoslavia d/	28.3	55.2	40.5	42.5	57.6	96.3	...	52.0 b/
Canada	-6.5	16.4	-6.7	30.7	29.3	...	...	...
United States	-4.7	8.3	2.4	17.3	14.9	5.6	...	7.0 b/
Japan	11.5	13.6	3.3	-5.0	5.4	26.8	...	8.9 b/

Source : ECE/ITD - Data bank of the Engineering Industries and Technology Section.
Year to year percentage changes and annual average increase 1986/1979 are calculated from Government replies to ECE questionnaires for the Annual Reviews of Engineering Industries and Automation.

a/ Based on values expressed in national currency at current prices.
b/ 1985/1979.
c/ Including precision instruments (ISIC, Rev.2 : 385).
d/ Social product including turnover taxes.

Table III.41 Index of total number of persons engaged in transport equipment industries (ISIC, Rev.2:384) in ECE member countries and Japan, 1979-1986

(1980 = 100)

COUNTRY	1979	1980	1981	1982	1983	1984	1985	1986
Austria a/	97	100	96	95	92	97	...	...
Belgium	100	100	95	89	90	86	86	...
Bulgaria	103	100	134	135	138	140	141	136
Cyprus	109	100	86	61	56	56	60	58
Czechoslovakia a/ b/	100	100	101	102	103	104	104	105
Denmark	98	100	100	106	97	92	100	92
Finland	93	100	103	105	107	102	102	93
France	103	100	95	94	92	87	83	...
German Dem.Rep. c/	...	...	...	...	...	...	...	...
Germany,Fed.Rep. of	98	100	99	99	98	98	101	105
Greece	...	100	106	...	...	...	...	...
Hungary	104	100	97	95	84	84	91	92
Ireland	99	100	93	90	80	63	54	...
Italy	100	100	96	93	87	77	...	...
Luxembourg d/	104	100	102	106	108	...	...	...
Malta	101	100	90	66	70	...	...	...
Netherlands	101	100	99	95	86	84	83	...
Norway	99	100	96	92	76	72	66	...
Poland	100	100	99	91	89	88	86	85
Portugal	108	100	103	100	94	86	...	...
Romania	...	...	...	...	...	...	...	...
Spain	95	100	94	93	93	88	...	...
Sweden	100	100	98	96	92	94	96	98
Switzerland	98	100	102	103	88	86	96	...
Turkey	104	100	104	104	104	109	...	...
USSR	...	...	...	...	...	...	...	...
United Kingdom e/	106	100	89	82	75	68	65	62
Yugoslavia	103	100	104	107	110	112	113	...
Canada	106	100	100	89	89	99	...	...
United States	109	100	100	91	92	100	104	104
Japan	99	100	104	103	102	104	109	...

Source : ECE/ITD - Data bank of the Engineering Industries and Technology Section.
Government replies to ECE questionnaires for the Annual Reviews of Engineering Industries and Automation.

a/ Including homeworkers.
b/ State national industry only.
c/ Included with metal products (ISIC, Rev.2 : 381).
d/ Including precision instruments (ISIC, Rev.2 : 385).
e/ At June of each year.

Table III.42 Total number of persons engaged in transport equipment industries (ISIC, Rev.2:384) in ECE member countries and Japan, 1979-1986

(Percentage change)

COUNTRY	1980/1979	1981/1980	1982/1981	1983/1982	1984/1983	1985/1984	1986/1985	1986/1979
Austria a/	2.8	-3.9	-1.6	-2.5	5.5	...	...	...
Belgium	0.4	-5.0	-5.9	0.8	-4.3	0.1	...	-2.4 b/
Bulgaria	-2.8	33.8	1.1	2.0	1.5	0.5	-3.0	4.1
Cyprus	-7.9	-13.8	-29.7	-7.7	-0.5	8.4	-3.6	-8.5
Czechoslovakia a/ c/	0.4	0.8	1.2	0.8	0.8	0.4	1.2	0.8
Denmark	2.2	0.4	5.8	-9.0	-4.5	8.7	-8.8	-0.9
Finland	7.6	2.5	2.5	1.7	-4.3	-0.5	-8.5	0.0
France	-2.9	-5.0	-0.9	-1.7	-5.4	-5.1	...	-3.5 b/
German Dem.Rep. d/	...	...	...	...	...	...	...	...
Germany,Fed.Rep. of	1.8	-1.2	0.5	-1.5	0.4	3.2	3.6	0.9
Greece	...	6.4	...	...	...	...	...	...
Hungary	-3.4	-3.4	-2.0	-11.8	1.0	8.0	1.1	-1.7
Ireland	0.8	-6.5	-3.5	-11.7	-20.4	-15.4	...	-9.7 b/
Italy	0.0	-3.8	-3.7	-6.2	-11.8	...	...	...
Luxembourg e/	-3.5	1.7	4.0	1.9	...	...	...	...
Malta	-1.3	-10.1	-26.5	6.6	...	...	...	...
Netherlands	-0.6	-0.8	-4.6	-8.8	-2.4	-0.8	...	-3.1 b/
Norway	0.9	-4.4	-3.5	-18.1	-5.0	-8.4	...	-6.6 b/
Poland	0.0	-1.3	-7.5	-2.5	-1.2	-2.0	-1.2	-2.3
Portugal	-7.4	2.9	-3.0	-5.5	-8.5	...	...	...
Romania	...	...	...	...	...	...	...	...
Spain	5.7	-6.2	-0.4	-0.6	-5.0	...	...	...
Sweden	-0.3	-1.9	-1.9	-4.7	2.2	2.5	2.0	-0.3
Switzerland	1.9	1.9	1.2	-14.7	-2.2	11.0	...	-0.4 b/
Turkey	-3.8	4.2	-0.2	0.0	4.6	...	...	...
USSR	...	...	...	...	...	...	...	...
United Kingdom f/	-5.9	-10.6	-8.7	-7.9	-9.5	-4.1	-4.4	-7.3
Yugoslavia	-2.6	4.1	3.3	2.5	1.2	1.2	...	1.6 b/
Canada	-5.8	0.0	-10.6	-0.6	11.3	...	...	...
United States	-8.5	-0.1	-8.6	0.7	9.1	3.4	0.7	-0.7
Japan	1.2	3.7	-0.8	-1.1	2.6	4.2	...	1.6 b/

Source : ECE/ITD - Data bank of the Engineering Industries and Technology Section.
Year to year percentage changes and annual average increase 1986/1979 are calculated from Government replies to ECE questionnaires for the Annual Reviews of Engineering Industries and Automation.

a/ Including homeworkers.
b/ 1985/1979.
c/ State national industry only.
d/ Included with metal products (ISIC, Rev.2 381).
e/ Including precision instruments (ISIC, Rev.2 : 385).
f/ At June of each year.

Table III.43 Index of productivity in transport equipment industries (ISIC, Rev.2: 384) in ECE member countries and Japan, 1979-1986 a/

(1980 = 100)

COUNTRY	1979	1980	1981	1982	1983	1984	1985	1986
Austria	83	100	102	97	112	113	...	...
Belgium	106	100	108	127	135	130	139	...
Bulgaria	83	100	77	80	86	92	97	101
Cyprus	83	100	186	244	266	286	235	213
Czechoslovakia	95	100	103	105	108	114	119	123
Denmark b/	103	100	114	119	114	125	124	140
Finland	96	100	102	113	111	111	115	118
France	101	100	100	102	107	108	108	...
German Dem.Rep.	...	...	...	...	...	...	...	...
Germany,Fed.Rep. of c/	103	100	109	110	113	110	119	121
Greece	...	100	97	...	...	...	...	...
Hungary	100	100	107	115	130	132	125	129
Ireland	119	100	109	105	114	147	153	...
Italy	88	100	107	107	119	133	...	...
Luxembourg	...	...	...	...	...	...	...	...
Malta	98	100	122	151	136	...	...	...
Netherlands	95	100	98	115	123	125	122	...
Norway	100	100	98	103	109	107	114	...
Poland	100	100	87	94	102	109	116	123
Portugal	71	100	107	106	84	73	...	...
Romania	...	...	...	...	...	...	...	...
Spain d/	105	100	99	103	112	113	...	...
Sweden	109	100	104	112	120	121	125	130
Switzerland	...	100	...	...	...	...	...	...
Turkey e/	...	...	100	123	156	159	...	...
USSR	...	...	...	...	...	...	...	...
United Kingdom	99	100	103	111	119	127	139	141
Yugoslavia	97	100	98	96	104	120	122	...
Canada	115	100	101	102	117	125	...	...
United States	102	100	98	99	111	116	120	124
Japan	84	100	102	97	93	100	99	...

Source : ECE/ITD - Data bank of the Engineering Industries and Technology Section. Calculated on the basis of indices of industrial production and indices of total number of persons engaged. See tables III.39 and III.41.

a/ Productivity expressed in terms of labour productivity.
b/ Indices of industrial production exclude shipbuilding and repairing.
c/ Indices of industrial production exclude manufacture of aircraft.
d/ From 1983 onwards, data are not comparable with previous years.
e/ 1981=100.

Table III.44 Gross fixed capital formation in transport equipment industries (ISIC, Rev.2:384) in ECE member countries and Japan, 1979-1986

(Percentage change) a/

COUNTRY	1980/1979	1981/1980	1982/1981	1983/1982	1984/1983	1985/1984	1986/1985	1986/1979
Austria	64.0	192.8	-24.5	-70.8	-13.1	...	...	...
Belgium	...	...	...	...	...	...	...	...
Bulgaria b/	50.8	16.8	59.4	27.2	-18.8	10.7	...	21.5 c/
Cyprus	-59.5	75.6	-79.3	88.7	0.0	100.0	-75.0	-24.5
Czechoslovakia	...	...	...	...	...	...	...	...
Denmark	134.9	-2.0	22.3	13.7	-3.7	16.7	...	23.8 c/
Finland	14.4	-11.9	46.5	28.7	-6.9	54.6	-16.9	12.5
France	26.7	-15.0	7.2	-2.2	37.6	23.7	...	11.5 c/
German Dem.Rep. d/	.	.	.	.	.	.	...	.
Germany,Fed.Rep. of	19.7	...	...	-2.8	...	...	...	...
Greece	...	35.5	...	...	...	...	...	...
Hungary	-12.1	1.1	7.6	1.3	-29.2	-0.1	34.6	-1.1
Ireland	-9.6	34.7	-17.8	-22.9	...	...	...	...
Italy	13.9	12.8	62.6	-5.7	2.9	...	...	...
Luxembourg	...	...	...	...	...	...	...	...
Malta	-89.9	-5.7	-60.6	173.1	...	...	...	...
Netherlands	68.0	-13.9	15.0	-27.7	55.1	15.7	4.9	12.4
Norway	35.8	20.5	-18.4	-19.7	-19.7	-7.0	...	-3.6 c/
Poland	-13.0	-14.8	...	31.1	18.6	13.3	...	...
Portugal	43.4	52.6	85.5	-39.1	-25.6	...	...	...
Romania	...	...	...	...	...	...	...	...
Spain	-4.5	-1.8	2.2	16.2	203.4	...	...	...
Sweden	31.7	10.5	6.4	12.9	49.9	21.5	19.4	21.0
Switzerland	...	...	...	...	...	...	...	...
Turkey	40.8	62.0	150.6	-24.4	260.3	...	...	...
USSR	...	...	...	...	...	...	...	...
United Kingdom	-3.3	-13.8	-10.4	5.8	17.6	5.6	1.9	0.0
Yugoslavia	44.7	4.1	16.2	46.9	35.2	85.1	...	36.4 c/
Canada	91.5	-3.8	-43.1	11.5	-14.1	...	...	...
United States	18.5	1.0	-17.3	-25.3	42.8	20.3	-0.2	3.4
Japan	54.7	24.4	3.7	-13.6	-5.7	28.0	...	13.0 c/

Source : ECE/ITD - Data bank of the Engineering Industries and Technology Section. Year to year percentage changes and annual average increase 1986/1979 are calculated from Government replies to ECE questionnaires for the Annual Reviews of Engineering Industries and Automation.

a/ Based on values expressed in national currency at current prices.
b/ Excluding land.
c/ 1985/1979.
d/ Included with metal products (ISIC, Rev.2 : 381).

Figure III.8 (a-d)

Development of production, employment and productivity indices in the transport equipment industries (ISIC 384) of four selected (standard) countries, 1979-1986 [a]

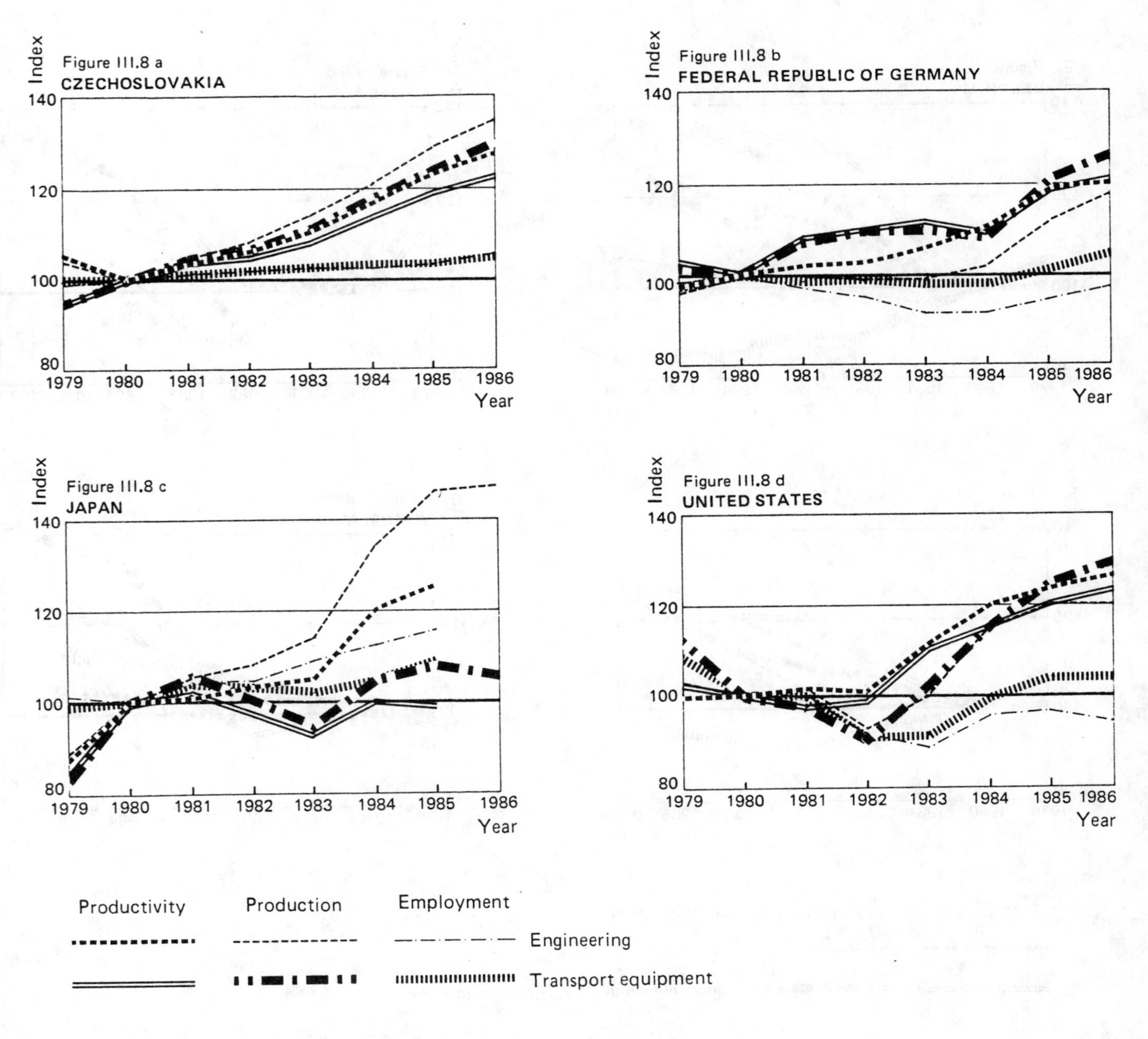

a Productivity expressed in terms of labour productivity.
1980 = 100.

Figure III.9 (a-d)

Development of production, employment and productivity indices in the transport equipment industries (ISIC 384) of four selected (additional) countries, 1979-1986 [a]

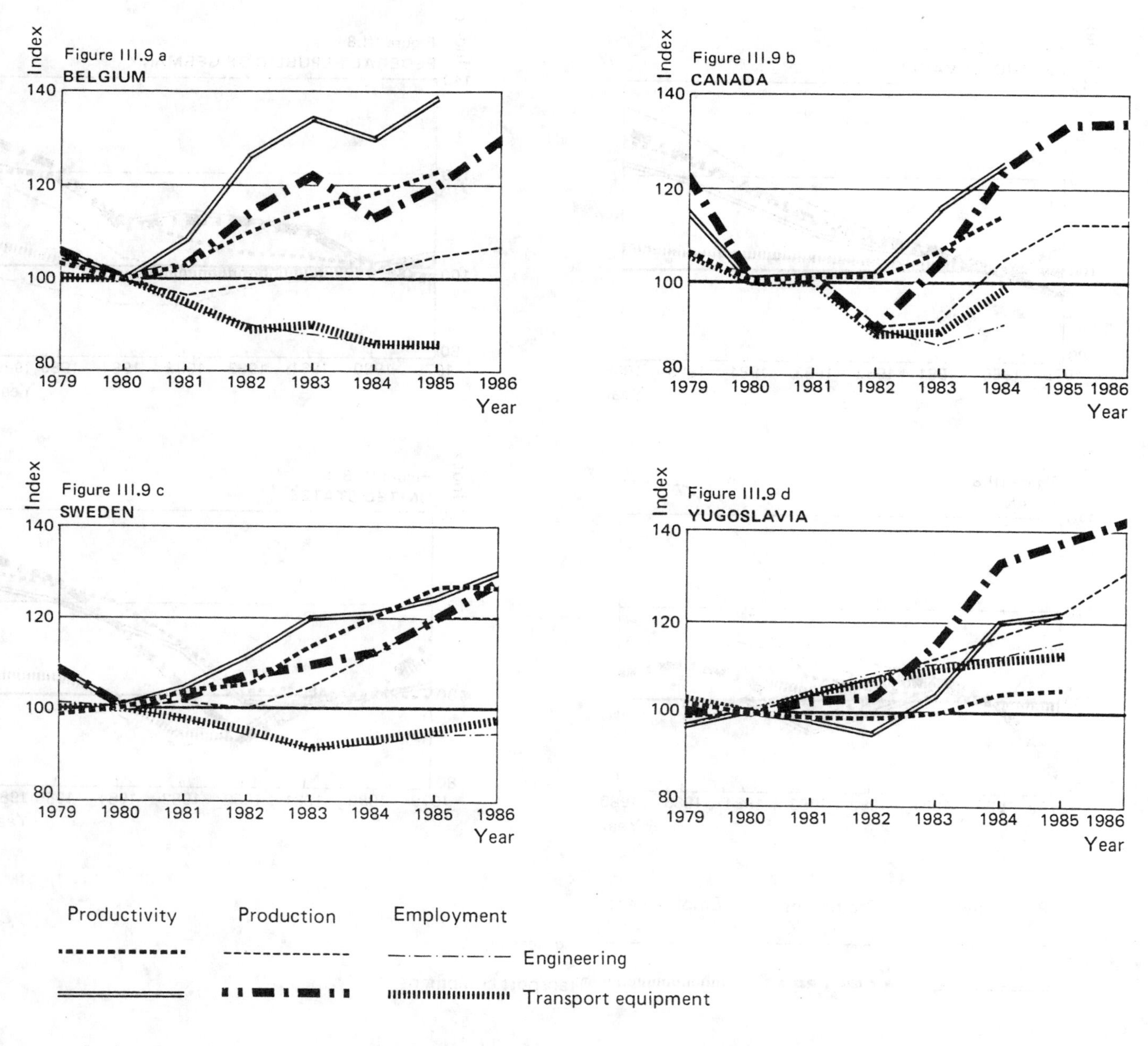

a Productivity expressed in terms of labour productivity.
1980 = 100.

Table III.45 Share of value added of transport equipment industries (ISIC, Rev.2:384) in engineering industries (ISIC, Rev.2:38) in ECE member countries and Japan, 1979-1986

(Percentage) a/

COUNTRY	1979	1980	1981	1982	1983	1984	1985	1986
Austria	13.7	13.2	13.2	11.8	15.3	16.4	...	...
Belgium	...	...	...	...	...	...	...	...
Bulgaria	...	...	...	...	...	...	...	...
Cyprus	16.6	17.9	13.9	8.3	8.6	7.6	7.4	...
Czechoslovakia	24.6	24.4	23.8	24.2	24.5	24.1	23.4	22.8
Denmark	16.1	15.8	17.7	19.3	19.9	17.4	15.6	...
Finland	22.5	21.3	20.7	22.1	22.3	21.4	21.6	18.5
France	32.3	32.2	31.2	30.9	31.8	31.0	31.2	...
German Dem.Rep.	...	...	...	...	...	...	...	...
Germany,Fed.Rep. of	26.2	24.3	25.3	25.9	26.7	26.6	26.7	...
Greece	...	36.3	37.9	...	...	...	...	...
Hungary	23.8	22.9	22.6	22.9	21.6	21.9	21.7	21.5
Ireland	14.4	12.8	11.8	10.6	8.3	...	...	...
Italy	28.6	28.7	28.7	29.1	24.8	24.8	...	...
Luxembourg b/	11.7	11.0	10.5	11.5	12.8	...	...	...
Malta	16.0	11.0	14.9	7.4	8.0	...	...	...
Netherlands c/	18.3	19.2	18.2	19.0	18.7	15.2	15.4	...
Norway	33.1	33.6	31.8	31.8	28.1	25.8	21.6	...
Poland	27.3	27.5	25.5	27.7	27.0	26.1	25.0	24.0
Portugal	36.7	33.9	33.8	26.9	25.9	22.2	...	...
Romania	...	...	...	...	...	...	...	...
Spain	30.3	30.1	28.3	31.1	30.8	28.8	...	...
Sweden	28.0	25.4	24.5	26.3	26.6	25.9	26.6	30.7
Switzerland	...	...	...	...	...	...	...	...
Turkey	27.9	28.2	30.3	32.2	34.0	30.6	...	...
USSR	...	...	...	...	...	...	...	...
United Kingdom	27.6	26.4	27.2	26.3	26.8	25.0	25.0	24.8
Yugoslavia d/	21.2	20.4	21.4	21.8	22.3	22.1	21.3	...
Canada	35.8	31.5	32.1	32.1	38.8	41.3	...	...
United States	25.1	22.9	22.6	23.3	26.0	25.7	26.8	...
Japan	23.5	23.6	24.0	23.5	23.0	21.7	23.1	...

Source : ECE/ITD - Data bank of the Engineering Industries and Technology Section.
Government replies to ECE questionnaires for the Annual Reviews of Engineering Industries and Automation.

a/ Based on values expressed in national currency at current prices.
b/ Transport equipment (ISIC, Rev.2 : 384) includes precision instruments (ISIC, Rev.2 : 385).
c/ In 1979, engineering industries (ISIC, Rev.2 : 38) exclude precision instruments (ISIC, Rev.2 : 385).
d/ Social product including turnover taxes.

Table III.46 Share of total number of persons engaged in transport equipment industries (ISIC, Rev.2:384) in total number engaged in engineering industries (ISIC, Rev.2:38) in ECE member countries and Japan, 1979-1986

(Percentage)

COUNTRY	1979	1980	1981	1982	1983	1984	1985	1986
Austria a/	13.0	13.1	12.9	13.2	13.2	13.8	...	...
Belgium	25.8	26.1	26.3	26.0	26.8	26.2	26.5	...
Bulgaria	15.9	15.2	20.2	20.1	20.1	19.8	19.6	19.6
Cyprus	16.9	14.5	11.8	8.1	7.6	7.3	7.3	7.3
Czechoslovakia a/ b/	22.4	22.4	22.3	22.4	22.3	22.4	22.3	22.2
Denmark	17.8	18.1	19.2	20.3	18.6	16.7	16.5	15.3
Finland	23.0	23.0	23.4	23.7	24.5	23.9	23.8	22.9
France	31.6	31.0	30.3	30.3	30.4	29.8	29.3	...
German Dem.Rep. c/	...	...	...	...	...	...	...	...
Germany,Fed.Rep. of	22.8	22.9	23.0	23.8	24.3	24.5	24.5	24.5
Greece	...	34.8	36.2	...	...	...	...	...
Hungary	20.4	20.3	20.1	20.0	18.5	18.9	19.8	20.1
Ireland	20.1	19.4	18.1	17.4	16.2	13.3	11.6	...
Italy	31.2	31.5	31.6	32.3	29.8	26.5	...	...
Luxembourg d/	15.6	14.1	14.4	15.4	15.8	...	...	...
Malta	14.5	14.8	13.1	10.1	11.2	...	...	...
Netherlands	20.0	20.0	20.5	20.2	19.6	19.1	18.4	...
Norway	35.8	36.1	34.6	33.9	30.3	29.0	25.9	...
Poland	24.0	24.1	24.3	24.0	24.0	24.0	23.9	23.7
Portugal	34.5	31.5	31.3	30.3	29.9	29.4	...	...
Romania	...	...	...	...	...	...	...	...
Spain	28.6	30.8	30.9	32.1	32.6	33.1	...	...
Sweden	27.7	27.9	27.8	28.3	27.8	27.9	28.3	28.8
Switzerland	4.7	4.6	4.7	4.8	4.3	4.4	4.7	...
Turkey	30.4	29.2	28.9	27.6	29.4	29.8	...	...
USSR	...	...	...	...	...	...	...	...
United Kingdom e/	28.8	28.8	29.0	28.7	28.1	24.9	24.2	23.9
Yugoslavia	22.4	21.2	21.0	20.9	21.0	20.9	20.6	...
Canada	31.6	29.9	29.6	29.6	30.9	32.5	...	...
United States	22.9	21.6	21.5	21.3	22.2	22.5	23.1	23.6
Japan	19.7	20.1	19.8	19.7	18.8	18.6	18.8	...

Source : ECE/ITD - Data bank of the Engineering Industries and Technology Section.
Government replies to ECE questionnaires for the Annual Reviews of Engineering Industries and Automation.

a/ Including homeworkers.
b/ State national industry only.
c/ Transport equipment (ISIC, Rev.2 : 384) is included with metal products (ISIC, Rev.2 : 381).
d/ Transport equipment (ISIC, Rev.2 : 384) includes precision instruments (ISIC, Rev.2 : 385).
e/ At June of each year.

Table III.47 Share of gross fixed capital formation of transport equipment industries (ISIC, Rev.2:384) in engineering industries (ISIC, Rev.2:38) in ECE member countries and Japan, 1979-1986

(Percentage) a/

COUNTRY	1979	1980	1981	1982	1983	1984	1985	1986
Austria	18.5	24.2	45.1	42.1	16.6	14.5	...	...
Belgium	...	...	...	...	...	...	...	...
Bulgaria b/	11.7	7.4	15.0	18.2	14.2	13.4	12.5	...
Cyprus	20.0	10.4	13.3	3.6	8.0	5.1	6.3	1.8
Czechoslovakia	...	...	...	...	...	...	...	...
Denmark	7.4	13.9	14.6	15.6	16.6	12.3	9.8	...
Finland	31.2	23.0	18.9	22.2	27.8	23.7	30.8	28.7
France	35.7	37.0	31.8	32.1	30.5	33.7	34.8	...
German Dem.Rep. c/	.	.	.	.	.	.	.	...
Germany,Fed.Rep. of	35.3	36.1	...	41.0	38.5	...	...	...
Greece	...	58.8	65.9	...	...	...	...	...
Hungary	26.7	27.7	32.1	32.8	32.6	23.9	27.9	36.4
Ireland	10.8	7.1	8.6	5.7	5.8	...	...	...
Italy	34.7	33.2	33.0	44.0	34.3	31.7	...	...
Luxembourg	...	...	...	...	...	...	...	...
Malta	18.9	6.2	3.1	0.8	1.1	...	...	...
Netherlands	15.1	21.3	20.7	23.6	18.0	20.2	17.4	16.2
Norway	24.1	32.8	34.0	27.7	26.9	17.4	12.1	...
Poland	15.1	16.4	17.7	17.7	19.8	21.0	23.0	...
Portugal	33.1	33.7	36.4	44.3	34.3	24.5	...	...
Romania	...	...	...	...	...	...	...	...
Spain	35.1	36.5	31.8	34.0	32.4	60.3	...	...
Sweden	30.1	30.8	30.4	33.9	33.7	36.9	37.0	38.4
Switzerland	...	...	...	...	...	...	...	...
Turkey	19.3	35.9	24.8	43.5	22.4	43.6	...	...
USSR	...	...	...	...	...	...	...	...
United Kingdom	38.7	38.7	37.5	34.5	34.2	32.4	32.4	31.0
Yugoslavia	22.4	25.1	21.6	20.5	24.3	23.2	23.7	...
Canada	48.2	55.5	52.8	42.3	45.4	35.7	...	...
United States	42.4	42.9	40.9	36.7	29.0	32.6	35.7	37.8
Japan	33.6	36.3	36.5	35.0	31.8	24.2	26.0	...

Source : ECE/ITD - Data bank of the Engineering Industries and Technology Section.
Government replies to ECE questionnaires for the Annual Reviews of Engineering Industries and Automation.

a/ Based on values expressed in national currency at current prices.
b/ Excluding land.
c/ Transport equipment (ISIC, Rev.2 : 384) is included with metal products (ISIC, Rev.2 : 381).

Table III.48

Production of passenger cars and commercial vehicles in nine selected countries, 1982-1985 [a]

(Millions of units)

Country	Passenger cars				Commercial vehicles			
	1982	*1983*	*1984*	*1985*	*1982*	*1983*	*1984*	*1985*
United States	5.074	6.781	7.622	8.140	1.906	2.416	3.076	3.397
Japan	6.336	7.156	7.073	7.645	3.857	3.966	4.037	4.711
Germany, Federal Republic of	3.771	3.875	3.783	4.165	0.311	0.303	0.264	0.295
France	3.086	2.960	2.910	2.784	0.446	0.457	0.424	0.460
Italy	1.296	1.386	1.439	1.354	0.156	0.179	0.160	0.185
USSR	1.307	1.315	1.327	1.332	0.866[b]	...	...	...
Spain	0.944	1.136	1.174	1.250	0.102	0.031	0.032	...
Canada	0.808	0.971	1.033	1.075	0.468	0.554	0.809	0.856
United Kingdom	0.888	1.045	0.910	1.048	0.269	0.245	0.224	0.263
Total world [c]	26.389	29.599	30.272	31.895	9.396	9.108	10.080	11.404
Percentage share of nine selected countries	89.1	90.0	90.1	90.3	89.2	89.5	89.6	89.2

Source: *Annual Review of Engineering Industries and Automation 1985*, vol. I.UN/ECE publication, Sales No.E.86.II.E.30.

a Countries listed in decreasing sequence of 1985 passenger-car production volume.

b Excluding wheeled tractors.

c Figures represent total of reporting countries.

Table III.49

Production of transport equipment (ISIC 384) by geographic region, 1975-1985

Region	*Value in billions of constant 1975 US dollars* [a]			*Index*	
	1975	*1980*	*1985*	*1980/1975*	*1985/1980*
Total world	345.9	441.9	561.2	127.8	127.0
of which:					
Developed market economies	217.4	256.3	295.8	117.9	115.4
of which:					
North America	79.7	94.4	107.6	118.4	114.0
Europe	97.4	111.3	111.5	114.3	100.2
Japan and others	40.3	50.6	76.7	125.6	151.6
Centrally planned economies	105.0	153.9	203.7	146.6	132.4
of which:					
European	99.8	147.1	194.8	147.4	132.4
Asian	5.2	6.8	8.9	130.9	130.9
Developing countries	23.5	31.7	61.7	134.9	194.6
of which:					
African	1.1	2.3	1.9	209.1	82.6
Asian	6.8	8.6	17.1	126.5	198.8
Latin America	15.6	20.8	42.7	133.3	205.3

Source: As for table III.11.
a See table II.1.

Table III.50

Share of geographic regions in world production of transport equipment, 1975-1985

(Percentage)

Region	*1975*	*1980*	*1985*
Total world	100.0	100.0	100.0
of which:			
Developed market economies	62.8	58.0	52.7
of which:			
North America	23.0	21.4	19.2
Europe	28.2	25.2	19.9
Japan and others	11.6	11.4	13.7
Centrally planned economies	30.4	34.8	36.3
of which:			
European	28.9	33.3	34.7
Asian	1.5	1.5	1.6
Developing countries	6.8	7.2	11.0
of which:			
African	0.3	0.5	0.3
Asian	2.0	2.0	3.1
Latin America	4.5	4.7	7.6
ECE region	80.1	79.9	73.8

Source: As for table III.11.

III.6 MANUFACTURE OF PRECISION INSTRUMENTS

ISIC major group 385 (precision instruments) covers the following types of instruments and devices:

- Scientific and laboratory apparatus;
- Measuring and controlling equipment;
- Photographic and optical goods; and
- Watches and clocks.

It includes a wide range of instruments for industrial use, but also for research and laboratory purposes, non-industrial services and personal use. The rapid innovation rate and product design is strongly influenced by recent developments in microelectronics, availability of new materials and increased sophistication of measurement and other methods. The high intellectual input probably explains why the ECE region has retained the leading position in this industry (compared with ISIC major groups 381-384).

Tables III.51-III.56 describe the development of production, value added, employment, productivity and gross fixed capital formation indicators characterizing the precision instrument industry in 1979-1986. The largest production growth rates were achieved in Japan with an index of industrial production of 161 in 1986 (1980 = 100) and in several relatively smaller countries, such as Finland (207 in 1986), Ireland (189 in 1985), Malta (172 in 1984), Yugoslavia (165 in 1986), Hungary (142 in 1986) and Norway (141 in 1986). This tendency indicates the attractive opportunity offered by precision instrument technologies to countries which are not yet highly specialized in other engineering fields (e.g. in non-electrical machinery), but have at their disposal an experienced research and design basis as well as skilled manufacturing personnel.

Figures III.10 (a-d) illustrate the development of production, employment and productivity indices concerning precision instrument industries of four "standard" countries and figures III.11 (a-d) reveal the situation in four "additional" countries: Denmark, Hungary, Ireland and Norway. It will be noted that some countries have already reached high productivity in this labour-intensive engineering sector (e.g. Federal Republic of Germany, Hungary, Japan and Norway).

Tables III.57-III.59 show the share of precision instruments in total engineering production - in terms of value added, employment and capital formation. From the available data, it may be seen that the precision instrument industry is still developing and, regarding general quantitative measures, represents only a "marginal" part of engineering activities. However, several countries accounted for an important share regarding, for instance, employment in 1983: Malta accounted for 21.3% of total employment in engineering industries, Switzerland for 15.0%, Ireland 11.7%, Hungary 11.6%, the United States 8.8% etc. In terms of value added, Malta registered a 21.4% share in 1983, Ireland 13.1%, Hungary 12.4% and the United States 9.2%. As the United States is a "leading-edge" country in this field (in both quantitative and qualitative terms), it might be useful to illustrate its production structure, despite the fact that its statistical classification does not correspond fully with ISIC 385 (see table III.60).

Table III.61 shows the development of world watch production. In value terms, Switzerland accounted for some 40% of total world production during the whole period (1975-1986). This high share resulted from the fact that the Swiss watch industry had the monopoly of the manufacture of high-quality and luxury watches.

Tables III.62 and III.63 present the geographic distribution of precision instrument production between 1975 and 1985. The European centrally planned economies increased their share in total world production from 61.7% in 1975 to 64.4% in 1985. North America's 1985 production attained a volume of nearly $US 40 billion (1985/1980 index 137.4). While the production share of developing countries remains marginal, and they still depend heavily on imports (according to the UNIDO data base, their imports amounted to nearly $10 billion in 1983), several countries have already achieved significant progress, e.g. Brazil, Hong Kong, Mexico, Republic of Korea and others. The ECE region's share in total world production remained the highest, contrary to its declining share in other major groups of ISIC 38: 91.5% share in 1975 and 88.2% in 1985.

Table III.51 Index of industrial production in precision instruments industries (ISIC, Rev.2:385) in ECE member countries and Japan, 1979-1986

(1980=100)

COUNTRY	1979	1980	1981	1982	1983	1984	1985	1986
Austria	...	...	...	...	...	...	...	...
Belgium	77	100	104	107	119	121	122	112
Bulgaria a/	.	.	.	.	.	.	.	.
Cyprus	92	100	108	...	...	...	...	...
Czechoslovakia	95	100	104	108	112	114	114	119
Denmark	91	100	109	117	111	132	140	138
Finland	90	100	122	131	135	175	206	207
France	100	100	102	102	101	105	110	...
German Dem.Rep. b/	.	.	.	.	.	.	.	...
Germany,Fed.Rep. of	97	100	95	89	87	90	99	102
Greece	85	100	92	92	94	103	78	64
Hungary	102	100	113	119	127	127	135	142
Ireland	115	100	122	128	159	186	189	...
Italy	87	100	95	72	64	78	91	91
Luxembourg b/	.	.	.	.	.	.	.	.
Malta	42	100	109	152	167	172	...	...
Netherlands	96	100	99	99	99	109	111	103
Norway	133	100	110	113	125	130	135	141
Poland	98	100	94	91	101	111	119	133
Portugal	...	...	...	...	...	...	...	...
Romania a/	.	.	.	.	.	.	.	.
Spain	62	100	81	95	95	93	98	102
Sweden	81	100	98	98	103	109	137	128
Switzerland	...	100	...	...	...	...	...	...
Turkey	...	...	...	...	...	...	...	...
USSR	...	100	...	...	...	...	...	...
United Kingdom	103	100	100	94	95	101	109	112
Yugoslavia	110	100	100	106	107	130	144	165
Canada	91	100	96	92	86	95	102	...
United States	97	100	104	103	101	112	114	116
Japan	78	100	111	103	111	130	152	161

Source : ECE/ITD - Data bank of the Engineering Industries and Technology Section.
Government replies to ECE questionnaires for the Annual Reviews of Engineering Industries and Automation.

a/ Included with non-electrical machinery (ISIC, Rev.2 : 382).
b/ Included with electrical machinery (ISIC, Rev.2 : 383).

Table III.52 Value added in precision instruments industries (ISIC, Rev.2:385) in ECE member countries and Japan, 1979-1986

(Percentage change) a/

COUNTRY	1980/1979	1981/1980	1982/1981	1983/1982	1984/1983	1985/1984	1986/1985	1986/1979
Austria	-18.8	-2.4	4.9	-2.4	21.7	...	...	...
Belgium	...	...	...	...	...	...	...	...
Bulgaria	...	...	...	...	...	...	...	...
Cyprus	...	...	...	...	...	...	...	...
Czechoslovakia	-2.2	-15.7	10.6	0.0	-2.3	-5.7	-12.2	-4.2
Denmark	10.1	16.0	14.4	16.9	16.2	10.9	...	14.1 b/
Finland	14.5	37.5	9.3	17.2	33.6	6.1	0.6	16.3
France	2.7	6.6	14.8	9.7	13.1	9.7	...	9.4 b/
German Dem.Rep.	...	...	...	...	...	...	...	...
Germany,Fed.Rep. of	14.3	2.2	-5.9	-2.5	-0.7	14.3	...	3.3 b/
Greece	...	4.2	...	...	...	...	...	...
Hungary	-13.9	13.9	10.2	4.9	6.5	16.1	14.3	7.0
Ireland	13.2	32.2	30.5	29.5	...	...	...	...
Italy	19.5	16.5	16.7	-60.9	51.6	...	...	...
Luxembourg c/	...	...	...	...	...	...	...	...
Malta	-17.9	6.9	23.4	5.8	...	...	...	...
Netherlands	...	8.5	9.8	-21.4	4.8	5.4	...	...
Norway	-4.6	11.4	4.4	14.6	18.6	29.1	...	11.8 b/
Poland	1.9	31.5	...	9.8	29.5	27.2	...	...
Portugal	55.2	3.3	101.9	24.7	51.1	...	...	...
Romania	...	...	...	...	...	...	...	...
Spain	-7.0	2.7	6.0	13.1	2.5	...	...	...
Sweden	31.7	6.3	6.9	21.3	8.1	42.7	-3.5	15.2
Switzerland	...	...	...	...	...	...	...	...
Turkey	50.4	131.2	91.3	2.8	27.7	...	...	...
USSR	...	...	...	...	...	...	...	...
United Kingdom	-25.8	-1.1	21.3	-1.8	12.5	12.7	7.0	2.5
Yugoslavia d/	49.1	53.4	33.4	51.0	72.7	88.0	...	57.0 b/
Canada	18.2	15.4	-23.3	10.1	11.8	...	...	...
United States	13.5	12.8	6.9	4.7	13.1	1.0	...	8.6 b/
Japan	14.2	10.8	-0.4	-1.1	2.5	18.7	...	7.2 b/

Source : ECE/ITD - Data bank of the Engineering Industries and Technology Section.
Year to year percentage changes and annual average increase 1986/1979 are calculated from Government replies to ECE questionnaires for the Annual Reviews of Engineering Industries and Automation.

a/ Based on values expressed in national currency at current prices.
b/ 1985/1979.
c/ Included with transport equipment (ISIC, Rev.2 : 384).
d/ Social product including turnover taxes.

Table III.53 Index of total number of persons engaged in precision instruments industries (ISIC, Rev.2:385) in ECE member countries and Japan, 1979-1986

(1980 = 100)

COUNTRY	1979	1980	1981	1982	1983	1984	1985	1986
Austria a/	117	100	68	72	70	74	...	...
Belgium	98	100	96	93	87	89	93	...
Bulgaria b/	.	.	.	.	.	.	.	.
Cyprus	...	...	...	...	...	...	...	...
Czechoslovakia a/ c/	113	100	100	100	100	100	94	100
Denmark	96	100	95	97	103	113	123	128
Finland	112	100	105	107	105	114	133	128
France	101	100	95	93	91	89	89	...
German Dem.Rep. d/	.	.	.	.	.	.	.	...
Germany,Fed.Rep. of	98	100	96	93	84	83	86	89
Greece	...	100	75	...	...	...	...	...
Hungary	103	100	99	97	93	89	91	93
Ireland	100	100	103	113	115	108	105	...
Italy	105	100	98	91	45	42	...	...
Luxembourg e/	.	.	.	.	.	...	...	...
Malta	94	100	124	134	135	...	...	...
Netherlands	110	100	101	81	76	78	82	...
Norway	108	100	92	92	92	92	100	...
Poland	102	100	98	92	90	94	90	90
Portugal	93	100	107	104	104	100	...	...
Romania	...	...	...	...	...	...	...	...
Spain	130	100	100	90	89	78	...	...
Sweden	94	100	101	106	101	99	103	104
Switzerland	100	100	99	87	80	72	76	...
Turkey	86	100	143	150	136	143	...	...
USSR	...	...	...	...	...	...	...	...
United Kingdom f/	160	100	91	89	80	97	98	99
Yugoslavia	110	100	110	120	130	140	140	140
Canada	96	100	108	76	76	76	...	...
United States	97	100	103	101	97	100	102	101
Japan	98	100	106	99	100	97	97	...

Source : ECE/ITD - Data bank of the Engineering Industries and Technology Section.
Government replies to ECE questionnaires for the Annual Reviews of Engineering Industries and Automation.

a/ Including homeworkers.
b/ Included with non-electrical machinery (ISIC, Rev.2 : 382).
c/ State national industry only.
d/ Included with electrical machinery (ISIC, Rev.2 : 383).
e/ Included with transport equipment (ISIC, Rev.2 : 384).
f/ At June of each year.

Table III.54 Total number of persons engaged in precision instruments industries (ISIC, Rev.2:385) in ECE member countries and Japan, 1979-1986

(Percentage change)

COUNTRY	1980/1979	1981/1980	1982/1981	1983/1982	1984/1983	1985/1984	1986/1985	1986/1979
Austria a/	-14.5	-31.9	6.3	-2.9	6.1	...	...	...
Belgium	1.9	-3.7	-3.8	-6.0	2.1	4.2	...	-1.0 b/
Bulgaria c/	.	.	.	.	.	.	.	.
Cyprus	...	...	...	...	...	...	...	...
Czechoslovakia a/ d/	-11.1	0.0	0.0	0.0	0.0	-6.3	6.7	-1.7
Denmark	4.4	-5.3	2.2	6.6	9.3	9.4	3.4	4.2
Finland	-10.4	4.7	2.2	-2.2	8.9	16.3	-3.5	2.0
France	-1.1	-4.5	-2.4	-2.4	-2.5	0.0	...	-2.2 b/
German Dem.Rep. e/	.	.	.	.	.	...	...	.
Germany,Fed.Rep. of	1.8	-3.6	-3.7	-9.7	-1.4	4.3	2.8	-1.5
Greece	...	-25.0	...	...	...	...	...	...
Hungary	-2.5	-1.4	-1.6	-3.7	-4.8	2.1	1.9	-1.4
Ireland	0.0	3.2	9.4	1.4	-5.6	-3.0	...	0.8 b/
Italy	-4.5	-1.6	-7.9	-50.0	-6.9	...	...	...
Luxembourg f/	...	...	.	.	...	...	...	...
Malta	6.6	23.5	8.1	1.1	...	...	...	...
Netherlands	-9.2	1.1	-20.0	-5.6	1.5	5.8	...	-4.8 b/
Norway	-7.1	-7.7	0.0	0.0	0.0	8.3	...	-1.2 b/
Poland	-2.0	-2.0	-6.1	-2.2	4.4	-4.3	0.0	-1.8
Portugal	8.0	7.4	-3.4	0.0	-3.6	...	...	...
Romania	...	...	...	...	...	...	...	...
Spain	-23.1	0.0	-10.0	-1.6	-12.1	...	...	...
Sweden	5.8	0.9	4.5	-4.3	-1.8	3.7	0.9	1.3
Switzerland	0.3	-1.1	-12.5	-7.8	-9.9	5.7	...	-4.4 b/
Turkey	16.7	42.9	5.0	-9.5	5.3	...	...	...
USSR	...	...	...	...	...	...	...	...
United Kingdom g/	-37.7	-8.9	-2.2	-10.0	20.9	1.0	1.1	-6.7
Yugoslavia	-9.1	10.0	9.1	8.3	7.7	0.0	0.0	3.5
Canada	4.2	8.0	-29.6	0.0	0.0	...	...	...
United States	2.9	2.7	-2.1	-3.2	3.2	1.3	-0.8	0.5
Japan	1.7	5.6	-6.1	0.4	-2.6	-0.4	...	-0.3 b/

Source : ECE/ITD - Data bank of the Engineering Industries and Technology Section.
Year to year percentage changes and annual average increase 1986/1979 are calculated from Government replies to ECE questionnaires for the Annual Reviews of Engineering Industries and Automation.

a/ Including homeworkers.
b/ 1985/1979.
c/ Included with non-electrical machinery (ISIC, Rev.2 : 382).
d/ State national industry only.
e/ Included with electrical machinery (ISIC, Rev.2 : 383).
f/ Included with transport equipment (ISIC, Rev.2 : 384).
g/ At June of each year.

Table III.55 Index of productivity in precision instruments industries (ISIC, Rev.2:385) in ECE member countries and Japan, 1979-1986 a/

(1980 = 100)

COUNTRY	1979	1980	1981	1982	1983	1984	1985	1986
Austria	...	...	...	...	...	...	...	...
Belgium	78	100	108	115	137	136	132	...
Bulgaria b/	.	.	.	.	.		.	...
Cyprus	...	...	...	...	...	...	...	...
Czechoslovakia	84	100	104	108	112	114	122	119
Denmark	95	100	115	121	108	117	113	108
Finland	81	100	116	122	129	153	156	162
France	99	100	107	109	111	118	124	...
German Dem.Rep.	...	...	...	...	...	...	...	...
Germany,Fed.Rep. of	99	100	99	95	103	109	115	115
Greece	...	100	123	...	...	...	...	...
Hungary	99	100	114	123	136	143	149	153
Ireland	115	100	118	113	139	172	180	...
Italy	83	100	96	79	141	184	...	...
Luxembourg	...	...	...	...	...	...	...	...
Malta	45	100	88	114	124	...	...	...
Netherlands	87	100	98	122	130	141	135	...
Norway	124	100	119	122	135	141	135	...
Poland	96	100	96	99	112	118	132	148
Portugal	...	...	...	...	...	...	...	...
Romania	...	...	...	...	...	...	...	...
Spain	47	100	81	105	107	119	...	...
Sweden	86	100	97	93	102	110	134	124
Switzerland	...	100	...	...	...	...	...	...
Turkey	...	...	...	...	...	...	...	...
USSR	...	...	...	...	...	...	...	...
United Kingdom	64	100	110	106	119	105	112	113
Yugoslavia	100	100	91	88	82	93	103	118
Canada	95	100	89	121	113	125	...	...
United States	100	100	101	102	104	112	112	115
Japan	79	100	105	104	111	134	157	...

Source : ECE/ITD - Data bank of the Engineering Industries and Technology Section. Calculated on the basis of indices of industrial production and indices of total number of persons engaged. See tables III.51 and III.53.

a/ Productivity expressed in terms of labour productivity.
b/ Included with non-electrical machinery (ISIC, Rev.2 : 382).

Table III.56 Gross fixed capital formation in precision instruments industries (ISIC, Rev.2:385) in ECE member countries and Japan, 1979-1986

(Percentage change) a/

COUNTRY	1980/1979	1981/1980	1982/1981	1983/1982	1984/1983	1985/1984	1986/1985	1986/1979
Austria	4.8	-13.6	5.3	-18.0	20.1	...	...	...
Belgium	...	...	...	...	...	...	...	...
Bulgaria b/ c/	.	.	.	.	.	.	...	.
Cyprus	...	...	...	...	...	...	...	...
Czechoslovakia	...	...	...	...	...	...	...	...
Denmark	2.9	5.7	23.4	12.4	36.4	59.0	...	21.9 d/
Finland	88.8	-6.5	76.2	-17.0	23.0	59.7	-9.2	24.4
France c/	.	.	.	.	.	.	...	.
German Dem.Rep. e/	.	.	.	.	.	.	...	.
Germany,Fed.Rep. of	22.5	...	...	10.6	4.5	24.3	...	8.0 d/
Greece	...	-43.4	...	...	...	...	...	...
Hungary	-11.2	-2.1	35.2	10.0	1.2	-22.4	-19.1	-2.8
Ireland	143.8	-35.4	92.9	-37.9	...	...	...	...
Italy	51.4	13.4	-14.3	-52.7	-14.6	...	...	...
Luxembourg	...	...	...	...	...	...	...	...
Malta	-23.3	145.5	24.9	-51.5	...	...	...	...
Netherlands	-8.5	-12.3	7.0	-49.2	19.4	54.1	21.1	-0.4
Norway	-11.5	-23.3	17.0	210.7	-59.0	31.7	...	4.9 d/
Poland	8.4	-11.5	...	4.4	25.6	-32.3	...	...
Portugal	290.8	-43.1	104.0	4.9	27.2	...	...	...
Romania	...	...	...	...	...	...	...	...
Spain	-16.7	0.0	100.0	-25.3	43.9	...	...	...
Sweden	57.0	-7.3	22.6	7.8	7.2	43.6	28.6	21.1
Switzerland	...	...	...	...	...	...	...	...
Turkey	-55.0	677.8	205.7	16.8	16.0	...	...	...
USSR	...	...	...	...	...	...	...	...
United Kingdom f/	-41.6	-18.6	22.9	18.6	...	...	...	...
Yugoslavia	-10.4	110.1	51.0	8.7	-20.6	121.4	...	32.6 d/
Canada	79.2	34.9	-12.1	29.4	-3.0	...	...	...
United States	...	...	...	...	...	...	...	...
Japan	27.6	28.8	-6.8	-8.7	-2.2	33.1	...	10.5 d/

Source : ECE/ITD - Data bank of the Engineering Industries and Technology Section. Year to year percentage changes and annual average increase 1986/1979 are calculated from Government replies to ECE questionnaires for the Annual Reviews of Engineering Industries and Automation.

a/ Based on values expressed in national currency at current prices.
b/ Excluding land.
c/ Included with non-electrical machinery (ISIC, Rev.2 : 382).
d/ 1985/1979.
e/ Included with electrical machinery (ISIC, Rev.2 : 383).
f/ From 1984 onwards, included with electrical machinery (ISIC, Rev.2 : 383).

Figure III.10 (a-d)

Development of production, employment and productivity indices in precision instruments industries (ISIC 385) of four selected (standard) countries, 1979-1986 [a]

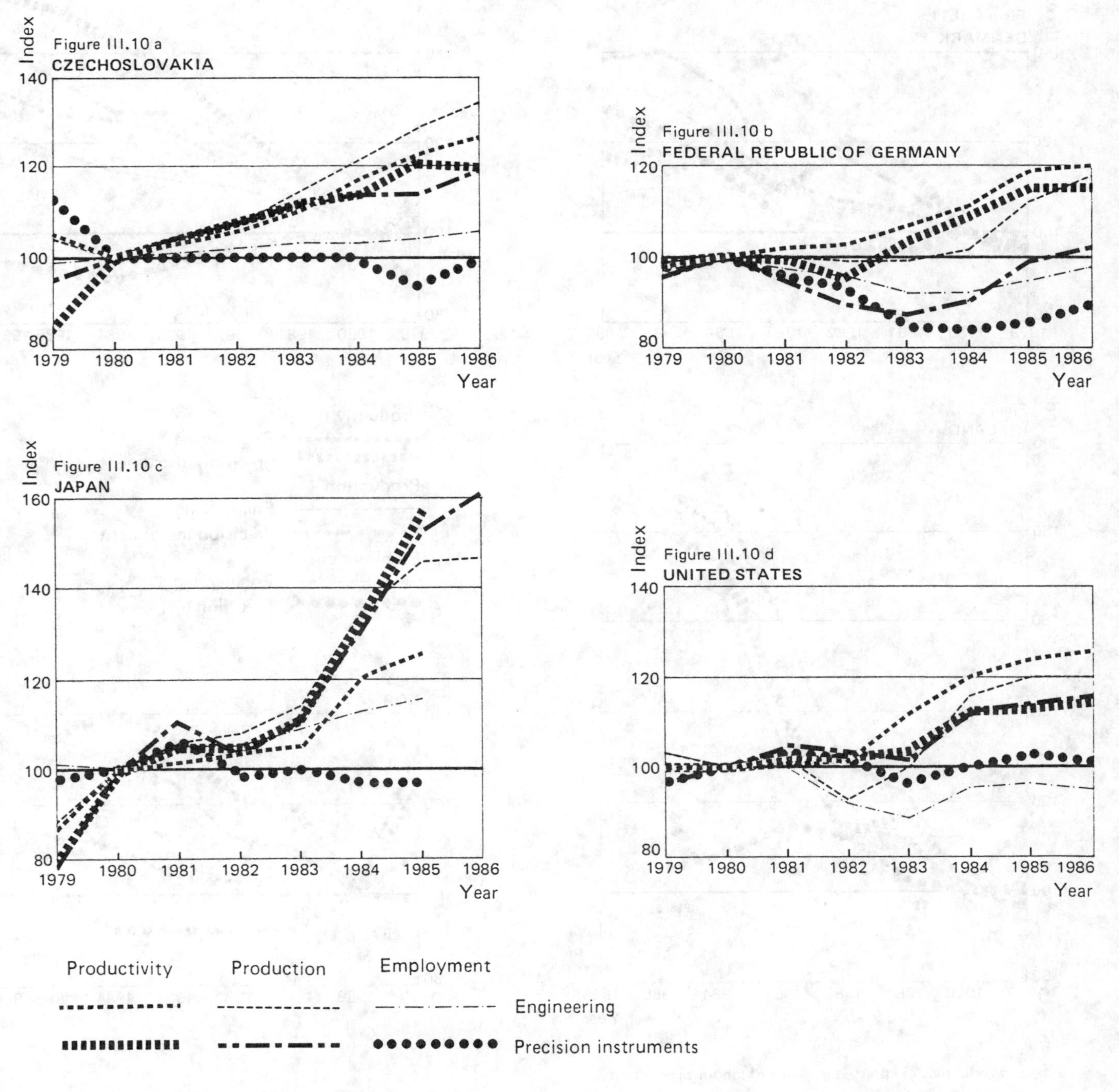

a Productivity expressed in terms of labour productivity. 1980 = 100.

Figure III.11 (a-d)

Development of production, employment and productivity indices in precision instruments industries (ISIC 385) of four selected (additional) countries, 1979-1986 [a]

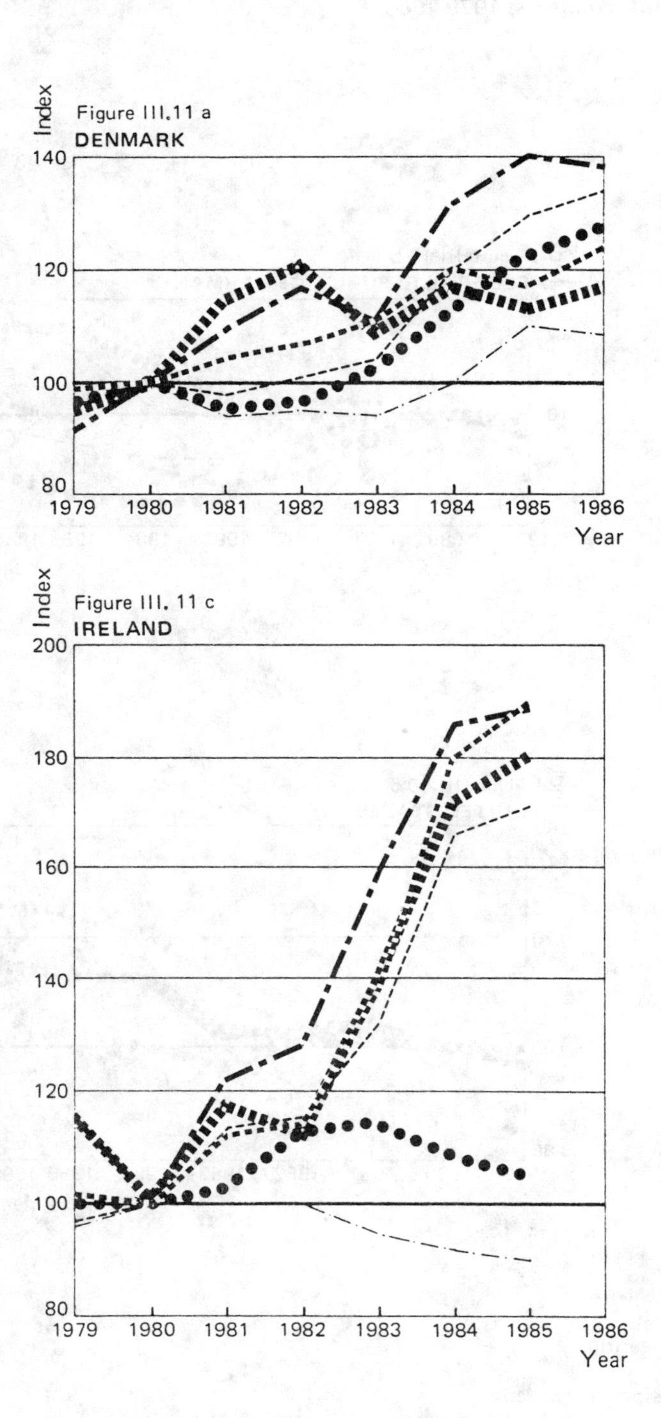

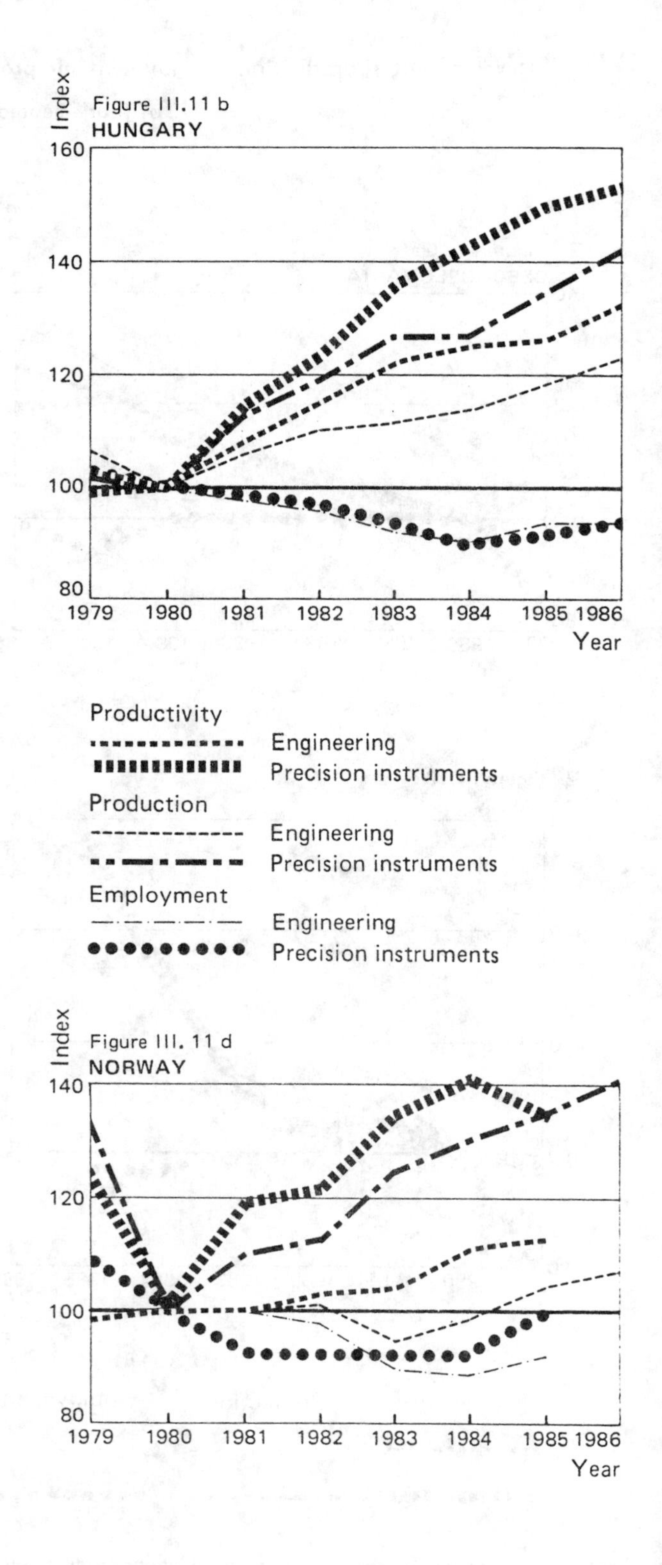

a Productivity expressed in terms of labour productivity.
1980 = 100.

Table III.57 Share of value added of precision instruments industries (ISIC, Rev.2:385) in engineering industries (ISIC, Rev.2:38) in ECE member countries and Japan, 1979-1986

(Percentage) a/

COUNTRY	1979	1980	1981	1982	1983	1984	1985	1986
Austria	3.2	2.4	2.2	2.1	2.0	2.3	...	...
Belgium	...	...	...	...	...	...	...	...
Bulgaria	...	...	...	...	...	...	...	...
Cyprus	0.0	0.0	0.0	0.0	0.0	0.0	0.0	...
Czechoslovakia	1.5	1.4	1.4	1.5	1.4	1.3	1.2	1.1
Denmark	7.0	6.8	7.4	7.4	7.7	8.1	8.0	...
Finland	2.9	2.9	3.3	3.1	3.4	4.2	3.9	3.9
France	2.9	2.7	2.5	2.5	2.6	2.7	2.8	...
German Dem.Rep.	...	...	...	...	...	...	...	...
Germany,Fed.Rep. of	3.7	3.8	3.7	3.4	3.2	3.0	3.2	...
Greece	...	0.6	0.5	...	...	...	...	...
Hungary	11.4	12.8	12.9	12.6	12.4	12.1	12.3	12.7
Ireland	12.8	11.4	12.1	13.1	13.1	...	...	...
Italy	5.8	5.7	5.6	5.9	2.4	3.2	...	...
Luxembourg b/	.	.	.	.	.	...	...	...
Malta	24.7	20.3	19.0	20.0	21.4	...	...	...
Netherlands c/	...	2.5	2.7	2.8	2.1	2.1	2.2	...
Norway	1.2	1.1	1.0	1.0	1.2	1.3	1.5	...
Poland	2.8	2.8	3.6	2.7	2.5	2.7	2.8	3.1
Portugal	1.0	1.2	1.0	1.8	1.8	2.4	...	...
Romania	...	...	...	...	...	...	...	...
Spain	1.6	1.3	1.3	1.2	1.2	1.3	...	...
Sweden	2.3	2.8	2.7	2.8	2.9	2.7	3.2	3.1
Switzerland	...	...	...	...	...	...	...	...
Turkey	0.5	0.4	0.6	0.8	0.6	0.5	...	...
USSR	...	...	...	...	...	...	...	...
United Kingdom	4.8	3.3	3.2	3.7	3.4	3.5	3.5	3.6
Yugoslavia d/	1.3	1.4	1.5	1.4	1.6	1.7	1.6	...
Canada	3.2	3.6	3.6	2.9	3.0	2.8	...	...
United States	7.7	8.3	8.6	9.2	9.2	9.0	8.9	...
Japan	4.0	4.1	4.1	3.8	3.9	3.6	3.6	...

Source : ECE/ITD - Data bank of the Engineering Industries and Technology Section.
Government replies to ECE questionnaires for the Annual Reviews of Engineering Industries and Automation.

a/ Based on values expressed in national currency at current prices.
b/ Precision instruments (ISIC, Rev.2 : 385) are included with transport equipment (ISIC, Rev.2 : 384).
c/ In 1979, engineering industries (ISIC, Rev.2 : 38) exclude precision instruments (ISIC, Rev.2 : 385).
d/ Social product including turnover taxes.

Table III.58 Share of total number of persons engaged in precision instruments industries (ISIC, Rev.2:385) in total number engaged in engineering industries (ISIC, Rev.2:38) in ECE member countries and Japan, 1979-1986

(Percentage)

COUNTRY	1979	1980	1981	1982	1983	1984	1985	1986
Austria a/	4.4	3.7	2.6	2.8	2.8	3.0	...	...
Belgium	1.7	1.8	1.8	1.8	1.8	1.8	2.0	...
Bulgaria b/	.	.	.	.	.	.	.	.
Cyprus	0.0	0.0	0.0	0.0	0.0	0.0	0.0	0.0
Czechoslovakia a/ c/	1.7	1.5	1.4	1.4	1.4	1.4	1.3	1.4
Denmark	6.0	6.2	6.2	6.4	6.8	7.0	7.0	7.4
Finland	3.0	2.5	2.6	2.6	2.6	2.9	3.4	3.4
France	4.0	4.0	4.0	3.9	3.9	3.9	4.1	...
German Dem.Rep. d/	.	.	.	.	.	.	.	.
Germany,Fed.Rep. of	4.4	4.5	4.4	4.3	4.1	4.0	4.1	4.0
Greece	...	1.0	0.7	...	...	...	...	...
Hungary	11.3	11.3	11.5	11.4	11.6	11.2	11.1	11.3
Ireland	10.2	9.8	10.0	11.0	11.7	11.4	11.4	...
Italy	5.0	4.8	5.0	4.8	2.4	2.2	...	...
Luxembourg e/	.	.	.	.	.	...	...	...
Malta	13.4	14.8	18.0	20.4	21.3	...	...	...
Netherlands	2.7	2.5	2.6	2.1	2.1	2.2	2.2	...
Norway	1.1	1.0	1.0	1.0	1.1	1.1	1.1	...
Poland	3.1	3.1	3.1	3.1	3.1	3.3	3.2	3.2
Portugal	1.8	1.9	2.0	1.9	2.0	2.0	...	...
Romania	...	...	...	...	...	...	...	...
Spain	1.6	1.3	1.4	1.3	1.3	1.2	...	...
Sweden	2.4	2.6	2.6	2.8	2.8	2.7	2.8	2.8
Switzerland	18.3	17.9	17.6	15.5	15.0	14.1	14.5	...
Turkey	0.7	0.9	1.2	1.2	1.1	1.1	...	...
USSR	...	...	...	...	...	...	...	...
United Kingdom f/	5.3	3.5	3.6	3.8	3.6	4.3	4.4	4.6
Yugoslavia	1.6	1.4	1.5	1.6	1.7	1.8	1.7	...
Canada	4.0	4.2	4.5	3.5	3.7	3.5	...	...
United States	7.6	8.1	8.3	8.8	8.8	8.4	8.5	8.5
Japan	5.3	5.4	5.4	5.1	4.9	4.6	4.5	...

Source : ECE/ITD - Data bank of the Engineering Industries and Technology Section.
Government replies to ECE questionnaires for the Annual Reviews of Engineering Industries and Automation.

a/ Including homeworkers.
b/ Precision instruments (ISIC, Rev.2 : 385) are included with non-electrical machinery (ISIC, Rev.2 : 382).
c/ State national industry only.
d/ Precision instruments (ISIC, Rev.2 : 385) are included with electrical machinery (ISIC, Rev.2 : 383).
e/ Precision instruments (ISIC, Rev.2 : 385) are included with transport equipment (ISIC, Rev.2 : 384).
f/ At June of each year.

Table III.59 Share of gross fixed capital formation of precision instruments industries (ISIC, Rev.2:385) in engineering industries (ISIC, Rev.2:38) in ECE member countries and Japan, 1979-1986

(Percentage) a/

COUNTRY	1979	1980	1981	1982	1983	1984	1985	1986
Austria	2.4	2.0	1.1	1.4	1.6	1.9	...	...
Belgium	...	...	...	...	...	...	...	...
Bulgaria b/ c/	.	.	.	.	.	.	.	...
Cyprus	0.0	0.0	0.0	0.0	0.0	0.0	0.0	0.0
Czechoslovakia	...	...	...	...	...	...	...	...
Denmark	6.9	5.7	6.4	7.0	7.3	7.7	8.4	...
Finland	2.5	3.0	2.6	3.7	3.0	3.4	4.6	4.6
France c/	.	.	.	.	.	.	.	...
German Dem.Rep. d/	.	.	.	.	.	.	.	...
Germany,Fed.Rep. of	2.8	2.9	...	2.5	2.6	2.8	2.8	...
Greece	...	1.7	0.8	...	...	...	...	...
Hungary	6.2	6.6	7.3	9.4	10.2	10.7	9.7	7.6
Ireland	10.4	18.4	10.8	16.7	13.6	...	...	...
Italy	5.3	6.8	6.8	4.7	1.9	1.4	...	...
Luxembourg	...	...	...	...	...	...	...	...
Malta	6.3	15.8	20.6	16.1	4.3	...	...	...
Netherlands	3.7	2.8	2.8	3.0	1.6	1.4	1.6	1.7
Norway	1.0	0.9	0.6	0.7	2.6	0.9	0.9	...
Poland	1.4	2.0	2.2	3.2	2.8	3.2	2.1	...
Portugal	0.6	1.8	0.7	1.0	1.3	1.5	...	...
Romania	...	...	...	...	...	...	...	...
Spain	0.7	0.7	0.6	1.2	0.8	0.7	...	...
Sweden	1.9	2.3	1.9	2.5	2.3	1.8	2.2	2.4
Switzerland	...	...	...	...	...	...	...	...
Turkey	0.1	0.1	0.2	0.5	0.4	0.3	...	...
USSR	...	...	...	...	...	...	...	...
United Kingdom e/	4.1	2.5	2.3	2.9	3.2	.	.	.
Yugoslavia	1.1	0.8	1.3	1.6	1.4	0.8	1.0	...
Canada	1.8	1.9	2.5	3.1	3.9	3.5	...	...
United States	...	...	...	...	...	...	...	...
Japan	4.7	4.2	4.4	3.8	3.6	2.9	3.2	...

Source : ECE/ITD - Data bank of the Engineering Industries and Technology Section.
Government replies to ECE questionnaires for the Annual Reviews of Engineering Industries and Automation.

a/ Based on values expressed in national currency at current prices.
b/ Excluding land.
c/ Precision instruments (ISIC, Rev.2 : 385) are included with non-electrical machinery (ISIC, Rev.2 : 382).
d/ Precision instruments (ISIC, Rev.2 : 385) are included with electrical machinery (ISIC, Rev.2 : 383).
e/ From 1984 onwards, precision instruments (ISIC, Rev.2 : 385) are included with electrical machinery (ISIC, Rev.2 : 383).

Table III.60

Production of selected precision instruments in the United States, 1975-1985

Type of product	*Value of shipments in billions of 1982 US dollars (industry data)* 1975	1980	1985	*Index* 1980/1975	1985/1980
Engineering and scientific instruments	2.8	3.2	3.6	114.3	112.5
Optical devices and lenses	1.1	2.9	5.4	263.6	186.2
Environmental controls	1.3	1.8	1.8	138.5	100.0
Controlling instruments	2.4	4.0	4.0	166.7	100.0
Fluid meters and devices	0.7	1.0	0.8	142.9	80.0
Instruments to measure electricity	4.3	7.0	7.2	162.8	102.9
Surgical and medical instruments	2.5	3.2	4.8	128.0	150.0
Photographic equipment and supplies	12.0	16.4	17.3	136.7	105.5

Source: U.S. Industrial Outlook 1988, Department of Commerce, Washington, D.C.

Table III.61

World market for watches, 1975-1986

Year	*World* Units	Swiss francs	*Switzerland* Units	Value	*Japan* Units	Value	*Hong Kong* Units	Value
	(millions)		*(percentage share of world total)*					
1975	218	8 250	34	40	14	19	3	1
1980	300	9 500	29	39	29	30	24	11
1985	451	11 000	13	40	39	36	21	5
1986	550	12 000	14	40	n.a	n.a	n.a	n.a

Source: Annual Review of Engineering Industries and Automation 1986, vol.I (based on *Financial Times*, 28 April 1987. Watchmaking Industry: Switzerland on the advance, Bulletin Credit Suisse, 1-87).

Table III.62

Production of precision instruments (ISIC 385) by geographic region, 1975-1985

Region	*Value in billions of constant 1975 US dollars* [a]			*Index*	
	1975	*1980*	*1985*	*1980/1975*	*1985/1980*
Total world	124.2	183.0	238.4	147.3	130.3
of which:					
Developed market economies	42.8	64.3	77.3	150.2	120.2
of which:					
North America	22.1	28.6	39.3	129.4	137.4
Europe	14.8	16.5	17.4	111.5	105.5
Japan and others	5.9	19.2	20.6	325.4	107.3
Centrally planned economies	79.8	117.7	158.7	147.5	134.8
of which:					
European	76.7	113.8	153.5	148.4	134.9
Asian	3.1	3.9	5.2	125.8	133.3
Developing countries	1.6	1.0	2.4	62.5	240.0
of which:					
African	-	-	-		
Asian	0.7	0.7	1.9	100.0	271.4
Latin America	0.9	0.3	0.5	33.3	166.7

Source: As for table III.11.

a See table II.1.

Table III.63

Share of geographic regions in world production of precision instruments, 1975-1985

(Percentage)

Region	*1975*	*1980*	*1985*
Total world	100.0	100.0	100.0
of which:			
Developed market economies	34.5	35.1	32.4
of which:			
North America	17.8	15.6	16.5
Europe	11.9	9.0	7.3
Japan and others	4.8	10.5	8.6
Centrally planned economies	64.2	64.3	66.6
of which:			
European	61.7	62.2	64.4
Asian	2.5	2.1	2.2
Developing countries	1.3	0.6	1.0
of which:			
African	-	-	-
Asian	0.6	0.4	0.8
Latin America	0.7	0.2	0.2
ECE region	91.5	86.8	88.2

Source: As for table III.11.

REFERENCES

III.1 See also: international Labour Organisation, Metal Trades Committee, *11th session - General Report* ILO, Geneva 1983.
Technologies mentioned represent only a selection of emerging, mainly information technologies, significantly influencing the structure of manpower (as studied by ILO). It is obvious that the introduction of new materials, as well as recent developments in laser and plasma technology nuclear engineering etc., have also contributed significantly to the innovation of manufacturing approaches.

III.2 *Recent Trends in Flexible Manufacturing.* UN/ECE publication, Sales No. E.85.II.E.35..

III.3 World crude-steel production in 1979-1986 remained at the annual level of 700-750 million tonnes (747 in 1979, 718 in 1980, 719 in 1985 and 715 million tonnes in 1986). The share of the ECE region in world steel production declined from some 70% in 1979 to less than 65% in 1986. For details, see UN/ECE publications *The Steel Market in 1981* (Sales No. E.82.II.E.16) and *in 1986* (Sales No. E.87.II.E.32).

III.4 See also *Annual Review of Engineering Industries and Automation 1986,* vol.I, pp.19-20. UN/ECE publication, Sales No. E.88.II.E.16.

III.5 For detailed analysis see the recently published ECE study *Software for Industrial Automation.* UN/ECE publication, Sales No. E.87.II.E.19.

III.6 See also *Annual Review of Engineering Industries and Automation 1986,* vol.I, pp.21-51. The number of NC-machine tools, industrial robots and FMS (FMU, FMC) installed indicates the general level of manufacturing automation in individual industrialized countries.

III.7 Figure III.5, as published in the *Annual Review* for 1986, is based on data collected and compiled, on a regular basis, by the *American Machinist* (February issues, 1973-1987). The four largest producers (Japan, Federal Republic of Germany, Soviet Union and United States) are also the dominant machine-tool consumers.

III.8 For technical programmes, conclusions and recommendations, see Reports of ECE Seminars - documents *ENG.AUT/SEM.3/4* (Seminar on Flexible Manufacturing Systems: Design and Applications, Sofia, September 1984) and *ENG.AUT/SEM.5/4* (Seminar on Industrial Robotics '86 - International Experience, Developments and Applications, Brno, February 1986).
Relevant studies: *Production and Use of Industrial Robots.* UN/ECE publication, Sales No. E.84.II.E.33; *Recent Trends in Flexible Manufacturing.* UN/ECE publication, Sales No. E.85.II.E.35.

III.9 *Trends in the Electrical and Electronics Industries.* ECE document ECE/ENG.AUT/31, Geneva 1988. This internal study is divided into three parts (chapters): (I) A review of the present state of the electrical and electronics industries, (II) Characteristics of the electrical and electronics industries and (III) Some aspects of international co-operation. Chapter III includes detailed information on the standardization activities of the International Electrotechnical Commission (IEC), which complements the information on the activities of the International Organization for Standardization (ISO), as described in subchapter IV.7 of the present study.

III.10 *The Telecommunication Industry: Growth and Structural Change.* UN/ECE publication, Sales No. E.87.II.E.35.

III.11 *Digital Imaging in Health Care.* UN/ECE publication, Sales No. E.86.II.E.29.

III.12 Report of the Seminar on Automation Means in Preventive Medicine '87, held at Piestany (Czechoslovakia) from 28 September to 2 October 1987 (ENG.AUT/SEM.6/3). The Seminar made several recommendations for further ECE engagement in selected fields of medical engineering, including the study of rehabilitation engineering (taken in a broad sense to include technical aids for both the handicapped and the old). See also Report of the eighth session of the Working Party on Engineering Industries and Automation, held in February 1988 (ECE/ENG.AUT/34, annex I).

III.13 *Annual Review of Engineering Industries and Automation 1986,* vol.I. UN/ECE publication, Sales No. E.88.II.E.16, chapter II.D - Manufacture of transport equipment.

III.14 See individual issues of the *Annual Review* (covering 1979-1986). The Working Party has, in several projects, examined developments in the transport equipment industry: e.g. studies *Development of Airborne Equipment to Intensify World Food Production.* UN/ECE publication, Sales No. E.81.E.24; *Techno-Economic Aspects of the International Division of Labour in the Automotive Industry.* UN/ECE publication, Sales No. E.83.II.E.14; and Seminar "Shipbuilding 2000 Maritime Conference BALTEXPO '88" Gdansk, Poland, September 1988 (Report document ENG.AUT/SEM.7/3).

CHAPTER IV WORLD TRADE AND OTHER FORMS OF INTERNATIONAL CO-OPERATION IN ENGINEERING INDUSTRIES

IV.1. LONG-TERM TRENDS IN WORLD TRADE IN ENGINEERING PRODUCTS

The contemporary engineering industries constitute a huge and highly diversified "macro-sector" composed of hundreds of groups and subgroups (types of manufacturing). The number of products manufactured is enormous, approaching some half a million different types of items, from bolts and nuts to large computer mainframes or spacecraft, to say nothing of the various dimensions and other modifications within the item types.

It is quite evident and natural that no country alone can satisfy its entire domestic demand for all engineering goods. Neither can it, in most cases, achieve the economy of scale to ensure economically justified batches of engineering products. Therefore, export (and import) orientation is an intrinsic feature of the engineering industry.

In the post-war period, international trade in machinery and equipment progressed, on average, faster than total world trade. During 1965-1985, world exports of engineering goods increased by a factor of 12-13, compared with 10-11 for total merchandise exports. While the share of engineering goods in cumulative exports remained practically constant at 30%, in some periods, variations of finished goods, raw materials and energy prices caused noticeable fluctuations (from 33% in 1970 to 27% in 1980 and 32% in 1985 when oil prices dropped again).[IV.1]

The situation in the engineering goods market has also been influenced by the cyclical development of many economies, the increasing debt of developing countries and the fluctuating financial climate.

In this respect, one cannot overestimate the role of technological progress as a major factor in the sectoral and regional restructuring of the engineering goods market.

In general, the influence of all the above-mentioned factors on world trade in the 1980s may be judged as negative.

In 1980-1985, world exports of machinery and equipment increased from $US 511 billion to 591 billion (at current prices), that is, by 15.7%, compared with a twofold increase in 1975-1980. Actually, they even shrank if measured at constant prices, especially in such traditional segments as exports of agricultural machinery, tractors, metalworking equipment etc.

This negative trend has been offset in part by growing exports of high-technology products, primarily computers (from $US 12.9 billion in 1980 to 26.0 billion in 1985) and electronic components (from $US 9.6 billion to 19.9 billion, respectively). Quite a considerable increase has been achieved in telecommunication equipment and in the electronics industries in general.

The competitive positions of various countries and regions have changed as a result of all these events. Exports of engineering goods of some west European countries (the Federal Republic of Germany, France, Italy and the United Kingdom) decreased by 10-20% in 1980-1985 (in value terms), as well as in their share in the world market. The same is to some extent applicable to the United States.

In contrast, Japan has, since 1981, maintained its position as the leading world exporter of machinery and equipment. Japanese exports of engineering goods have jumped from $76 billion in 1980 to 146 billion in 1986. Nowadays, more than 20% (by value) of all such goods exported in the world are "made in Japan". Japanese computer exports (basically personal computers and peripheral devices to the American and European mainframes) have rocketed from $542 million in 1980 to 4.6 billion in 1985), lagging behind the United States only.

Some newly industrialized and developing countries have also strengthened their competitiveness as exporters of engineering goods.

Since 1963, ECE publishes regularly annual *Bulletins of Statistics on World Trade in Engineering Products*. These publications provide basic information on export data, as reported by individual countries, on the basis of Standard International Trade Classification, Revision 2 (SITC, Rev.2).[IV.2]

The structure and development of exports (and imports) of individual ECE member countries is further analysed in subchapter IV.2 (for 1979-1987). In this introductory subchapter, an attempt is made to illustrate some long-term trends in world trade (including non-ECE countries) in 1961-1985 arising from the development of engineering imports. This subchapter is based on a study recently prepared by the ECE General Economic Analysis Division (GEAD).[IV.3]

The following review of global trends in imports of engineering goods is an extension of a recent secretariat study which focused upon the machinery and equipment imports of the European centrally planned economies. Data presented in that study showed that the volume of the Soviet Union's machinery purchases rose throughout the 1970s and into the early 1980s. This reflected roughly similar increases in the import volumes of these goods from the country's socialist and non-socialist (largely western) trade partners. By contrast, the total imports of these goods into the east European countries peaked in the mid-1970s and declined thereafter. The volume of these countries' imports from the socialist economies fell slightly during this period, but picked up again in the mid-1980s. Imports from the non-socialist area evolved even more unfavourably, falling by some 40% in volume from 1977 to 1984. While the magnitude of these imports from the non-socialist area was substantially less than that from the socialist countries, these flows had a disproportionate significance for the introduction of advanced technology. These trends, among others, reflected the relative stagnation of domestic investment in eastern Europe and contributed to the aging of these countries' stock of fixed capital, a phenomenon also present in the Soviet Union. [IV.4]

Noting these developments, it appears important to determine how imports of machinery and equipment into the eastern ECE countries compare with those of other countries and country groups. [IV.5] Had the European centrally planned economies somehow fallen behind their major competitors in this respect? Without an examination of the inflows of engineering goods into other regions, the question could not be confidently answered since certain other countries (e.g., those of Latin America) also reduced imports when external adjustments became necessary.

In the geographic breakdown adopted in this subchapter, the world has been divided into three major country groups: the developed market economies, the centrally planned economies of Europe, and the developing countries. The further breakdown of major country groupings shown in the tables aims at capturing formal trade associations or other geographic classifications adopted by the United Nations for the presentation of trade statistics. Selected data for individual countries are provided in table IV.1. [IV.6]

Volume estimates of engineering imports of the developed and developing market economies were obtained by deflating the value series by indices of machinery and equipment export prices. [IV.7] Approximating the engineering goods imports of developing areas from exports to these countries from the developed market economies and the Soviet Union results in some underestimating of the total value of such purchases. Not reflected in these estimates are the developing countries' intra-trade and their imports from eastern Europe. [IV.8] An idea of the magnitude of the resulting error can be obtained from table IV.2, which shows GATT estimates of the value of engineering goods imports by origin and destination in 1984. Bearing in mind that the geographic and product definitions differ somewhat from those used in this study (the GATT data in table IV.2 include trade in all vehicles and those consumer durables classified as engineering goods), the GATT data show the engineering goods exports of the centrally planned economies to the developing countries and the intra-trade of the developing countries combined to account for some 18% of total engineering goods imports of the developing countries in 1984. In fact, the estimates presented in this subchapter cover a part of this potential shortfall because they include data for total engineering goods imports of a limited reporting group of developing countries and Soviet exports to the others.

The definition of engineering goods used in this subchapter corresponds broadly to that adopted by the ECE Working Party on Engineering Industries and Automation, except that passenger automobiles (SITC 7321) have been *excluded*. The reason for this modification stems from the aim to achieve coverage for engineering products which is more consistent with the concept of productive capital investments. None the less, the engineering goods aggregate used here still contains some products which do not meet this criterion: goods destined largely for final consumption, such as household appliances and consumer electronics and certain components. Although some imported components serve to upgrade existing machinery or as intermediate inputs for machinery produced for domestic use, other components, the relative importance of which is unknown, have no impact on fixed investment.

In the international trade statistics of eastern Europe and the Soviet Union, exchanges with socialist countries (denominated in roubles) must be analysed separately from those with non-socialist countries (in convertible currencies). The statistical and analytical distinction between the two trading areas is important because of the differences in trade regimes and the sharp divergencies of price movements prevailing in these two markets from the early 1970s on. In order to present data for the total of eastern imports from both currency areas, all eastern trade data have been converted into United States dollars. The procedure involves converting data for non-socialist trade of these countries into dollars by means of the appropriate official exchange rates between the domestic currency (or foreign trade accounting unit) and the dollar. Data for trade with socialist countries have been converted from domestic currencies into transferable roubles (TR), and then to dollars by applying the 1980 *Soviet* official rouble/dollar exchange rate.

Estimates of the stock of imported capital (engineering) goods at the end of 1985 were obtained by cumulating the annual imports of these goods (at constant 1980 dollars) for the period 1961-1985. These figures were then used to calculate an estimate of the age structure of imported capital, defined here as the shares of the total imported capital stock at end 1985 imported during the five-year periods 1961-1965, 1966-1970, 1971-1975, 1976-1980 and 1981-1985.

Underlying these estimates is the assumption that none of the engineering goods imported in 1961-1985 have been retired, which is equivalent to positing a 25-year lifetime for equipment. While that may appear rather long, a recent study of machinery and equipment in several western countries [IV.9] concludes that service lives of over 20 years are not uncommon. [IV.10] The service life of equipment appears to be even longer in eastern Europe and the Soviet Union. [IV.11]

In summary, an attempt has been made to achieve world coverage for engineering goods imports in volume terms. This has required the utilization of a host of international and national data sources along with a number of key assumptions, especially as regards world price developments of engineering goods and the end-use of these products. Thus the estimates concerned should be treated as a rough approximation.

The *developed market economies* are the largest importing area. In constant 1980 dollars, their imports of engineering goods cumulated to almost $1,600 billion during the five-year period 1981-1985, accounting for some 58% of global imports (see tables IV.3 and IV.4 and figures IV.1 and IV.2). Within this grouping, the imports of the *west European* countries have been almost twice as large as those of *North America.* Since the mid-1970s, the shares of these two subregions have diverged, that of western Europe declining steadily while North America's climbed steeply in 1981-1985. *Japan's* share of global engineering goods imports remained unchanged, at less than 2%.[IV.12]

Eastern Europe and the Soviet Union together have accounted for some 11-13% of global engineering goods imports, a share which has shown a long-term tendency to decline. *Eastern Europe's* share peaked in the first half of the 1970s and has fallen since, owing to the fall in the volume of its imports. The experience of the *Soviet Union* has been the opposite.

On an overall basis the *ECE region* has consistently accounted for two thirds or more of global imports of engineering goods.

The *developing countries'* share of world engineering goods imports rose markedly in the past decade. This reflects the increasing importance of *Asia* as an importing region. In the latest five-year period, this region accounted for over 17% of the global market. Within Asia, the largest importers are the *Pacific Rim* and *Asia Middle East.* The shares of both had risen until the early 1980s, when that of the Middle East peaked while the Pacific Rim's share continued to move upward. Developments in these two regions stand in marked contrast to those in *other Asia,* which includes such countries as India and Pakistan. As a market, the imports of this group have only doubled in volume terms since the early 1960s, and their share in global machinery imports declined until the late 1970s, after which it moved upward.

Latin America's presence in the market for engineering goods has also diminished in relative terms, although the volume of its imports has more than tripled since the 1961-1965 quinquennium. At that time, its purchases were comparable to those of Asia (10% of the global market), but in 1981-1985 its share slid to only 6%.

Of the developing country areas, *Africa's* imports of engineering goods have consistently been the smallest and accounted for an unchanging 5-6% of the world market. In the most recent period, the *North African* countries have been the continent's most active importers.

The impressive long-term rate of growth of engineering goods imports of the Pacific Rim and Asia Middle East is demonstrated by the fact that all other regions shown in table IV.3 which had registered the same or lower levels of imports in the first half of the 1960s had been surpassed in this respect by the 1980s, often by large amounts. The Pacific Rim's share of the world market also exceeded those of eastern Europe and the Soviet Union (taken separately) for the first time in the 1981-1985 period (table IV.4).

In table IV.5, the 25 largest importers of engineering goods have been ranked according to their cumulated purchases during 1981-1985. In addition to the most industrialized countries of the ECE region, the list includes several oil exporters and most of the countries which constitute the Pacific Rim. (Similar data for all countries are included in table IV.1.)

A deceleration in growth during the past decade has been one of the major features in the development of global engineering goods trade. Growth rates averaged around 12% in volume in the 10-year period 1966-1975, but fell to 7% in 1976-1980 and then to only about 4% in the first half of the 1980s (table IV.6 and figure IV.3).

It was the period around the first oil price shock that gave rise to developments which had a profound impact on both the global economy and engineering goods trade. In the five years prior to the quadrupling of world oil prices at the end of 1973, the volume of engineering goods trade had on average expanded at almost 13% annually, generally with only small differences in the growth rates registered by the major regions.

In 1974, the year immediately following the oil price shock, the pace of world engineering import growth continued at a rate of some 15% (the same as in 1973) but then slowed to 8% in 1975-1976. A salient feature of this period is the emergence of considerable differences in regional growth rates. On the one hand, engineering goods imports into the *developed market economies* slowed sharply, those of the *European Economic Community, North America* and *Japan* actually having turned negative in 1975, as western economies reacted to the oil price rise by implementing restrictive macro-economic policies.[IV.13] On the other hand, imports into the *developing countries* accelerated to rates over 30% per annum in 1974-1975, while those into the *European centrally planned economies* continued to expand at a lower but still high rate.

The ballooning revenues of *Asia Middle Eastern* oil exporters enabled them to raise engineering goods imports by 60% in volume in 1974 and then to almost double them in 1975. Somewhat smaller but still very large increases were registered by *North African* countries, most of which also benefited from higher fuel prices. This also was the case for the *Soviet Union,* although the import increases were not as large. By contrast, *non-oil-producing developing countries* and *eastern Europe* had suffered terms-of-trade losses which reduced their import capacity. The financial liquidity in international markets created by the OPEC surpluses, reinforced by the depressed demand for private credit in the developed market economies, greatly facilitated the financing of productive investment and other purchases of the rest of the world. This took the form of borrowing both by governments and by transnational corporations, the aim of the latter often being to boost foreign direct investment. Conditions were particularly attractive in the mid-1970s, since real interest rates were low or even negative. In addition, engineering goods suppliers, faced with weak demand in the developed market economies, resorted to aggressive export promotion, notably through the increased use of government-financed or guaranteed export credit facilities. On an overall basis, stagnating demand in the developed market economies was to a large extent offset by the swelling of engineering goods purchases in the rest of the world.

With the exception of 1980 and 1984, volume growth of world engineering goods imports slowed during the *1976-1985 period* and in 1982-1983 actually turned negative. Initially, the deceleration was due mainly to the combined effect of markedly lower growth of imports into the developing countries and eastern Europe and the Soviet Union. Developed market economies' imports expanded only somewhat more slowly than in the preceding quinquennium. However, by the early 1980s the import growth of all major regions had slowed, except in North America.

As regards the *developing countries*, the slow-down in engineering goods import growth was the outcome of several major interacting international economic developments. Already in the 1970s some countries undertook measures to adjust to rising indebtedness, which generally required import growth to be held down. Even though international borrowing continued at a brisk pace, funds were also increasingly being allocated to debt servicing, the burden of which was raised by the sharp rise in interest rates at the end of the decade. As a result, *Latin America's* volume of imports advanced only marginally in the second half of the 1970s. *Africa's* import growth held up well only until 1977, after which the volume declined markedly.

Around the turn of the decade, the world economy experienced several shocks which, on the whole, had further slowing effects on the engineering goods trade (table IV.4). With the advent of the second round of oil price increases in 1979, oil bills swelled and further downward pressure was placed on other imports, including those of engineering goods. Interest rates rose further, while the western recession and the accompanying drop in import demand curtailed export revenues. Moreover, with the onset of the debt crisis in 1982, commercial bank lending to the developing countries was sharply reduced, which contributed to a shift towards a net outflow of financial resources. Finally, direct foreign investment flows into the developing countries fell sharply.

These pressures compelled the countries of both *Latin America* and *Africa* to cut back on their engineering goods imports in the 1981-1985 period. Latin American imports were particularly hard hit. There had been a strong upturn in engineering goods import growth rates in 1980-1981, but in the wake of the debt crisis they dropped by over 40% in volume (in 1982-1983). This was followed by an upturn in 1984-1985, which was, however, insufficient to compensate for the previous drop.

The *east European* countries were also affected by many of the same factors. A combination of a generally cautious attitude toward the accumulation of debt, the priority of other commodity imports, and rising debt service obligations led them to disproportionately curtail engineering goods imports from the non-socialist countries already in the mid-1970s.[IV.14] In addition, external adjustment to relative price changes in eastern Europe's socialist trade and the resulting losses in their terms of trade was achieved in part by slowing down imports of engineering goods from *socialist* trade partners. Together, these two developments resulted in the cessation of growth in total engineering goods imports. These pressures continued into the 1980s, when they were intensified by the sharp curtailment in 1982 of international credits available to eastern Europe. As a consequence, imports of engineering goods from the non-socialist countries were reduced further, although there was some recovery in 1984-1985.

The *Soviet Union's* import capacity was not similarly constrained during most of this period. Its export revenues from the convertible currency area rose steeply into the early 1980s, owing to a combination of higher fuel export volumes and/or prices, and then levelled off. Also, gold sales provided a continuing source of revenue. The surge in imports of engineering goods in 1982-1983 - destined for the Urengoi gas pipeline - was none the less to a large extent financed on credit. As regards Soviet trade with the socialist countries, adjustment to gains in its terms of trade stemming from fuel price rises in intra-CMEA trade largely took the form of increased volumes of machinery and equipment imports from eastern Europe. In general, the pace of Soviet machinery imports first slackened to 5% annually in 1976-1980 and then picked up to 7% in 1981-1985.

Oil-producing developing countries' imports of engineering goods tended to follow the movement of oil prices. Imports shot up in the wake of the first round of oil price increases, as noted. However, as purchasing power dipped when oil prices eased downwards and the dollar depreciated in the second half of the 1970s, *Asia Middle East's* and *North Africa's* import growth rates plunged, actually turning negative in 1979. Import expansion recommenced after the second oil price rise, only to turn negative when prices dropped again. In 1983-1985, the volume of Asia Middle East's engineering goods imports contracted by over 40%.

The highest average growth rates of engineering goods imports throughout this period were achieved by the *Pacific Rim,* with an annual average of 15% in 1976-1980, lowering to some 6% in 1981-1985. The imports of this group of countries appeared relatively less affected by the factors discussed above. Their receipts were based largely upon the exports of manufactures, which expanded strongly and steadily throughout most of the period. From the mid-1970s on, *other*

Asia's imports grew almost as rapidly as those of the Pacific Rim, and actually expanded more quickly in 1981-1985. The latter period reflects the particularly fast growth registered by India.

The highest growth rate of engineering goods imports achieved by any group of countries in the 1981-1985 quinquennium was that of all the *centrally planned economies of Asia*. This reflects entirely the imports of the People's Republic of China, the other countries of that group being considerably smaller importers and managing much lower growth rates. China's imports of engineering goods tripled in the two years 1984-1985.

In the period 1981-1985, *western Europe's* average rate of import growth was more than halved compared with the preceding quinquennium, owing to the sharp contraction of imports into virtually all countries during the western economic recession in 1982-1983. Low imports in these years were offset by the strong recovery in 1984-1985.

North America was the only major region whose imports of engineering goods imports accelerated significantly in 1981-1985 (to nearly 14%) — largely because of the step-up in United States purchases. The pace of import growth into Canada also picked up, but to a markedly lesser extent. In 1984, the United States' engineering goods imports grew by nearly 40% in volume, owing to the convergence of high internal demand (gross fixed capital formation grew by 16%) and the strong appreciation of the United States currency.

As noted above, the world economy suffered several shocks during the past decade to which countries were forced to adjust, although to differing extents. In general, the burden of economic adjustment was borne by domestic investment. In the developed market economies and eastern Europe and the Soviet Union, the slowing or stagnant import growth of engineering goods reflected the slow-down or absolute decline in overall capital formation, which contributed to the aging of domestic stocks of fixed capital.[IV.15] In both cases, these adverse developments are believed to be partially responsible for the deceleration in productivity growth. Similarly, adjustments undertaken in the developing countries, especially in the aftermath of debt crises, often caused engineering goods imports to be scaled back disproportionately, as is illustrated by the fall in the share of these goods in total developing country imports.[IV.16]

Up to this point, the focus has been upon establishing the magnitudes and regional dynamics of world machinery imports. While these results are of interest in themselves, the geographic pattern of engineering goods imports may also be important for the development of countries' relative technological position and for their export performance. To the extent that this is in fact the case, the global changes in the import flows of engineering goods documented above could cause some realignments in export competitiveness in the ECE region with implications for the future development of east-west trade.

As already noted, the engineering goods trade is a prime vehicle for the international transfer of technology. The development of this trade, and more specifically of imports, is *one of the factors* which determines the technological characteristics of the domestic stock of fixed assets, the potential growth of new and improved products and reductions in production costs. Domestic policies toward science and technology and the development of total investment would seem to predominate, particularly among the industrialized countries. In this respect, two measures are presented which allow inter-country (inter-regional) comparisons of the recent developments in this regard. The assumption that the engineering goods imported in a given year are technologically superior to and more productive than those purchased in prior years is crucial to the interpretation of both indicators.

The average rate of *growth of engineering goods imports* during 1981-1985 gives some indication of the rate at which new foreign technology was recently introduced into domestic capital stocks. Thus those countries having registered the fastest growth rates of the *imported* engineering goods during this period are likely to have experienced the most rapid improvement in technological parameters of their *imported* capital stocks. North America and Asia (especially the Pacific Rim) stand out in this respect, while Latin America and Africa registered an absolute decline in volumes (see figure IV.4).

The other indicator, the *age (or vintage) structure of the gross stock of imported capital,* provides a measure of the degree of modernization of a country's productive apparatus (table IV.7). Inter-country (inter-regional) comparisons of age structures give an idea of the relative *rate* at which the technical characteristics of their imported capital stocks have changed. Thus a country with a relatively young imported capital stock, i.e., one in which assets five years old or less represent a comparatively high share, is presumed to have modernized more quickly than countries with low shares. It should be stressed, however, that similar age structures *do not* necessarily imply similar *levels* of imported technology. Undoubtedly there are considerable differences in the sophistication of technology being traded at any given time, countries purchasing technology appropriate to their levels of economic development. At the end of 1985, regions with the most favourable age structures measured in this way were Asia (especially the Pacific Rim) and North America, while Africa, Latin America and eastern Europe showed the oldest age structures (figure IV.5). Eastern Europe's non-socialist machinery imports show an even less favourable age structure (see table IV.1 and IV.7). This is, however, not the case for the Soviet Union, whose socialist and non-socialist engineering goods imports display very similar vintage structures.

Indirectly, an inflow of engineering goods can boost *exports* through the augmentation of domestic supply, exploitation of international specialization and superior foreign technologies, introduction of new product lines, and so on. Provided they are efficiently absorbed (reports of gross underutilization or even non-installation of imported capital goods occasionally surface), the new technology embodied in engineering goods imports improves the characteristics of the productive capital stock and thereby creates the potential for advancement in international competitiveness. However, the *end-use structure* of these imports is also important.

The long-term dynamics of *east-west trade* has often been of some concern. At first this concern focused upon the unfavourable commodity structure of *eastern exports,* which has been heavily weighted towards primary commodities. On the one hand, the scope for increasing the supply of many of these goods in the eastern countries is limited. On the other hand, world demand for primary goods has generally been less dynamic than that for manufactures. Accordingly, the role of manufactures in the east's total exports would have to be progressively raised if east-west trade is to maintain its share of world trade. However, contrary to the improvement required in this respect, findings documented in other ECE publications [IV.17] and elsewhere revealed a sluggishness in the eastern countries' exports of manufactures which was reflected in the erosion of their share of the western market since the mid-1970s.

The possibility of a further erosion in market position is raised by the low level of eastern Europe's *imports of machinery and equipment* during the past decade. [IV.18] A continuation of sluggish export revenues would in turn necessarily constrain eastern import growth in the long term and thus all east-west trade. Recourse to credits, of course, would offer a means of financing faster import growth, but this is only a limited option for a number of eastern countries.

While the central role is given here to engineering goods imports, it should be borne in mind that there are other, possibly more important, determinants of countries' export performance. These obviously include foreign demand, the volume of total domestic investment, management and international marketing practices, and all those factors which come under the rubric of export policies.

The impact of engineering goods imports on output and exports is also related to the degree of substitutability between domestically produced and imported engineering goods and to the share of the latter goods in total domestic investment in machinery and equipment. These shares presumably vary considerably between the large number of countries dealt with here. In many developing countries, imported engineering goods account for the preponderant share of the domestic stock of machinery and equipment, since domestic production of such products is too small to satisfy needs. On the other hand, large industrially advanced countries are in a position to satisfy a much larger share of their engineering goods needs through domestic production.

A limited exploratory analysis tends to provide empirical support for the hypothesized relationship between engineering goods imports and a country's total exports. The results also suggest that the differential rates of engineering goods imports between countries documented above could eventually cause some realignment in relative export performance. In figure IV.6 the average annual changes in the export volume of goods and services during 1970-1981 have been plotted against the growth of the volume of engineering goods imports during 1968-1979 (in selected countries). [IV.19]

The growth of the volume of world engineering goods imports slowed considerably during the past decade, from an average annual rate of 12% in 1971-1975 to only 4% in 1981-1985. The latter figure reflects an absolute decline in the volume of imports in 1982-1983, which was more than offset by the recovery in 1984-1985. With the exception of North America, there was a long-term tendency for the expansion of all major regions' engineering goods imports to slow down.

At the end of 1985, nearly 36% of the world's stocks of imported engineering goods was five years old or less. Nearly one half of the Pacific Rim's share of imported capital stocks fell into this category, a share somewhat above that of North America (40 %) and the Soviet Union (37%). At the other end of the scale are Latin America and eastern Europe, both with shares of less than 30%.

The global coverage of engineering goods trade presented here makes possible an examination of the question whether eastern Europe, having held down its purchases of these goods during the past decade, had fallen behind its main competitors in this respect. For this purpose two indicators, the volume growth rates of engineering goods imports in 1981-1985 and the share of imported capital stocks five years old or less, were adopted as measures of recent rates of technological advancement of the imported capital stocks. The indicators suggest that the technological characteristics of eastern Europe's stock of total imported capital goods developed in a fashion similar to those of Latin America, parts of Africa and of southern Europe. All three of these groups appear to have lagged in the improvement of their imported capital stocks relative to most of western Europe, the Soviet Union, Asia (especially the Pacific Rim and India) and North America. If only eastern Europe's imports of engineering goods from the *non-socialist* (largely western) countries are taken into consideration — they are believed to contain a higher share of technologically advanced products than those obtained from the socialist countries — the indicators suggest an even greater lag. This is, however, not the case for the Soviet Union, since its imports of engineering goods from the socialist and non-socialist areas developed at comparable rates in 1981-1985.

If engineering goods imports positively influence export growth, as some evidence presented here suggests, then the recent marked changes in the global pattern of engineering goods imports should eventually cause some realignment in regional export flows. Thus those regions which registered *relative* improvement in the technological characteristics of their capital stocks — Asia (especially the Pacific Rim), North America and the Soviet Union — could also be better placed to advance in world markets. Conversely, Latin America, eastern Europe, and parts of Africa and of southern Europe would have fallen behind in this respect. It is useful to reiterate at this point that there are other determinants of export growth not addressed here, which may well be considerably more important than engineering goods imports. At the same time, while additional imports of engineering goods do not necessarily translate into an additional supply of saleable goods, a relatively low level of these imports in recent years is likely to have hindered a country's technological advance and adversely affected its competitive position in world markets.

Attention may be drawn to the fact that the volume of east European imports of engineering good from the *non-socialist* countries started to recover in 1984 and accelerated to a growth rate of some 12% in 1986. [IV.20] Thus it is possible that in the past year or so eastern Europe's engineering goods imports developed more favourably than those of other groups. However, the level of machinery imports was still below that reached in the mid-1970s and it is doubtful if the recent surge will be sufficient to compensate for the low purchases during the previous decade. In 1986, engineering goods purchases appear to have benefited from the general relaxation of pressure on east European current accounts. [IV.21] In fact, judging from the sharp increase in the share of machinery and equipment imports in total east European imports from the west, it is likely that these goods received priority. This in turn may reflect a certain willingness or even urgency on the part of the east European countries to upgrade their stocks of western capital equipment. As regards the future, some east European countries are in a position to sustain the growth of engineering goods imports from the west, while others, by virtue of the recent deterioration in their convertible currency current account balance of payments, or other financial constraints, may have difficulties maintaining the dynamics of these purchases.

Table IV.1

World imports of engineering goods, by region and country: values, volume growth, age structure, population

	Population 1980	*Value (Billion US $)*		*Volume growth (Average annual percentage change)*					*Age structure of imported stock (Percentage shares, 1961-1985 = 100)*				
	(Millions)	*1985*	*1981-1985 (1980 $)*	*1962-1965*	*1966-1970*	*1971-1975*	*1976-1980*	*1981-1985*	*1961-1965*	*1966-1970*	*1971-1975*	*1976-1980*	*1981-1985*
Developed market economies....	784.6	305.1	1597.2	9.6	14.0	9.9	7.8	6.4	6.0	10.8	19.7	27.5	36.1
North America..........	251.7	113.7	526.5	13.1	17.0	7.6	9.4	13.9	4.7	11.1	17.3	25.8	41.2
Canada..........	24.0	35.0	131.3	10.9	11.7	8.5	3.0	4.6	7.7	14.9	21.2	26.8	29.4
United States of America...	227.7	78.7	395.2	16.2	22.6	7.0	13.9	17.6	3.1	9.0	15.2	25.2	47.5
Western Europe..........	416.1	181.7	1022.9	9.1	12.7	11.3	7.2	3.2	6.4	10.7	20.7	28.3	33.9
European Communities........	317.6	143.9	812.7	10.1	13.5	11.0	8.5	3.1	6.0	10.4	20.3	28.5	34.8
Belgium-Luxembourg.........	9.9	12.1	68.1	9.3	12.7	11.6	5.1	-0.4	7.2	11.6	23.1	29.5	28.6
Denmark..........	5.1	4.5	22.6	9.9	9.7	10.5	-0.9	6.0	8.6	14.0	23.3	27.4	26.6
France..........	53.9	24.0	145.0	12.2	15.9	11.6	8.7	1.2	5.3	10.5	21.0	28.9	34.2
Germany, Federal Republic of	61.6	33.1	188.6	13.3	17.0	10.3	11.9	3.8	4.6	8.6	18.5	29.6	38.7
Greece..........	9.6	1.9	12.5	12.5	17.0	6.8	6.2	-6.3	7.3	13.8	21.8	32.9	24.2
Ireland..........	3.4	3.1	16.5	7.2	9.6	8.7	14.9	5.2	6.9	9.7	16.8	27.6	38.9
Italy..........	56.4	14.6	81.8	0.6	18.3	10.9	11.0	2.8	6.9	10.0	20.3	27.4	35.5
Netherlands..........	14.1	14.1	75.4	7.0	11.3	9.6	4.9	3.0	8.2	12.2	21.7	27.9	30.1
Portugal..........	9.8	1.3	10.7	15.3	8.8	9.6	9.1	-4.2	7.1	12.1	23.8	25.7	31.2
Spain..........	37.5	6.3	34.0	22.2	6.7	15.7	2.0	4.8	6.9	12.4	23.1	25.7	31.9
United Kingdom..........	56.3	28.9	157.5	13.1	12.6	13.5	12.1	6.7	4.7	9.1	17.6	28.0	40.5
EFTA..........	31.3	31.8	176.0	7.1	9.7	11.4	4.2	3.0	8.1	11.8	22.0	27.0	31.2
Austria..........	7.6	5.7	31.9	4.5	10.1	11.8	10.8	0.8	7.0	9.9	20.7	30.0	32.4
Finland..........	4.8	3.6	21.2	5.0	8.9	16.3	2.0	2.1	9.0	11.0	23.0	25.2	31.7
Iceland..........	0.2	0.2	1.4	17.9	-0.7	26.7	3.1	2.3	9.9	12.3	24.1	25.2	28.5
Norway..........	4.1	5.2	31.7	10.6	8.8	14.7	-1.5	5.5	8.4	12.6	22.4	27.4	29.2
Sweden..........	8.3	9.2	46.6	10.6	9.9	12.2	1.5	4.9	8.0	12.7	22.6	26.7	30.0
Switzerland..........	6.3	7.8	43.3	4.1	12.5	5.5	12.1	2.3	8.2	11.7	21.2	25.7	33.2
Other Europe..........	67.2	6.0	34.1	3.3	15.0	17.2	-2.1	8.4	7.0	11.0	22.1	29.0	30.9
Faeroe Islands..........	0.0	0.1	0.4	...	...	306.2	10.9	13.6	-	0.1	10.4	40.4	49.1
Gibraltar..........	0.0	0.1	0.2	...	...	49.5	-0.7	20.1	-	2.5	31.0	24.6	41.9
Malta..........	0.4	0.2	1.1	4.4	13.4	2.0	15.5	2.4	10.1	13.1	14.0	26.2	36.6
Turkey..........	44.4	3.4	13.5	-0.8	10.1	24.6	-10.1	30.1	8.0	10.2	22.4	24.5	34.9
Yugoslavia..........	22.3	2.2	18.9	7.3	17.7	15.1	1.9	-2.4	6.4	11.6	22.3	31.6	28.1
Japan..........	116.8	9.7	47.8	-0.8	22.7	5.5	10.2	4.0	7.0	11.1	20.1	26.2	35.6
Other industrial market economies	50.3	16.4	94.6	16.9	6.3	10.1	3.5	0.3	9.1	13.7	21.5	24.5	31.2
Australia..........	14.7	8.3	42.7	15.8	4.7	7.8	5.1	7.5	9.6	14.4	17.5	24.4	34.1
New Zealand..........	3.1	1.8	8.7	10.3	3.0	12.2	0.4	5.6	11.5	13.5	21.0	23.8	30.2
Israel..........	3.9	2.1	10.4	9.8	15.7	30.1	-4.7	4.8	7.0	9.2	33.5	23.4	26.9
South Africa..........	28.6	4.2	32.8	23.8	8.1	10.5	7.6	-9.7	8.6	14.5	21.9	25.2	29.8
Centrally planned economies.....	377.6	77.4	305.3	13.3	10.8	13.7	2.3	4.1	6.9	11.4	20.0	29.0	32.8
Eastern Europe..........	112.1	32.0	130.5	13.4	12.8	13.5	-	0.8	6.8	12.4	22.8	29.9	28.0
Albania..........	2.7	0.1	0.3	163.0	15.4	19.2	16.0	12.9	4.0	11.7	15.8	17.5	51.1
Bulgaria..........	8.9	5.4	21.4	14.9	7.6	16.8	-2.1	7.4	7.3	13.5	20.6	25.6	33.0
Czechoslovakia..........	15.3	7.6	29.8	13.1	9.8	12.2	2.4	3.2	7.4	12.1	20.9	28.7	30.8
German Democratic Republic	16.7	6.4	28.8	17.0	22.9	10.3	-0.6	1.2	4.7	12.4	23.5	29.6	29.9
Hungary..........	10.7	3.0	12.8	7.4	14.7	14.7	3.6	-0.4	7.2	11.0	20.9	31.2	29.8
Poland..........	35.6	6.5	24.6	7.7	12.2	19.2	-2.4	-1.2	7.0	11.8	25.6	32.9	22.7
Romania..........	22.2	3.0	13.0	9.6	11.5	9.5	3.6	-4.3	8.6	14.2	23.5	30.6	23.1
Soviet Union..........	265.5	45.4	174.8	13.3	8.8	14.4	5.0	7.2	6.9	10.3	17.2	28.0	37.6
Developing countries..........	3188.8	136.8	762.1	5.2	10.7	18.7	7.0	1.1	6.1	9.3	16.7	30.1	37.7
Africa..........	441.6	19.4	133.4	12.6	9.2	20.5	1.9	-1.1	6.2	9.5	20.4	31.6	32.3
North Africa..........	109.0	9.5	68.1	1.9	15.5	26.0	3.2	3.6	5.1	7.2	15.8	32.0	39.9
Algeria..........	18.7	3.6	24.1	8.2	26.6	26.5	1.6	1.5	3.2	6.2	17.9	34.3	38.3
Egypt..........	42.1	2.3	15.7	-0.1	14.4	25.0	10.7	22.6	7.5	7.0	9.7	29.5	46.3
Libyan Arab Jamahiriya....	3.0	2.0	16.2	4.5	13.8	39.5	8.6	-1.0	3.3	6.4	15.5	30.2	44.6
Morocco..........	20.1	0.7	4.4	2.0	20.4	19.1	-1.7	1.5	7.5	12.2	18.9	33.8	27.6
Republic of Sudan..........	18.7	0.3	2.1	-10.9	15.5	20.4	1.4	-4.8	12.0	10.0	18.6	31.1	28.3
Tunisia..........	6.4	0.7	5.5	9.1	0.8	30.4	3.7	1.5	6.3	6.7	16.4	31.2	39.5

Table continued.

Table IV.1 *(continued)*

	Population 1980	Value (Billion US $)		Volume growth (Average annual percentage change)					Age structure of imported stock (Percentage shares, 1961-1985 = 100)				
	(Millions)	1985	1981-1985 (1980 $)	1962-1965	1966-1970	1971-1975	1976-1980	1981-1985	1961-1965	1966-1970	1971-1975	1976-1980	1981-1985
CEUCA	13.4	1.2	6.8	17.7	9.4	21.4	4.8	7.0	4.9	8.9	17.9	28.5	39.8
Republic of Cameroon	8.5	0.5	3.0	17.0	12.7	9.5	18.1	2.4	4.9	9.5	14.6	29.4	41.7
Central African Republic	2.3	-	0.2	13.7	16.7	3.7	14.9	9.2	10.3	20.6	26.1	20.4	22.6
Congo	1.5	0.2	1.7	12.2	4.9	20.9	11.8	23.7	6.2	9.5	16.4	21.6	46.2
Gabon	1.1	0.4	2.0	28.4	9.0	42.1	-5.1	9.3	3.3	6.4	22.1	32.8	35.4
ECOWAS	139.1	5.2	40.1	22.3	6.6	23.1	0.7	-7.2	6.5	9.5	24.2	34.3	25.5
Benin	3.4	0.1	0.5	6.4	11.7	17.6	18.3	18.5	6.1	8.8	17.3	28.1	39.7
Burkina Faso	6.2	0.1	0.3	10.8	10.2	18.8	9.3	2.5	7.1	9.1	18.1	33.3	32.4
Cape Verde	0.3	-	0.1	...	...	12.4	31.5	14.5	-	2.0	17.2	33.8	47.0
Ivory Coast	8.2	0.3	2.4	18.0	12.4	12.4	12.7	-16.2	6.5	11.3	19.9	40.7	21.5
Gambia	0.6	-	0.1	23.5	-10.5	30.2	19.2	-10.9	9.0	11.7	16.1	37.6	25.7
Ghana	11.5	0.2	1.2	19.3	-7.4	15.0	-0.1	1.3	20.8	16.3	19.7	26.7	16.5
Guinea-Bissau	0.8	-	0.1	18.4	14.8	-5.8	40.2	1.8	7.9	16.4	18.9	24.1	32.8
Liberia	1.9	1.5	10.7	35.0	10.8	23.3	-16.1	-1.3	6.4	12.3	35.9	28.9	16.5
Mali	7.1	0.1	0.4	9.7	5.4	19.9	10.9	5.4	10.6	9.7	17.5	31.3	30.9
Mauritania	1.6	0.1	0.6	9.7	4.7	22.2	-0.5	8.4	9.2	12.5	21.1	23.5	33.7
Niger	5.3	0.1	0.5	11.5	15.0	17.1	23.6	-5.8	4.4	7.6	14.3	41.5	32.3
Senegal	5.7	0.2	1.2	2.2	-0.2	18.4	6.6	-5.0	13.1	11.8	18.0	30.3	26.8
Sierra Leone	3.3	-	0.2	17.1	-3.1	1.6	17.5	-10.0	19.6	19.2	21.2	23.1	16.9
Togo	2.6	0.1	0.3	40.9	-0.7	32.8	20.4	-10.5	5.4	7.5	17.3	49.8	20.1
Other Africa	180.1	3.4	18.3	13.0	9.9	7.7	5.7	-0.8	8.6	15.6	23.8	24.7	27.3
Angola	7.7	0.6	3.0	10.0	15.6	-3.1	23.4	9.5	7.6	16.7	20.0	19.2	36.6
Burundi	4.1	-	0.2	29.9	-1.2	19.0	9.2	10.7	7.8	9.3	14.0	24.4	44.7
Chad	4.5	-	0.1	5.2	19.1	12.9	-14.9	41.5	9.6	15.6	24.8	30.2	19.9
Ethiopia	38.5	0.7	2.0	20.0	-2.4	3.7	15.2	23.1	10.6	14.0	14.2	22.0	39.3
Kenya	16.7	0.3	2.2	7.7	10.5	6.5	16.3	-8.0	10.1	15.8	20.1	30.9	23.1
Madagascar	8.7	0.1	0.7	11.5	8.8	6.0	21.9	-13.0	11.6	18.0	21.7	27.3	21.4
Malawi	6.1	0.1	0.3	...	26.9	14.7	12.6	-6.0	1.4	16.1	22.6	35.3	24.7
Mauritius	0.9	0.1	0.3	12.6	-1.8	38.2	-1.8	-1.4	8.9	8.9	25.2	35.1	22.0
Mozambique	12.1	0.1	0.8	12.1	18.7	-4.8	9.8	-6.0	9.8	22.8	35.0	14.8	17.6
Reunion	0.5	0.2	0.9	17.1	11.8	9.5	10.2	4.3	6.0	11.0	18.4	27.3	37.3
Rwanda	5.2	-	0.3	...	118.1	34.9	14.9	2.6	0.5	6.7	15.2	36.6	41.1
Somalia	4.0	0.1	0.6	17.4	0.6	48.5	12.4	-7.3	6.7	7.4	31.7	28.8	25.4
Uganda	13.1	0.1	0.4	36.2	4.4	13.4	37.4	-2.8	7.9	18.9	21.9	28.0	23.2
United Republic of Tanzania	18.6	0.3	1.7	...	32.4	9.5	15.7	-5.6	0.9	11.6	20.3	37.4	29.8
Zimbabwe	26.5	0.4	2.0	15.3	17.0	14.0	-7.1	-0.7	9.4	16.0	33.4	21.6	19.7
Zambia	5.8	0.2	1.3	-8.2	30.3	14.0	-4.8	-0.6	12.4	20.2	29.2	21.7	16.5
Asia	2390.7	85.7	476.4	7.7	10.2	22.3	9.6	4.1	4.7	7.7	14.1	30.1	43.5
Middle East	87.9	24.3	182.3	12.4	10.1	41.9	6.1	-2.4	2.7	5.2	13.4	36.4	42.3
Bahrain	0.4	0.4	2.9	13.4	23.1	15.1	13.3	1.4	3.6	8.9	15.3	34.9	37.3
Cyprus	0.6	0.4	2.1	11.3	13.7	5.5	26.2	1.6	5.9	10.4	21.1	27.6	35.0
Iran	38.4	3.5	22.2	21.1	20.3	39.7	-13.5	15.5	3.1	8.9	24.5	38.5	25.0
Iraq	13.2	3.5	32.8	3.4	2.1	77.7	9.9	-1.3	2.4	2.7	11.3	34.6	48.9
Jordan	2.9	0.7	5.2	11.2	2.0	41.9	21.1	6.1	3.9	5.2	10.0	30.3	50.7
Kuwait	1.4	2.2	13.7	8.4	12.9	28.4	11.6	3.3	4.0	8.0	11.1	33.0	44.0
Lebanon	2.7	0.3	3.0	11.3	5.0	25.3	47.8	-12.5	10.6	14.8	28.1	20.4	26.1
Oman	1.0	1.0	5.1	3.2	30.5	71.5	3.3	16.8	1.8	5.6	11.7	25.7	55.2
Qatar	0.2	0.4	3.0	14.8	14.4	41.6	11.7	2.4	3.5	6.8	12.3	41.1	36.3
Saudi Arabia	9.4	8.1	71.2	19.5	-0.0	50.7	27.1	-3.7	1.7	3.0	6.9	37.5	50.9
Democratic Yemen	2.0	0.1	0.8	14.0	-21.8	21.8	31.3	-4.3	19.2	10.7	7.7	29.8	32.6
Syrian Arab Republic	8.7	1.1	5.0	-0.6	15.0	40.6	9.5	-2.4	4.1	6.3	17.0	39.6	33.1
United Arab Emirates	1.0	2.0	13.7	...	...	95.3	13.3	-4.5	-	0.3	11.9	42.6	45.3
Yemen	6.0	0.3	1.7	89.3	38.4	31.0	46.2	-3.7	0.5	2.3	6.8	43.7	46.7
Pacific Rim	271.2	34.2	208.7	5.7	18.8	16.5	15.2	5.7	3.5	6.8	14.2	28.1	47.4
Indonesia	146.4	2.7	24.1	11.1	8.5	22.3	10.3	0.5	5.3	5.6	15.4	27.6	46.2
Hong Kong	5.0	5.8	37.0	5.9	21.2	9.1	24.8	8.3	3.7	6.4	12.3	27.3	50.4
Malaysia	13.7	3.3	23.9	-0.0	11.8	15.7	19.7	2.1	5.8	7.2	13.3	24.9	48.9
Philippines	48.3	1.5	9.3	6.8	10.6	14.2	4.1	-3.5	10.0	16.1	17.8	26.5	29.6
Singapore	2.4	6.9	44.4	66.9	28.5	20.1	17.6	4.2	1.7	4.8	14.9	28.4	50.3
Republic of Korea	38.1	8.8	42.4	-1.7	69.2	16.2	15.4	18.4	1.5	6.5	12.2	29.2	50.5
Other Pacific Rim	17.0	5.2	27.4	26.0	24.4	20.3	15.7	1.2	1.9	6.3	16.2	31.0	44.6

Table continued.

Table IV.1 *(continued)*

	Population 1980	*Value (Billion US $)*		*Volume growth (Average annual percentage change)*					*Age structure of imported stock (Percentage shares, 1961-1985 = 100)*				
	(Millions)	*1985*	*1981-1985 (1980 $)*	*1962-1965*	*1966-1970*	*1971-1975*	*1976-1980*	*1981-1985*	*1961-1965*	*1966-1970*	*1971-1975*	*1976-1980*	*1981-1985*
CPEs of Asia	1073.7	17.3	37.6	21.6	7.8	18.5	12.7	28.1	5.5	10.6	16.9	23.2	43.7
China	996.1	13.2	29.2	65.4	8.6	35.7	23.2	37.9	2.5	5.7	13.6	23.9	54.4
Democratic Kampuchea	6.4	0.1	0.2	-3.2	4.8	-17.1	29.3	58.4	26.6	32.9	18.2	3.5	18.8
Dem. People's Rep. of Korea	18.0	0.5	1.6	20.7	40.2	23.3	-0.6	1.2	4.9	13.8	35.5	20.2	25.6
Lao People's Dem.Rep.	3.7	0.1	0.3	24.3	4.3	-4.5	50.4	3.7	8.3	16.9	11.8	22.4	40.7
Mongolia	1.7	2.1	3.9	-1.3	14.8	16.2	8.4	4.2	6.3	10.7	15.5	30.9	36.6
Viet Nam	54.2	1.3	2.4	12.2	7.8	-6.3	6.2	2.4	15.7	26.7	22.9	17.3	17.5
OCEANIA	4.5	0.5	2.7	28.0	17.4	-2.0	7.4	-5.9	8.0	16.9	22.0	25.8	27.2
American Samoa	0.0	-	-	...	...	12.2	17.0	15.3	-	3.0	30.5	29.7	36.8
Fiji	0.6	0.1	0.5	14.5	12.5	9.7	8.3	-11.6	8.7	12.2	24.9	29.1	25.1
French Polynesia	0.1	0.1	0.4	...	3.1	2.3	15.4	3.8	5.9	16.5	17.4	26.3	33.9
New Caledonia	0.1	0.1	0.4	25.1	32.2	-11.2	-2.7	-4.3	11.0	26.8	26.2	19.7	16.3
Papua New Guinea	3.0	0.2	1.3	25.7	25.0	2.3	11.1	-8.3	5.8	14.2	20.5	27.3	32.3
Samoa	0.2	-	0.1	...	24.8	31.4	12.9	16.7	2.7	6.9	24.1	31.5	34.8
Solomon Islands	0.2	-	0.1	...	...	24.1	15.7	-4.9	-	3.3	23.4	36.8	36.5
Tonga	0.1	-	-	...	...	63.6	36.0	13.3	-	1.0	18.8	43.3	36.9
Vanuatu	0.1	-	0.1	22.0	-13.7	16.8	16.1	32.8	25.1	29.9	12.6	15.8	16.6
Other Asia	953.4	9.5	45.1	3.2	0.9	3.8	10.6	8.7	14.8	16.3	13.3	20.6	35.0
Afghanistan	16.0	0.1	0.3	22.8	5.5	-1.3	28.7	-1.7	12.0	14.5	13.5	37.4	22.6
Bangladesh	88.7	0.3	1.6	6.3	-4.0	95.5	28.5	-5.2	20.7	20.5	11.2	23.9	23.6
Brunei Barussalam	0.2	0.1	0.7	174.6	18.4	62.6	6.1	8.7	2.0	7.8	20.5	27.5	42.2
Burma	33.6	0.2	1.6	7.6	0.2	-6.8	39.2	-2.6	14.9	14.6	12.1	26.8	31.6
India	663.6	4.7	19.0	1.6	-9.0	5.1	6.8	26.1	20.5	14.7	12.0	15.3	37.5
Sri Lanka	14.8	0.3	1.8	-5.3	11.0	0.1	32.3	-1.0	12.9	17.4	9.0	22.2	38.5
Pakistan	83.6	1.3	6.7	3.9	9.5	5.1	13.0	4.3	13.3	22.0	11.9	21.8	31.0
Thailand	46.5	2.3	13.1	10.9	15.5	9.4	8.5	4.0	7.0	14.3	16.5	25.3	36.8
Latin America	356.5	31.7	152.3	-1.1	12.5	12.8	6.7	-3.5	9.0	12.6	19.4	29.2	29.8
LAIA	308.0	20.1	108.4	-4.3	12.6	13.3	8.8	-5.7	9.6	12.5	19.1	29.7	29.2
Argentina	28.2	1.4	9.7	-21.0	15.3	4.3	32.4	-18.0	17.5	13.1	14.9	29.0	25.5
Bolivia	5.6	0.1	0.8	7.7	10.4	23.8	-4.8	3.1	8.6	16.6	19.9	37.6	17.3
Brazil	121.3	3.3	16.3	-18.7	38.9	22.1	-4.6	-6.1	6.7	10.7	30.0	31.5	21.1
Chile	11.1	0.8	4.9	-6.9	18.9	-2.6	15.2	-4.1	13.3	20.0	16.8	25.5	24.4
Colombia	25.9	1.2	7.8	-3.2	16.5	0.5	13.8	-4.9	12.1	14.4	16.0	24.0	33.5
Ecuador	8.1	0.6	2.9	9.6	15.0	27.7	9.6	-9.2	5.3	9.6	17.5	40.0	27.5
Mexico	69.4	9.3	43.5	9.4	9.0	11.5	22.5	1.4	7.5	11.4	15.0	26.5	39.6
Paraguay	3.2	0.1	0.6	10.4	4.9	18.0	14.4	-10.3	7.9	12.4	18.1	34.5	27.1
Peru	17.3	0.6	3.9	0.6	0.5	20.3	1.7	-8.2	16.8	17.6	20.8	21.0	23.9
Uruguay	2.9	0.1	1.1	-17.4	15.9	4.7	29.8	-25.0	18.8	12.3	11.2	31.7	26.0
Venezuela	15.0	2.7	16.9	12.7	3.2	19.8	5.1	-3.7	8.3	12.0	17.5	36.3	25.9
CACM	20.1	1.0	4.9	13.6	7.3	9.1	2.6	-5.2	9.9	15.8	20.6	32.9	20.8
Costa Rica	2.3	0.2	1.0	10.4	13.5	5.8	7.4	-4.5	8.5	14.3	23.0	35.8	18.3
El Salvador	4.5	0.2	0.8	19.9	-0.1	17.0	-5.9	5.7	11.3	14.9	20.7	34.2	18.9
Guatemala	6.9	0.2	1.2	15.8	3.7	10.8	4.6	-15.1	10.4	15.1	19.1	35.4	20.0
Honduras	3.7	0.2	0.9	8.5	17.7	4.4	11.8	-9.5	8.5	17.6	18.0	33.0	22.9
Nicaragua	2.7	0.3	1.0	13.1	6.3	12.5	6.2	17.3	10.9	18.4	21.9	23.4	25.5
Other America	28.4	10.5	39.1	10.0	14.1	12.4	0.3	5.9	6.9	12.4	20.2	26.7	33.8
Bahamas	0.2	0.3	1.5	-1.8	9.9	-1.2	10.3	27.8	14.2	19.8	18.9	13.5	33.7
Barbados	0.3	0.1	0.7	11.5	22.6	-2.1	14.3	6.9	5.9	13.8	16.6	22.4	41.3
Bermuda	0.1	0.2	1.0	-17.0	31.7	13.0	17.0	25.8	12.3	15.9	16.4	32.3	23.2
Cuba	9.7	3.9	8.2	20.2	16.5	16.0	0.6	2.8	6.6	13.7	18.8	30.3	30.7
Dominican Republic	5.4	0.3	1.3	7.3	24.7	11.6	-0.4	-6.9	9.4	15.1	26.9	26.4	22.2
French Guiana	0.1	0.4	0.9	...	38.6	-7.7	24.0	46.3	1.5	14.3	8.7	17.8	57.7
Greenland	0.1	0.1	0.3	...	...	136.8	6.2	2.7	-	0.2	19.5	39.3	41.0
Guadeloupe	0.3	0.1	0.7	11.7	7.6	6.5	15.1	-8.4	9.7	13.0	17.3	29.9	30.1
Guyana	0.9	0.1	0.3	16.1	8.2	11.3	-3.9	-9.2	12.6	22.0	24.5	25.6	15.3
Haiti	5.1	0.2	0.7	4.9	15.1	24.3	6.8	5.5	4.8	6.2	17.5	31.9	39.5
Jamaica	2.1	0.2	1.1	8.6	20.2	-1.4	-12.7	7.1	12.0	26.9	28.2	13.7	19.1
Martinique	0.3	0.1	0.6	9.0	13.9	5.6	15.0	-5.6	7.6	12.6	18.6	29.1	32.2
Neth.Antil and Aruba	0.2	0.2	1.3	-4.8	38.4	-1.3	2.3	-4.6	7.7	19.6	22.5	24.4	25.8
Panama	2.0	0.1	0.4	32.3	-2.3	7.4	7.1	-10.5	14.1	18.4	19.0	27.7	20.7
Suriname	0.4	4.0	17.2	26.0	11.8	29.8	-1.3	16.8	3.8	7.2	21.1	26.4	41.5
Trinidad and Tobago	1.1	0.3	2.7	14.5	-1.9	11.0	23.6	-11.7	10.1	9.4	14.4	30.6	35.5
Virgin Islands	0.1	-	0.1	...	...	-2.0	5.3	20.6	-	5.9	34.1	26.1	33.9
Total	4401.3	535.8	2759.2	9.3	12.3	12.3	6.6	4.4	6.2	10.6	19.0	28.2	35.9

Table continued.

Table IV.1 *(concluded)*

	Population 1980	*Value (Billion US $)*		*Volume growth (Average annual percantage change)*					*Age structure of imported stock (Percentage shares, 1961-1985 = 100)*				
	(Millions)	*1985*	*1981-1985 (1980 $)*	*1962-1965*	*1966-1970*	*1971-1975*	*1976-1980*	*1981-1985*	*1961-1965*	*1966-1970*	*1971-1975*	*1976-1980*	*1981-1985*
Memorandum items:													
Southern Europe	123.6	15.2	89.6	12.6	11.1	13.3	1.4	2.8	7.0	12.2	22.7	28.1	30.0
ECE region	329.6	372.8	1854.7	10.6	13.2	10.8	6.7	6.0	6.1	10.9	19.7	27.8	35.5
Imports of engineering goods from non-socialist countries only													
Centrally planned economies	377.6	13.4	83.8	16.9	15.3	16.9	-2.1	1.1	5.8	12.1	21.7	30.4	30.0
Eastern Europe	112.1	4.9	29.2	20.8	18.5	14.6	-5.5	-2.0	5.4	13.8	27.1	31.0	22.8
Albania	2.7	0.1	0.3	163.0	15.4	19.2	16.0	12.9	4.0	11.7	15.8	17.5	51.1
Bulgaria	8.9	0.7	3.8	56.7	15.9	27.5	-9.6	14.8	5.5	16.8	20.8	22.7	34.2
Czechoslovakia	15.3	1.1	6.2	20.7	26.8	4.1	0.4	1.2	5.3	15.3	24.2	28.5	26.6
Democratic Republic of Germany	16.7	1.4	8.9	34.6	26.7	2.6	3.9	4.3	4.2	13.4	23.7	26.7	31.9
Hungary	10.7	0.5	3.0	20.9	12.8	15.7	3.8	-0.0	5.9	11.2	21.9	32.4	28.6
Poland	35.6	0.9	4.7	5.3	19.2	39.5	-12.9	-5.4	4.6	10.2	33.7	38.0	13.5
Romania	22.2	0.4	2.6	22.8	13.5	7.6	2.2	-13.0	7.9	18.6	29.6	31.4	12.5
Soviet Union	265.5	8.5	54.6	14.4	12.3	21.0	1.0	4.0	6.2	10.6	17.1	29.9	36.2

Sources: Data bank of the ECE General Economic Analysis Division. ECE secretariat estimates.

Table IV.2

World engineering products trade, by area, 1984

(Billions of US dollars, f.o.b.)

Origin/Destination	*Developed countries*	*Developing countries*	*Eastern trading area*	*World*
Developed countries	379.15	129.60	17.75	526.50
Developing areas	37.60	19.25	1.30	58.15
Eastern trading area	3.95	10.40	46.00	60.35
World	420.70	159.25	65.05	645.00

Source: GATT, *International Trade 1985-86*, Geneva 1986.

Note: In this table, the eastern trading area includes the People's Republic of China and the other centrally planned economies of Asia, in addition to eastern Europe and the Soviet Union. Engineering goods here includes all road motor vehicles.

Table IV.3

Value of world imports of engineering goods, 1961-1985

(Period cumulations, in billions of 1980 US dollars)

	1961-1965	*1966-1970*	*1971-1975*	*1976-1980*	*1981-1985*	*1961-1985*
Developed market economies	263.5	478.5	871.1	1216.2	1597.2	4426.4
North America	60.2	141.7	221.2	329.8	526.5	1279.3
Western Europe	193.9	321.8	623.0	851.3	1022.9	3012.9
European Community	140.6	243.3	474.8	667.1	812.7	2338.5
EFTA	45.6	66.4	123.9	152.2	176.0	564.0
Other Europe	7.7	12.2	24.3	32.0	34.1	110.4
Japan	9.4	14.9	26.9	35.1	47.8	134.2
Other industrial market economies	27.4	41.5	65.1	74.1	94.6	302.7
Centrally planned economies	63.8	106.1	186.1	269.7	305.3	931.1
Eastern Europe	31.6	58.0	106.3	139.4	130.5	465.8
Soviet Union	32.2	48.1	79.8	130.3	174.8	465.3
Developing countries	123.2	187.7	337.4	608.8	762.1	2019.2
Africa	25.5	39.3	84.0	130.1	133.4	412.3
North Africa	8.8	12.3	26.9	54.7	68.1	170.8
CEUCA	0.8	1.5	3.1	4.9	6.8	17.2
ECOWAS	10.1	15.0	38.1	53.9	40.1	157.3
Other Africa	5.7	10.4	15.9	6.6	18.3	67.0
Asia	51.7	84.0	154.3	329.8	476.4	1096.2
Middle East	11.8	22.4	57.8	156.8	182.3	431.1
Pacific Rim	15.4	29.8	62.6	123.8	208.7	440.3
Centrally planned economies	4.8	9.2	14.6	20.0	37.6	86.2
Oceania	0.8	1.7	2.2	2.6	2.7	10.0
Other Asia	19.0	21.0	17.1	26.5	45.1	128.6
Latin America	45.9	64.4	99.1	149.0	152.3	510.7
LAIA	35.6	46.4	70.9	110.4	108.4	371.8
CACM	2.3	3.7	4.8	7.7	4.9	23.3
Other America	8.0	14.3	23.4	30.8	39.1	115.6
Total world	477.9	813.7	1459.8	2168.8	2759.2	7679.4
Memorandum items:						
Eastern Europe (non-socialist imports)	6.9	17.8	34.7	39.7	29.2	128.3
Soviet Union (non-socialist imports)	9.3	16.0	25.8	45.2	54.6	150.9
ECE region	317.9	569.6	1030.3	1450.8	1854.7	5223.3
Southern Europe	21.0	36.3	67.7	84.0	89.6	298.7

Source: As for table IV.1.

Table IV.4

Structure of world engineering goods imports by region, 1961-1985

(Percentage shares of volume measured in 1980 US dollars)

	1961-1965	*1966-1970*	*1971-1975*	*1976-1980*	*1981-1985*	*1961-1985*
Developed market economies	55.1	58.8	59.7	56.1	57.9	57.6
North America	12.6	17.4	15.2	15.2	19.1	16.7
Western Europe	40.6	39.6	42.7	39.3	37.1	39.2
European Community	29.4	29.9	32.5	30.8	29.5	30.5
EFTA	9.5	8.2	8.5	7.0	6.4	7.3
Other Europe	1.6	1.5	1.7	1.5	1.2	1.4
Japan	2.0	1.8	1.8	1.6	1.7	1.7
Other industrial market economies	5.7	5.1	4.5	3.4	3.4	3.9
Centrally planned economies	13.4	13.0	12.7	12.4	11.1	12.1
Eastern Europe	6.6	7.1	7.3	6.4	4.7	6.1
Soviet Union	6.7	5.9	5.5	6.0	6.3	6.1
Developing countries	25.8	23.1	23.1	28.1	27.6	26.3
Africa	5.3	4.8	5.8	6.0	4.8	5.4
North Africa	1.8	1.5	1.8	2.5	2.5	2.2
CEUCA	0.2	0.2	0.2	0.2	0.2	0.2
ECOWAS	2.1	1.8	2.6	2.5	1.5	2.0
Other Africa	1.2	1.3	1.1	0.8	0.7	0.9
Asia	10.8	10.3	10.6	15.2	17.3	14.3
Middle East	2.5	2.8	4.0	7.2	6.6	5.6
Pacific Rim	3.2	3.7	4.3	5.7	7.6	5.7
Centrally planned economies	1.0	1.1	1.0	0.9	1.4	1.1
Oceania	0.2	0.2	0.2	0.1	0.1	0.1
Other Asia	4.0	2.6	1.2	1.2	1.6	1.7
Latin America	9.6	7.9	6.8	6.9	5.5	6.7
LAIA	7.5	5.7	4.9	5.1	3.9	4.8
CACM	0.5	0.5	0.3	0.4	0.2	0.3
Other America	1.7	1.8	1.6	1.4	1.4	1.5
Total above	100.0	100.0	100.0	100.0	100.0	100.0
Memorandum item:						
ECE region	66.5	70.0	70.6	66.9	67.3	68.0

Source: As for table IV.1.

Table IV.5

Twenty-five largest importers of engineering goods and their cumulated purchases during 1981-1985

(Billions of 1980 US dollars)

United States	395	
Germany, Federal Republic of	189	
Soviet Union	175	(55) [a]
United Kingdom	157	
France	145	
Canada	131	
Italy	82	
Netherlands	75	
Saudi Arabia	71	
Belgium	68	
Japan	48	
Sweden	47	
Singapore	44	
Australia	43	
Mexico	43	
Switzerland	43	
Republic of Korea	42	
Hong Kong	37	
Spain	34	
Iraq	33	
South Africa	33	
Austria	32	
Norway	32	
Czechoslovakia	30	(6) [a]
German Democratic Republic	29	(9) [a]

Source: As for table IV.1.

a Imports from non-socialist countries only are in parenthesis.

Figure IV. 1. Value of imports of engineering goods, by region, 1981 – 1985

(Billions of 1980 US dollars)

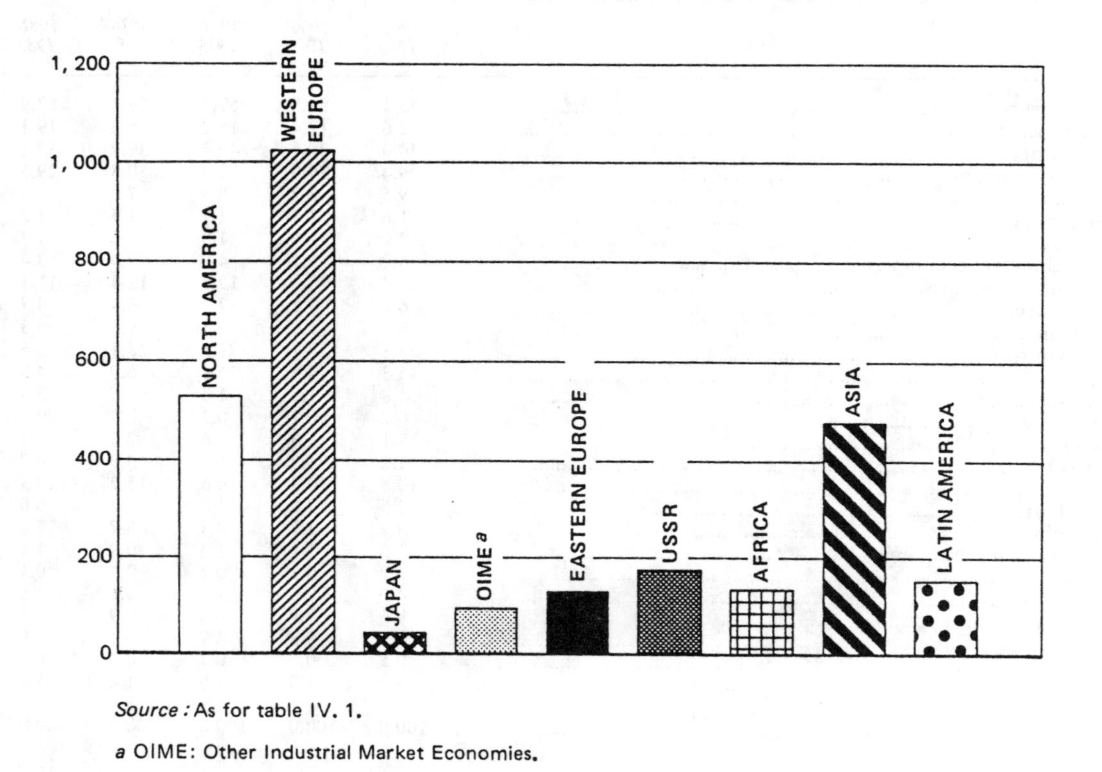

Source: As for table IV. 1.

a OIME: Other Industrial Market Economies.

Figure IV. 2. Share of imports of engineering goods, by region, 1981 – 1985

(Percentage)

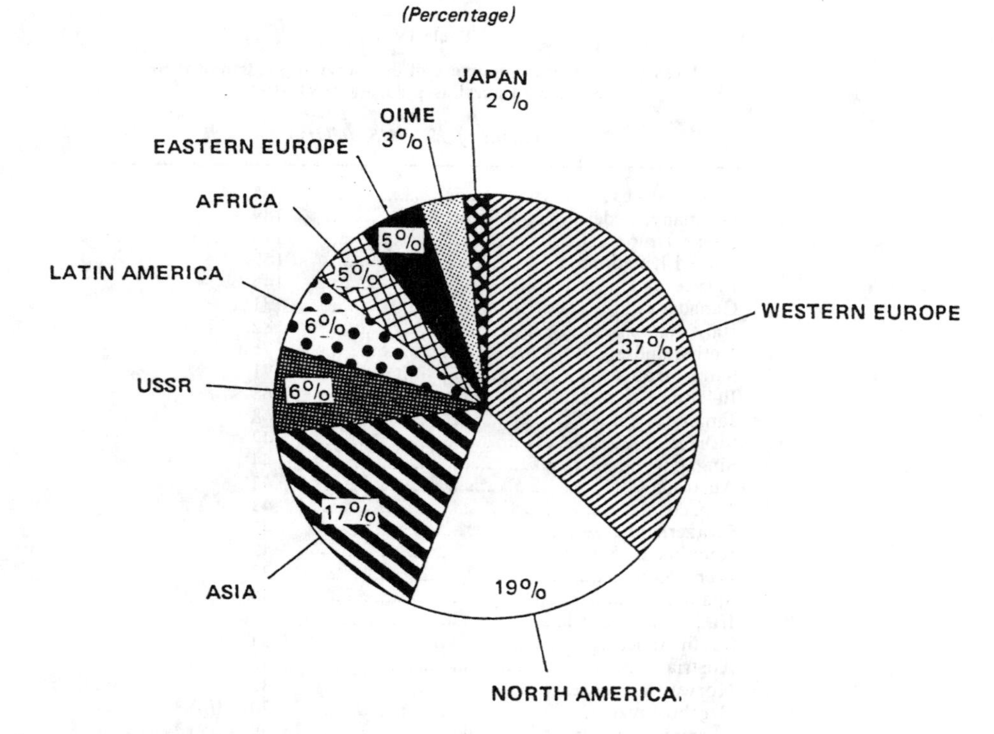

Source: As for table IV. 1.

Table IV.6

Growth of world import volume of engineering goods, 1962-1985

(Average annual percentage changes)

	1962-1965	*1966-1970*	*1971-1975*	*1976-1980*	*1981-1985*
Developed market economies	9.6	14.0	9.9	7.8	6.4
North America	13.1	17.0	7.6	9.4	13.9
Western Europe	9.1	12.7	11.3	7.2	3.2
European Community	10.1	13.5	11.0	8.5	3.1
EFTA	7.1	9.7	11.4	4.2	3.0
Other Europe	3.3	15.0	17.2	-2.1	8.4
Japan	-0.8	22.7	5.5	10.2	4.0
Other industrial market economies	16.9	6.3	10.1	3.5	0.3
Centrally planned economies	13.3	10.8	13.7	2.3	4.1
Eastern Europe	13.4	12.8	13.5	0.0	0.8
Soviet Union	13.3	8.8	14.4	5.0	7.2
Developing countries	5.2	10.7	18.7	7.0	1.1
Africa	12.6	9.2	20.5	1.9	-1.1
North Africa	1.9	15.5	26.0	3.2	3.6
CEUCA	17.7	9.4	21.4	4.8	7.0
ECOWAS	22.3	6.6	23.1	0.7	-7.2
Other Africa	13.0	9.9	7.7	5.7	-0.8
Asia	7.7	10.2	22.3	9.6	4.1
Middle East	12.4	10.1	41.9	6.1	-2.4
Pacific Rim	5.7	18.8	16.5	15.2	5.7
Centrally planned economies	21.6	7.8	18.5	12.7	28.1
Oceania	28.0	17.4	-2.0	7.4	-5.9
Other Asia	3.2	0.9	3.8	10.6	8.7
Latin America	-1.1	12.5	12.8	6.7	-3.5
LAIA	-4.3	12.6	13.3	8.8	-5.7
CACM	13.6	7.3	9.1	2.6	-5.2
Other America	10.0	14.1	12.4	0.3	5.9
Total above	9.3	12.3	12.3	6.6	4.4
Memorandum items:					
Eastern Europe (non-socialist imports)	20.8	18.5	14.6	-5.5	-2.0
Soviet Union (non-socialist imports)	14.4	12.3	21.0	1.0	4.0
ECE region	10.6	13.2	10.8	6.7	6.0
Southern Europe	12.6	11.1	13.3	1.4	2.8

Source: As for table IV.1.

Figure IV. 3. Growth of world engineering goods import volumes, 1971 – 1985

(Average annual percentage change)

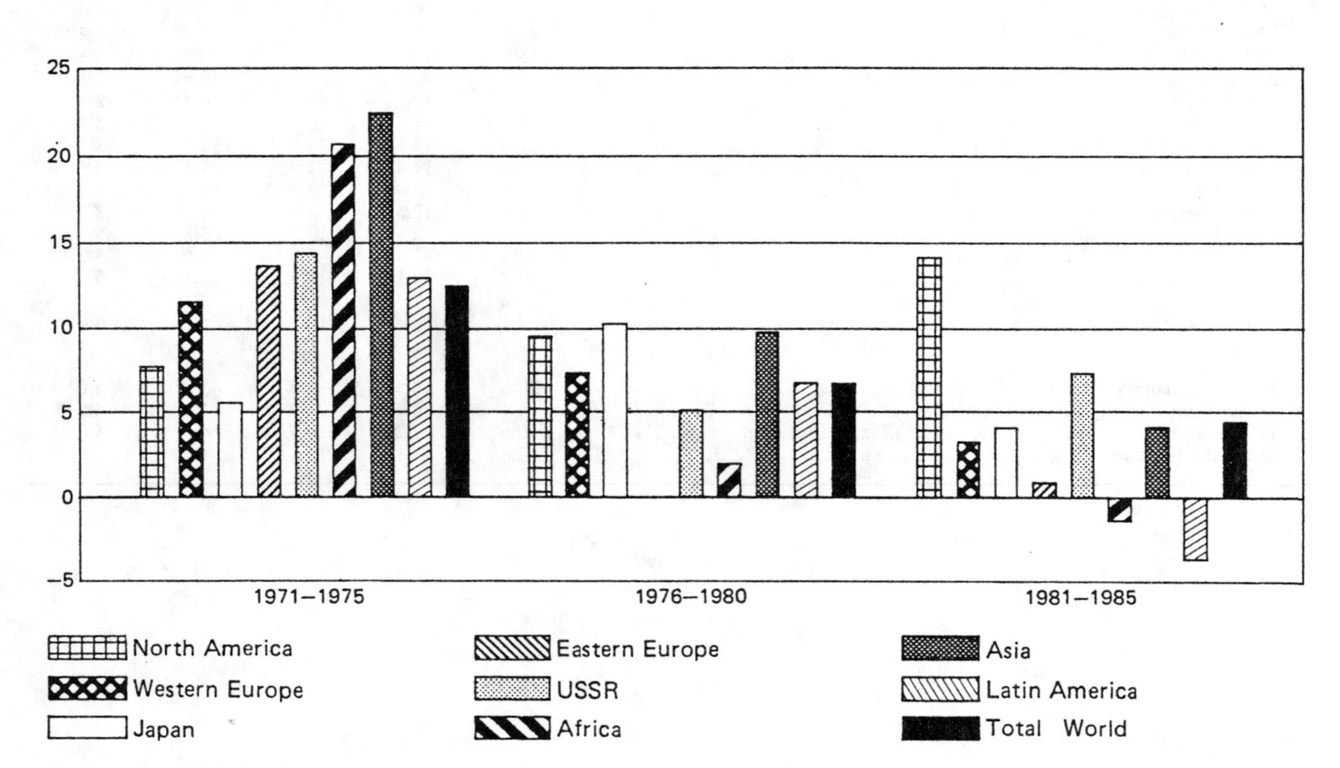

Source :As for table IV. 1.

Note: Growth rate for eastern Europe in 1976-1980 was zero.

Table IV.7

Stock of imported engineering goods, 1985: vintage structure by import period

(Percentage shares)

	1961-1965	*1966-1970*	*1971-1975*	*1976-1980*	*1981-1985*	*1961-1985*
Developed market economies	6.0	10.7	19.7	27.5	36.1	100.0
North America	4.6	11.1	17.3	25.8	41.2	100.0
Western Europe	6.4	10.7	20.7	28.3	33.9	100.0
European Community	6.0	10.4	20.3	28.5	34.8	100.0
EFTA	8.0	11.8	22.0	27.0	31.2	100.0
Other Europe	7.0	11.0	22.1	29.0	30.9	100.0
Japan	7.0	11.1	20.1	26.2	35.6	100.0
Other industrial market economies	9.1	13.7	21.5	24.5	31.2	100.0
Centrally planned economies	6.8	11.4	20.0	29.0	32.8	100.0
Eastern Europe	6.9	12.4	22.8	29.9	28.0	100.0
Soviet Union	6.9	10.3	17.2	28.0	37.6	100.0
Developing countries	6.2	9.3	16.7	30.1	37.7	100.0
Africa	6.2	9.5	20.4	31.6	32.3	100.0
North Africa	5.1	7.2	15.8	32.0	39.9	100.0
CEUCA	4.9	8.9	17.9	28.5	39.8	100.0
ECOWAS	6.5	9.5	24.2	34.3	25.5	100.0
Other Africa	8.6	15.6	23.8	24.7	27.3	100.0
Asia	4.6	7.7	14.1	30.1	43.5	100.0
Middle East	2.7	5.2	13.4	36.4	42.3	100.0
Pacific Rim	3.5	6.8	14.2	28.1	47.4	100.0
Centrally planned economies	5.6	10.6	16.9	23.2	43.7	100.0
Oceania	8.1	16.9	22.0	25.8	27.2	100.0
Other Asia	14.8	16.3	13.3	20.6	35.0	100.0
Latin America	9.0	12.6	19.4	29.2	29.8	100.0
LAIA	9.5	12.5	19.1	29.7	29.2	100.0
CACM	9.9	15.8	20.6	32.9	20.8	100.0
Other America	6.9	12.4	20.2	26.7	33.8	100.0
Total above	6.3	10.6	19.0	28.2	35.9	100.0
Memorandum items:						
Eastern Europe (non-socialist imports)	5.3	13.8	27.1	31.0	22.8	100.0
Soviet Union (non-socialist imports)	6.2	10.6	17.1	29.9	36.2	100.0
ECE region	6.1	10.9	19.7	27.8	35.5	100.0
Southern Europe	7.0	12.2	22.7	28.1	30.0	100.0

Source: As for table IV.1.

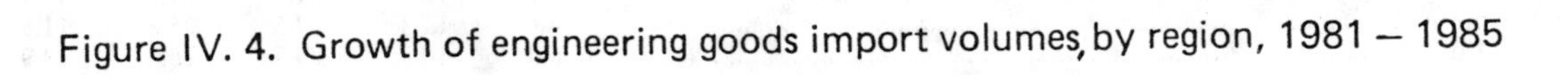

Figure IV. 4. Growth of engineering goods import volumes, by region, 1981 – 1985

(Average annual percentage change)

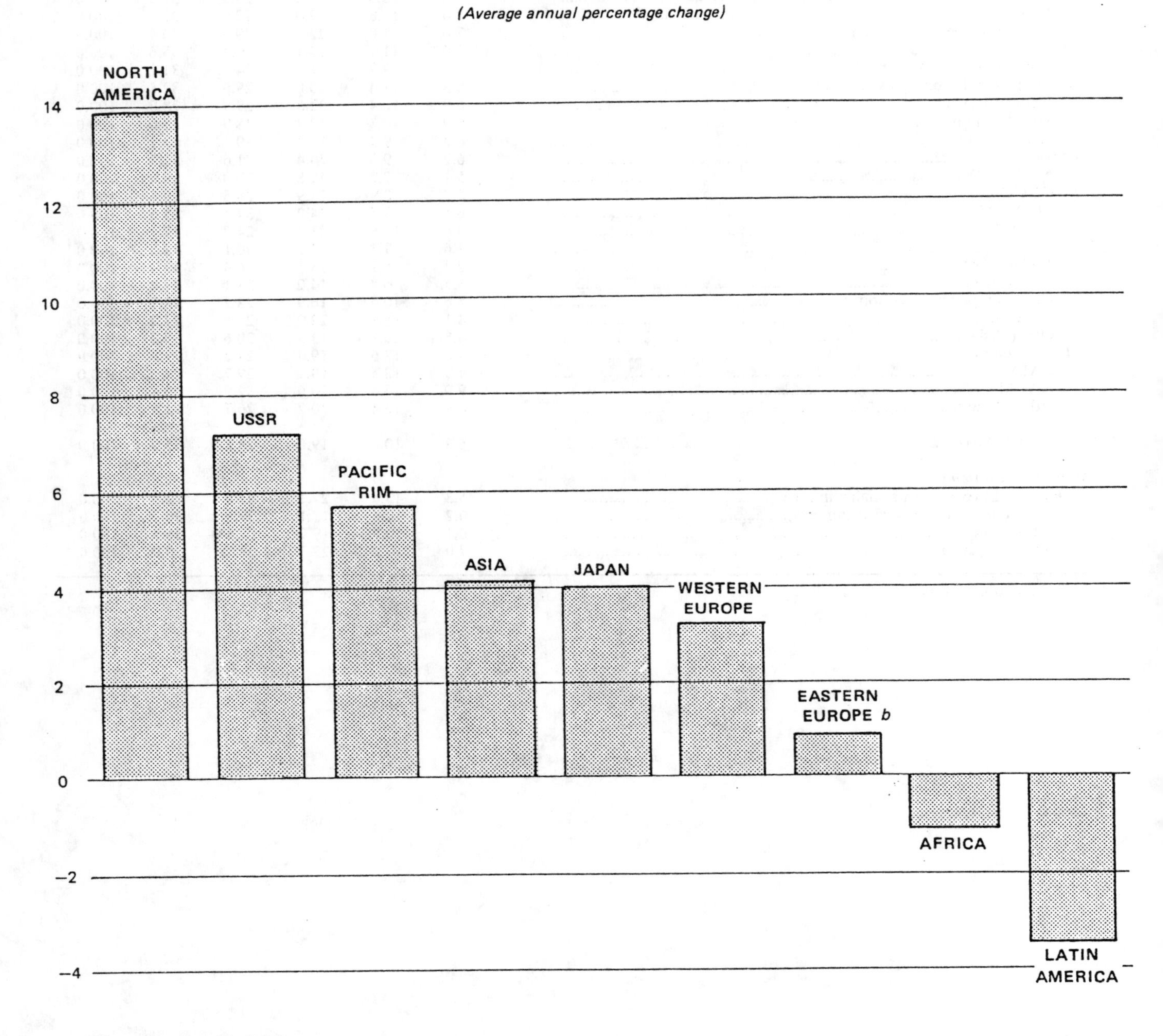

Source :As for table IV. 1

a Pacific Rim included in Asia.
b Excluding USSR

Figure IV. 5. Stocks of imported engineering goods, by region, 1985: share of stock five years old or less

(Percentage)

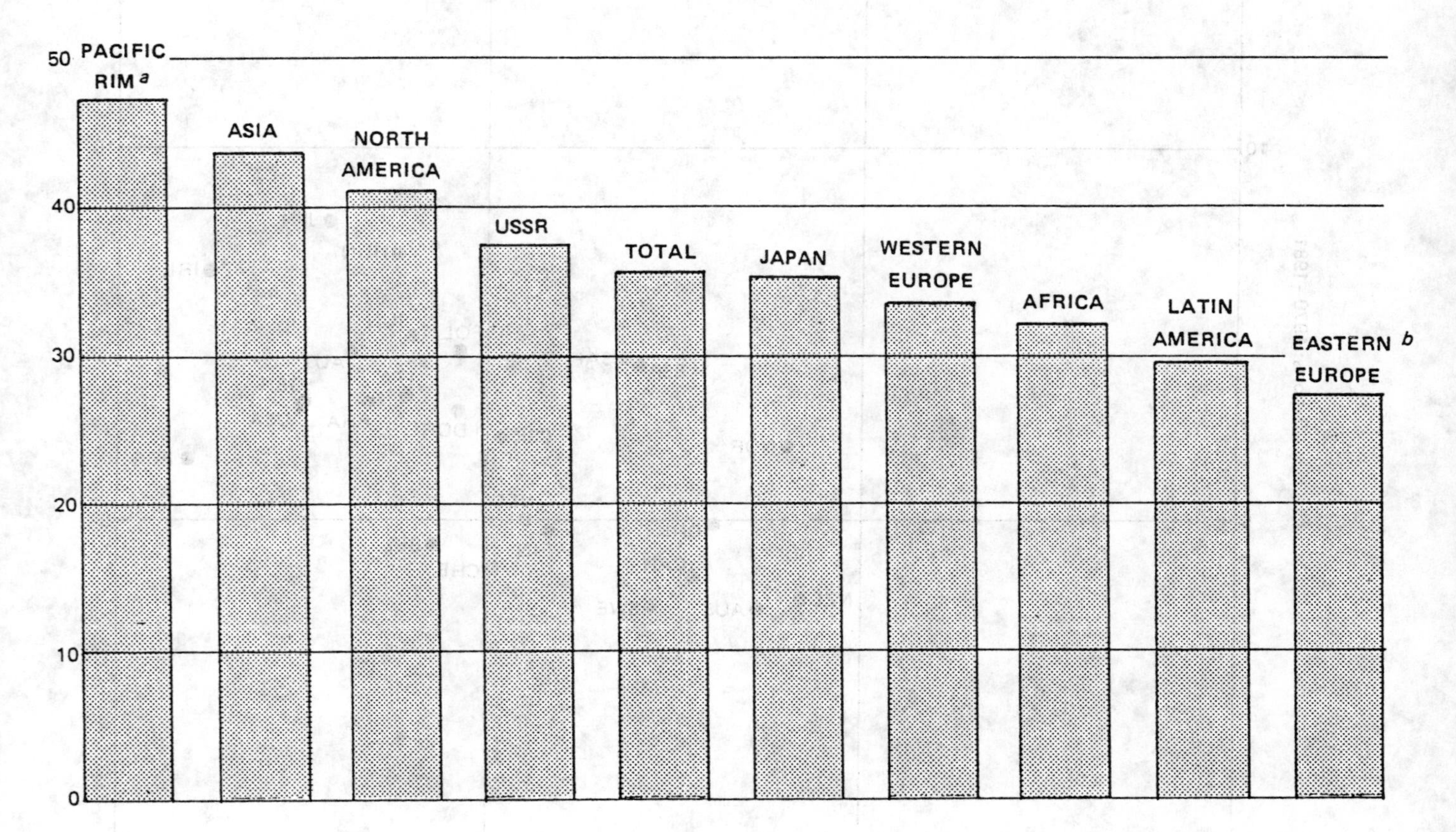

Source: As for table IV. 1

a Pacific Rim included in Asia.

b Excluding USSR.

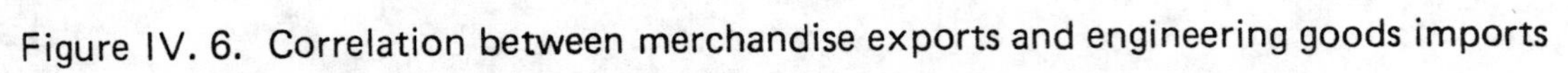

Figure IV. 6. Correlation between merchandise exports and engineering goods imports in selected countries, 1968 – 1981

(Volumes : average annual percentage change)

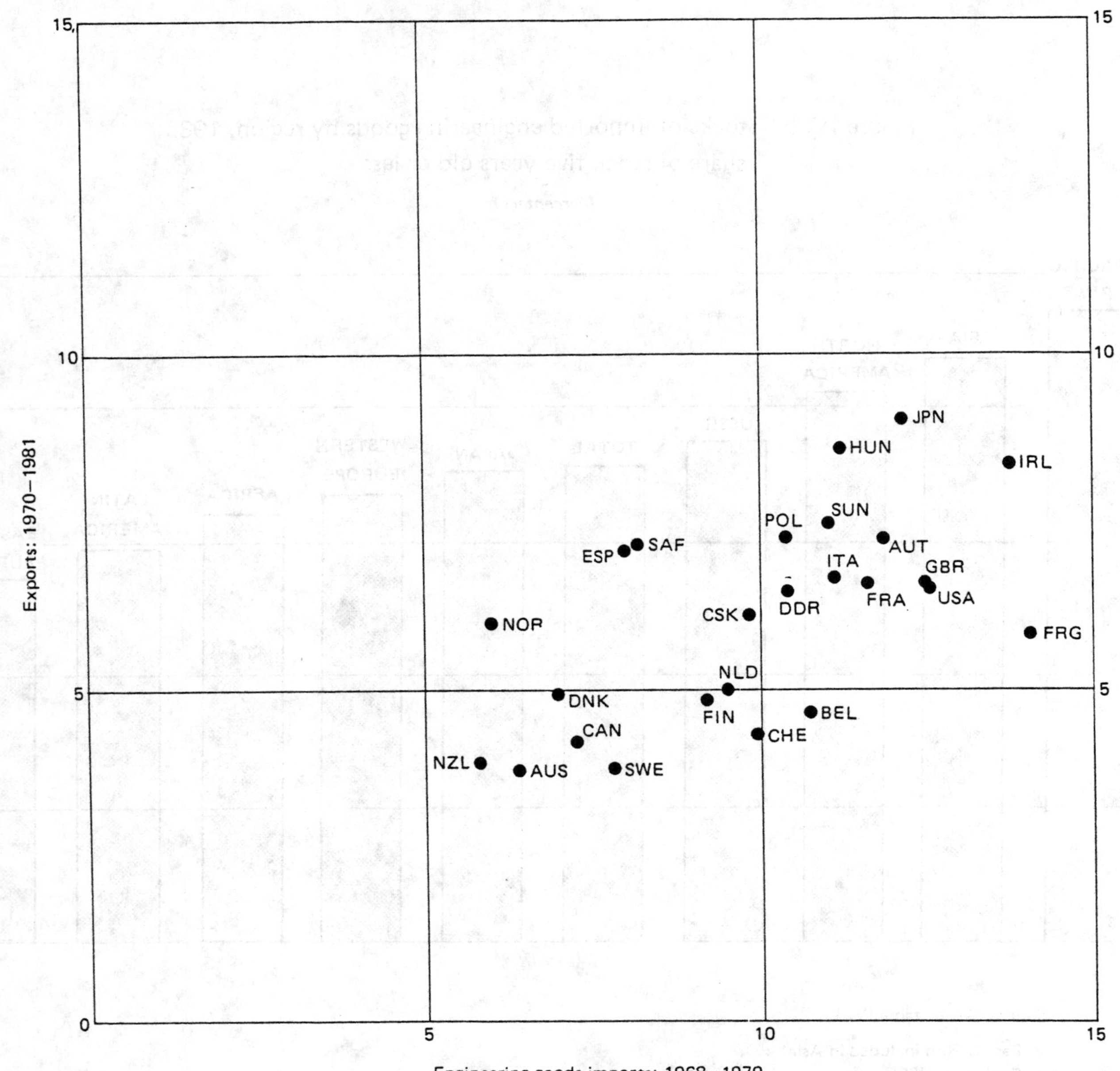

Source : As for table IV. 1. and, for exports, World Bank, *World Development Report 1983.*

Key:

AUS Australia
AUT Austria
BEL Belgium
CAN Canada
CSK Czechoslovakia
DNK Denmark
FIN Finland
FRA France
DDR German Democratic Republic
FRG Germany, Federal Republic of
HUN Hungary
IRL Ireland
ITA Italy
JPN Japan
NLD Netherlands
NZL New Zealand
NOR Norway
POL Poland
SAF South Africa
ESP Spain
SWE Sweden
CHE Switzerland
SUN Union of Soviet Socialist Republics
GBR United Kingdom
USA United States of America

IV.2. GEOGRAPHIC AND PRODUCT PATTERN OF REGIONAL TRADE IN ENGINEERING PRODUCTS

The role of engineering product exports, as a share of total exports, has grown systematically over the reviewed period [IV.22]. Globally, for the reporting countries it has increased from 26.9% of the total in 1979, to a record 31.6% in 1986, as can be seen from table IV.8 (despite a dip to 25.6% in 1980, picking up to 26.6% and 26.5 in 1981 and 1982). The total volume of machinery and transport equipment exports (SITC, Rev.2: Section 7) in this period rose from $430 billion in 1979 to $673 billion in 1986 (table IV.9). Table IV.10 shows the share of these exports by individual regions. The annual value of engineering exports in the 1979-1986 period went from $482 billion to $698 billion, as shown in table IV.11. Table IV.12 shows the respective percentage share of individual groups of exports in total exports.

The performance of the engineering sector in the first half of the 1979-1986 period must be seen against a background of decreasing world exports. The more robust recovery in the last four years of the period shows an acceleration based on general economic improvement, with the share of engineering exports going from 27.5% in 1983 to 28.3 in 1984, 29.5 in 1985 and 31.6 in 1986.

The performance of the regions in this export boom varied, however. Consistently, the export of engineering products played the most significant role (i.e. share in total exports) in the developed market economies, and the smallest in the developing countries, with the centrally planned economies of eastern Europe maintaining a steady intermediate position, as shown in table IV.8.

The developed market economies accounted for the largest share of engineering product exports as a share of global engineering exports, consistently between 86 and 89% in the review period, as shown in table IV.10. The centrally planned economies of eastern Europe provided a declining share of the total, garnering 8.2% of the total in 1986, down from the record high of 9.8% in 1979. Developing countries showed a general upward tendency in 1983-1985, ending 1986 with a minimum share of 2.7%.

The developed market economies also played the leading role when measured in terms of the share of engineering exports in each country's or region's total exports (see table IV.8). This share went from 35.2% in 1979 to a low of 34.7 in 1980, rising steadily thereafter to reach 40.8 in 1986. Countries which benefited most from the recovery of the world market (largely stimulated in the latter part of the review period by the recovery in the United States), were Japan, the United States, Canada and the Asian developing countries. In 1985, according to GATT estimates, trade in manufactured products was the only buoyant commodity trade flow, with a 6% growth, particularly for road motor vehicles, household appliances and specialized machinery.

The share of engineering exports in total exports in Japan has been consistently higher than in any other country reviewed, rising steadily from 53.7% in 1979 to 70% in 1986. Western Europe's exports of engineering products as a share of total exports shows a more varied pattern of significant swings with a low of 30.8% in 1980 and a high of 36.4% in 1981, ending 1986 at 34.5%. Smaller variations were experienced by the share of engineering exports of the United States and Canada in the review period, with engineering products accounting for 44.7% of total North American exports in 1986, up from a 37% low in 1980.

The share of engineering exports in total exports of the centrally planned economies of eastern Europe during the period started in 1979 with 28.3% and came close to the 30% level in 1986. In the remaining years of the period, the share of engineering exports in their total exports declined to 26.7% in 1980 and further to 25.1 in 1981, rebounding to 27.7 in 1982, dipping to 26.8 in 1984, and beginning the current recovery with 27.2% in 1984.

While the developing countries have sharply increased their share of engineering products in total exports, it continues to remain well below the growth level of the developed countries. This trend, however, is still assumed to be mostly powered by the Asian developing countries. The developing countries' share of engineering products in total exports went from 5% in 1979 to a record low of 2.4 in 1982, and reached a high of 5.4 in 1985.

In terms of groups of products, the structure of exports shows relatively little change, with non-electrical machinery and transport equipment vying for the largest share of engineering exports in all the years under review. Thus non-electrical machinery, which in 1985 accounted for 35% of total engineering exports, or $217 billion, was only marginally down from the 36.9% share reached in 1980 and over the low of 34.7% reached in 1984. Similarly, transport equipment, whose $209 billion exports in 1985 equalled 33.7% of the total, was the highest share of the review period, and only marginally up from the low of 31.8% posted in 1980.

The share of electrical equipment exports in total exports, the next largest category, reached 20.8% in 1986, for a value of $145 billion, compared with a low of 17.9%, or $86.5 billion, in 1979. Exports of precision instruments maintained a constant share of 6 to 6.4%, ending 1986 with 6.4%, or $45 billion. The export share of metal products, while accounting for more than 6% in the 1979-1982 period, has declined steadily since, reaching only 5.2%, or $36 billion, in 1986.

Although the 1979-1987 period shows significant changes in the volume of trade in engineering products over the years, the structure in terms of leading exporters and leading importers, as well as the relative significance of various categories of engineering products, show far fewer changes. The United States was the largest market in all categories, while Japan was the leading exporter for the period 1979-1986 (the Federal Republic of Germany took the lead in 1987). [IV.23]

As table IV.13 illustrates, United States imports of all engineering products in 1987 amounted to more than $200 billion, up from $182.4 billion in 1986. Although much smaller, the Federal Republic of Germany was the second largest single-country market, with $74.0 billion. The United Kingdom followed suit with $61.3 billion, followed by France with $57.6 billion, Canada with $52.1 billion, Italy $38.9 billion, the Netherlands $30.8 billion and the USSR $30.5 billion (all figures for 1987).

In terms of product groups, transport equipment was the single largest, with the United States buying $80.3 billion in 1987 and $75.9 billion in 1986. This was followed by Canada's $24.7 billion in 1987 and $23.8 billion in 1986 and the Federal Republic of Germany with $21.1 billion and $16.2 billion, respectively. The United Kingdom imported $17.3 billion in transport equipment in 1987 and $13.7 billion in 1986, followed by France with $15.4 billion and $11.4 billion and Italy with $12.4 billion and $8.8 billion respectively. In 1986, the USSR also reported high imports of transport equipment, amounting to $6.6 billion.

Electrical machinery supplies were another category of which the United States also took the largest amount, with $47.9 billion in 1987 and $43.3 billion in 1986. The Federal Republic of Germany was the second largest importer of electrical machinery, buying $17.5 billion in 1987 and $13.7 billion in 1986. The United Kingdom, a close third, purchased $13.3 billion in 1987 and $10.3 in 1986, followed by France's $11.5 billion and $8.7 billion respectively.

Non-electrical machinery purchases by the United States were also high: $54.8 billion in 1987 and $47.3 billion in 1986. The Federal Republic of Germany purchased $25.3 billion in 1987 and $20.5 billion in 1986, followed by the United Kingdom with $23.2 billion in 1987 and $18.1 billion in 1986. France represented a $22.9 billion dollar market in 1987 and $17.7 billion in 1986, while Canada bought $16.5 billion in 1987 and $14.9 billion in 1984. USSR imports amounted to $15.4 billion in 1985 ($16.1 billion in 1983). Of this, 1985 sales of automated data processing (ADP) equipment to the United States represented $4.1 billion, to the United Kingdom $3.3 billion, to the Federal Republic of Germany $2.9 billion, to Canada $2.4 billion and to France $2.3 billion, all significantly higher than in preceding years.

United States imports of precision instruments in 1987 amounted to $9.1 billion, up from $8.3 billion the preceding year. The comparable figures for the Federal Republic of Germany were $5.4 billion and $4.4 billion, and for the United Kingdom $4.4 billion and $3.7 billion. Controlling instruments were also a significant import for the USSR when it purchased $740 million worth of them in 1984. (The United States in the same year imported controlling instruments for some $1.4 billion).

On the export side, the figures in table IV.14 show Japan in the lead for the period 1979-1986, with total engineering exports in 1986 reaching a level of nearly $160 billion, up from some $130 billion in 1985. This compared with United States exports of $106 billion and $105 billion, the Federal Republic of Germany's $132 billion and $96 billion, followed by the United Kingdom's $43 billion and $37 billion and France's $47 billion and $37 billion, respectively. In 1987, the Federal Republic of Germany accounted for $161 billion, the United States for $122 billion, France for $57 billion and the United Kingdom for $54 billion. [IV.24]

Significant changes in the review period occurred in the metal goods sector, in which the United States exports dropped to under $3 billion in 1985-1986. The lead in metal goods exports was held by the Federal Republic of Germany, with sales of $9.3 billion in 1987 and $7.8 billion in 1986, followed by Japan's $4 billion in 1986 and Italy's $4.5 billion in 1987 and $3.8 billion in 1986. France posted metal goods exports of $3.3 billion in 1987, down from $3.4 billion in 1980, while the United Kingdom exported $2.5 billion worth of metal goods in 1987.

The leaders in non-electrical machinery exports were the Federal Republic of Germany with $57.6 billion in 1987; the United States with $47.4 billion (1987); Japan with $42.1 billion (1986); the United Kingdom with $23.7 billion (1987); Italy with $22.6 billion (1987); and France with $18.7 billion (1987).

In exports of electrical machinery, Japan far outperformed the rest, reaching $44.8 billion in 1986, followed by the Federal Republic of Germany with $25.7 billion (1987), the United States with $22.4 billion (1987), France with $10.8 billion (1987) and the United Kingdom with $9.4 billion (1987).

Foreign sales of transport equipment showed greater balance among the first three market leaders, with Japan hitting $59.4 billion in 1986 exports; the Federal Republic of Germany $57.2 billion (1987) and the United States $40.5 billion (1987). Other significant transport equipment exporters included Canada with $25.3 billion, France with more than $21 billion, the United Kingdom with more than $14 billion and Italy with more than $10 billion (all in 1987).

The same three countries took the lead in exports of precision instruments, with the Federal Republic of Germany accounting for $9.4 billion in 1987, Japan $8.6 billion in 1986 and the United States $8.4 billion (1987). Switzerland recorded exports of $5.3 billion, the United Kingdom $4.3 billion and France $3.4 billion (all in 1987).

In terms of more specific engineering products, the Federal Republic of Germany was the world's largest machine-tool exporter in 1984. Although in 1985 this position - according to estimates - was taken over by Japan, the export ratio remained notably higher in the Federal Republic of Germany than in Japan (61% against 40%). The success of the Federal Republic of Germany engineering industries on foreign markets in recent years occurred at a time of a growing world market for those products (favourable investment climate, growing consumer demand) and a relatively low-valued deutsche mark.

The relatively high value of the dollar in 1984-1985 in fact benefited most of the United States' foreign competitors (in particular Japan). For example, the United States increased its imports of machine tools to $1.7 billion in 1985, whilst its exports remained at a very modest level ($450 million). Similar developments occurred in other branches of the engineering industry. As a result, with the depreciation of the dollar in mid-1985 and the continued strengthening of the recovery in western Europe, the United States adverse trade balance is expected to improve after the review period.

The powerful export-oriented engineering industries of Japan (66% of all Japanese exports being engineering products) profited extensively from the recovery in the United States. Japan's machine-tool exports increased by 38.8% in 1984 and a further 20% in 1985, totalling $2.2 billion. With this strong export drive, Japan became the world's largest machine-tool exporter in 1985, compared with its position as a net machine-tool importer before 1972. In 1985, imports accounted for only about 6% of machine-tool consumption in Japan. Japan is also the world's leading exporter of electrical machinery, with particular strength in consumer electronics and telecommunication equipment markets.

Data presented above (e.g. in tables IV-1, IV.13 and IV.14) have already indicated significant deviations in the development of international trade in engineering products, by individual countries and regions. East-west trade (within the ECE region) remained stagnant during the period under consideration. Economic co-operation between east and west "was hampered considerably by protectionist trends, introvertedness and self-sufficiency, and the absence of a common monetary system ...". [IV.25] Recent changes in east-west trade and financial flows are analysed in greater detail in the *Economic Bulletin for Europe.* [IV.26] Regarding the structure of eastern exports to the west, Soviet Union exports continued to be fuel-oriented (60-70% share of total exports), whilst other east European countries' exports remained manufactures-oriented (more than 30% share of semi-manufactures and nearly 30% share of engineering and other consumer goods in total exports). On the other hand, eastern imports from the west (including those of the Soviet Union) continued to be heavily machinery-oriented (some 80% share of manufactures in total imports). In this respect, an increased proportion was recorded in imports of automated technologies, e.g. NC-metalworking machinery, welding and assembly robots (work-places), material-handling equipment. [IV.27]

East-west business opportunities and trade prospects were also discussed at the ECE Symposium, held at Thessaloniki (Greece) in September 1986. One of the Symposium workshops was fully devoted to east-west co-operation in the field of engineering industries' trade policies. Numerous examples of successful trade co-operation, including licensing, joint ventures and joint participation in third-country projects, were presented. Among the many existing problems (e.g. complicated procedures, payment conditions, limitations in technology transfer, lack of common data bases and associated communications), the need to develop and maintain mutually agreed standards for new engineering technologies and products was stressed (see also subchapter IV.7). [IV.28]

Table IV.8

Share of exports of engineering products [a] in total exports of selected regions, 1979-1986

(Percentage based on values in US dollars at current f.o.b. prices)

Region	*1979*	*1980*	*1981*	*1982*	*1983*	*1984*	*1985*	*1986* *
Developed market economies	35.2	34.7	36.0	36.8	37.0	37.7	39.3	40.8
of which:								
Western Europe	31.9	30.8	36.4	32.8	32.0	31.5	32.9	34.5
North America	38.5	37.0	39.6	39.5	40.0	40.9	43.7	44.7
Japan	53.7	54.9	56.8	61.4	63.8	66.6	67.8	70.0
Centrally planned economies of eastern Europe	28.3	26.7	25.1	27.1	26.8	27.2	28.8	29.8
Others [b]	5.0	4.9	5.6	2.4	4.5	5.4	4.6	3.8
Total world	26.9	25.6	26.6	26.5	27.5	28.3	29.5	31.6

a In terms of SITC, Rev.2: Section 7 - Machinery and transport equipment.

b Total world minus developed market economies and centrally planned economies.

Table IV.9 World exports of engineering products a/ by region of origin, 1979-1986

(Billions of US dollars at current f.o.b. prices)

Region of origin	1979	1980	1981	1982	1983	1984	1985	1986*	Index 1986/ 1979
Developed market economies	375.2	436.0	446.4	432.4	429.5	465.9	498.1	600.0	159.9
of which:									
Western Europe	230.2	259.5	238.1	236.6	226.2	228.3	248.1	319.0	138.6
North America	85.6	99.5	113.4	108.8	107.8	122.2	128.5	131.0	153.0
Japan	58.5	75.9	93.7	85.2	93.7	113.3	119.2	146.4	250.3
Centrally planned economies of eastern Europe	41.9	45.7	46.7	44.7	46.8	47.7	50.1	55.0	131.3
Others b/	12.5	16.1	18.5	12.3	21.5	26.6	22.6	18.0	144.0
Total world	429.6	497.8	511.6	489.4	497.8	540.6	570.8	673.0	156.7

Source : Bulletins of Statistics on World Trade in Engineering Products. UN/ECE publication.

a/ In terms of SITC, Rev.2 : Section 7 - Machinery and transport equipment.
b/ Total world minus developed market economies and centrally planned economies.

Table IV.10 World exports of engineering products: a/ share of individual regions, 1979-1986

(Percentage)

Region of origin	1979	1980	1981	1982	1983	1984	1985	1986*
Developed market economies	87.3	87.6	87.3	88.4	86.3	86.2	87.3	89.1
of which:								
Western Europe	53.6	52.1	46.5	48.3	45.4	42.2	43.5	47.4
North America	19.9	20.0	22.2	22.2	21.7	22.6	22.5	19.5
Japan	13.6	15.2	18.3	17.4	18.8	21.0	20.9	21.8
Centrally planned economies of eastern Europe	9.8	9.2	9.1	9.1	9.4	8.8	8.8	8.2
Others b/	2.9	3.2	3.6	2.5	4.3	4.9	3.9	2.7
Total world	100.0	100.0	100.0	100.0	100.0	100.0	100.0	100.0

Source and footnotes as for table IV.9.

Table IV.11

Total exports of metal and engineering products [a] of main exporters, by group of products, 1979 -1986

(Thousands of millions of US dollars at current f.o.b. prices)

Group of products	*1979*	*1980*	*1981*	*1982*	*1983*	*1984*	*1985*	*1986* [b]
Metal products	32.1	37.7	36.7	34.4	31.8	32.8	32.2	36.0
Non-electrical machinery	177.6	201.1	196.5	191.4	191.3	202.3	212.9	240.0
Electrical machinery	86.5	99.1	102.2	98.5	104.2	120.0	121.7	145.0
Transport equipment	157.2	173.3	178.1	173.7	178.3	193.0	207.5	232.0
Precision instruments	29.0	34.4	34.8	32.7	32.5	35.4	38.6	45.0
Total of above	482.4	545.6	548.3	530.7	538.1	583.5	612.9	698.0

Sources: Bulletins of Statistics on World Trade in Engineering Products, 1980-1986. United Nations Comtrade Data Base.

a In terms of SITC, Rev.2: Division 69, Section 7, Division 87 and Groups 881, 884 and 885. Footnote to table III.11.

b

Group of products	SITC, Rev.2 code
Metal products	69
Non-electrical machinery	71 (except 716) + 72 + 73 + 74 + 75
Electrical machinery	716 + 76 + 77
Transport equipment	78 + 79
Precision instruments	87 + 881 + 884 + 885

Table IV.12

Total exports of metal and engineering products of main exporters, by group of products, 1979-1986

(Percentage)

Group of products	*1979*	*1980*	*1981*	*1982*	*1983*	*1984*	*1985*	*1986* [b]
Metal products	6.7	6.8	6.7	6.5	5.9	5.6	5.2	5.2
Non-electrical machinery	36.8	36.9	35.8	36.1	35.6	34.7	34.7	34.4
Electrical machinery	17.9	18.2	18.6	18.5	19.4	20.6	19.9	20.8
Transport equipment	32.6	31.8	32.5	32.7	33.1	33.1	33.9	33.2
Precision instruments	6.0	6.3	6.4	6.2	6.0	6.1	6.3	6.4
Total of above	100.0	100.0	100.0	100.0	100.0	100.0	100.0	100.0

Sources: As for table IV.11.

Table IV.13 Imports of metal and engineering products of selected countries, by group of products, 1979-1987

(Millions of US dollars at current prices, f.o.b.)

Country		Group of products					
		Metal products	Non-electrical machinery	Electrical machinery	Transport equipment	Precision instruments	Total engineering products
Austria	1979	731.9	2371.2	1568.0	2071.8	475.5	7218.4
	1980	879.9	2873.9	1810.1	2305.5	548.9	8418.2
	1981	699.4	2547.2	1468.9	1742.8	466.3	6924.6
	1982	650.3	2305.0	1466.8	1661.8	446.5	6530.4
	1983	656.9	2317.6	1474.1	1972.1	449.4	6870.1
	1984	638.2	2390.7	1465.4	1639.7	460.2	6594.2
	1985	664.0	2705.8	1571.8	1938.4	499.9	7379.9
	1986	957.7	3943.4	2278.8	2851.7	733.2	10764.8
	1987	1178.1	4909.6	2999.5	3442.0	899.6	13428.9
Belgium	1979	1407.6	4754.1	2628.7	7356.8	691.0	16838.1
	1980	1661.8	5495.7	2997.4	7552.3	809.2	18516.4
	1981	1312.0	4233.8	2431.5	6572.9	687.7	15237.8
	1982	1189.2	4255.0	2039.5	6173.1	648.3	14305.0
	1983	1043.4	4017.2	1963.5	6169.5	607.7	13801.2
	1984	1011.8	4363.2	1916.1	5572.0	604.9	13468.1
	1985	1116.9	4787.3	2149.2	6039.1	662.6	14755.2
	1986	1519.2	6679.4	3299.7	9149.6	926.9	21574.7
	1987	1835.4	8584.0	3833.6	12013.7	1125.0	27391.6
Bulgaria	1979	71.1	1079.7	332.8	1361.3	...	3344.6
	1980	...	...	324.3	1335.0	213.8	3610.2
	1981	...	...	355.8	1408.8	250.7	3826.8
	1982	131.9	1762.9	451.2	1563.5	264.6	4234.6
	1983	192.3	1792.8	460.1	1779.0	292.4	4518.3
	1984	...	...	...	...	...	4580.8
	1985	...	...	...	...	...	4775.7
	1986	...	...	...	...	...	5969.4
	1987	...	...	...	...	...	...
Cyprus	1979	35.2	96.2	61.1	78.8	17.0	288.3
	1980	44.3	99.3	73.5	97.5	22.2	336.8
	1981	38.2	76.6	76.4	63.2	16.3	270.8
	1982	35.3	86.5	101.9	86.3	19.2	329.2
	1983	34.1	81.5	94.9	99.7	27.4	337.7
	1984	34.4	97.9	73.8	186.2	30.5	422.9
	1985	32.8	88.5	92.6	145.1	23.0	382.0
	1986	37.2	98.4	107.6	105.4	22.5	371.1
	1987	...	...	...	...	...	...
Czechoslovakia	1979	108.8	3414.0	1036.2	730.8	208.2	5498.0
	1980	118.5	3488.3	813.3	1138.1	209.0	5767.2
	1981	83.4	3282.3	877.7	654.8	212.0	5110.3
	1982	166.7	3549.2	619.6	598.4	216.5	5151.4
	1983	160.2	3796.4	659.5	645.6	219.3	5500.5
	1984	171.6	4049.1	703.6	690.9	234.5	5850.9
	1985	182.2	4064.9	732.5	665.8	262.2	5908.2
	1986	231.0	5066.0	930.0	784.5	320.4	7331.9
	1987	...	...	...	...	...	...
Denmark	1979	468.7	1952.6	1088.7	1313.2	282.7	5105.9
	1980	480.9	1893.5	1075.7	1008.0	301.8	4759.8
	1981	395.5	1619.1	890.9	1149.8	242.9	4298.2
	1982	385.6	1698.0	898.8	1014.2	246.3	4242.9
	1983	505.3	1719.2	887.4	1041.7	246.5	4400.2
	1984	409.5	1853.2	983.9	1127.4	264.0	4638.1
	1985	449.8	2216.6	1130.0	1429.4	297.4	5523.2
	1986	634.0	3231.3	1581.4	2388.8	436.5	8271.9
	1987	762.6	3609.2	1864.0	2104.5	497.7	8838.1
Finland	1979	258.4	1433.3	711.0	924.3	215.8	3542.7
	1980	366.1	2054.8	903.9	1195.8	288.2	4808.9
	1981	305.5	1900.2	782.1	1158.8	257.2	4403.7
	1982	292.8	1835.2	811.6	1123.9	257.1	4320.5
	1983	276.0	1773.9	822.8	1155.5	261.1	4289.3
	1984	281.6	1781.0	891.6	1020.9	251.3	4226.5
	1985	299.8	1895.5	972.4	1067.9	282.3	4518.1
	1986	394.4	2508.0	1302.9	1635.7	380.6	6221.5
	1987	533.4	3324.1	1755.1	2195.9	495.3	8303.7
France	1979	2215.7	11187.6	5386.1	7441.0	2090.5	28320.9
	1980	2696.6	13768.1	6500.1	8545.4	2448.7	33958.9
	1981	2275.2	12500.4	5952.2	8170.3	2236.4	31134.5
	1982	2212.9	12862.7	5965.4	8710.1	2189.4	31940.6
	1983	1988.3	11917.5	5505.1	8122.3	2101.3	29634.5
	1984	1924.0	11660.5	5699.6	7139.8	2022.7	28446.6
	1985	2125.9	12972.6	6058.6	7760.7	2235.2	31153.0
	1986	2973.2	17708.4	8719.1	11366.5	3019.0	43786.1
	1987	3818.3	22883.3	11538.2	15441.2	3875.7	57556.8

Table continued/

Table IV.13 (continued)

Country		Group of products					
		Metal products	Non-electrical machinery	Electrical machinery	Transport equipment	Precision instruments	Total engineering products
German Democratic Republic	1979	...	...	...	...	...	...
	1980	...	...	...	...	...	...
	1981	...	...	...	...	...	...
	1982	...	...	...	...	...	...
	1983	...	...	...	...	...	...
	1984	...	...	...	...	...	...
	1985	...	...	...	...	...	...
	1986	...	...	...	...	...	...
	1987	...	...	...	...	...	...
Germany, Federal Republic of	1979	2800.1	11602.1	8113.4	11081.0	2879.3	36476.0
	1980	3410.7	13646.7	9701.6	11692.6	3416.0	41867.6
	1981	2754.6	11912.2	8620.8	11702.3	2982.1	37972.1
	1982	2593.3	11338.4	8354.4	11856.9	2786.9	36930.0
	1983	2574.5	12303.9	8700.1	12311.8	2724.0	38614.3
	1984	2538.3	12807.2	9371.5	11262.1	2759.7	38738.8
	1985	2672.7	14856.1	9936.1	11483.6	3086.0	42034.4
	1986	3822.2	20485.3	13729.5	16235.2	4413.5	58685.7
	1987	4699.7	25273.1	17537.1	21087.6	5403.9	74001.4
Greece	1979	158.7	1015.3	394.8	2262.5	117.9	3949.3
	1980	169.1	1016.7	490.1	2279.9	126.8	4082.6
	1981	187.9	864.9	382.4	1208.2	115.9	2759.3
	1982	169.6	876.0	414.3	1267.7	96.6	2824.2
	1983	154.6	859.0	381.6	1120.4	91.9	2607.4
	1984	148.0	853.0	365.5	1251.6	101.6	2719.6
	1985	144.0	800.0	412.2	1189.5	121.1	2666.8
	1986	168.5	994.2	591.0	1333.6	158.7	3246.0
	1987	...	...	...	...	...	...
Hungary	1979	183.8	1495.0	555.9	755.3	126.8	3116.9
	1980	168.2	1412.6	530.8	756.5	126.5	2994.6
	1981	181.2	1375.2	520.6	667.1	132.2	2876.3
	1982	165.6	1419.3	487.6	647.8	128.8	2849.0
	1983	141.9	1248.3	482.7	577.4	118.0	2568.3
	1984	119.8	1145.8	433.5	527.7	100.8	2327.6
	1985	129.9	1180.5	501.6	551.0	110.7	2473.8
	1986	162.2	1413.9	574.2	722.3	135.3	3007.9
	1987	...	...	...	...	...	...
Ireland	1979	370.5	1374.5	587.9	875.0	134.9	3342.9
	1980	444.2	1463.4	743.7	820.7	162.9	3634.9
	1981	374.6	1439.2	765.6	706.8	149.0	3435.3
	1982	315.9	1408.7	727.1	530.3	166.8	3148.9
	1983	264.4	1439.6	699.3	505.2	176.9	3085.3
	1984	268.4	1733.9	794.7	483.3	191.8	3472.2
	1985	271.7	1837.2	814.5	489.2	212.5	3625.1
	1986	334.8	2116.4	904.3	617.7	236.5	4209.7
	1987	380.8	2614.0	1188.3	758.4	263.3	5204.9
Italy	1979	925.8	5599.7	3509.1	5107.2	1268.5	16410.3
	1980	1245.5	7625.4	4475.7	7953.1	1701.8	23001.4
	1981	1045.7	6786.4	3681.8	7245.5	1609.9	20369.4
	1982	949.8	6100.1	3631.3	6458.4	1549.3	18688.9
	1983	848.1	5598.6	3431.6	5107.1	1479.3	16464.7
	1984	866.8	6359.8	3945.9	5839.5	1543.6	18555.7
	1985	947.7	7341.6	4377.5	6518.8	1713.5	20899.1
	1986	1320.2	9760.3	6050.9	8849.6	2382.0	28363.0
	1987	1685.0	13305.2	8304.1	12398.5	3220.3	38913.2
Netherlands	1979	1637.6	5335.0	4111.0	5632.6	1131.2	17847.5
	1980	1864.7	6015.1	4629.1	4563.2	1269.4	18341.5
	1981	1499.9	5072.0	3627.8	3705.5	1161.1	15066.4
	1982	1417.1	5134.1	3341.0	3675.8	1038.6	14606.7
	1983	1352.0	5527.5	3392.3	4097.0	1040.4	15409.1
	1984	1427.7	6094.6	3647.6	3791.4	1046.0	16007.3
	1985	1532.7	6947.3	3943.9	4107.1	1178.9	17709.9
	1986	2160.6	9506.5	5504.4	6423.3	1654.2	25249.0
	1987	2658.4	11555.9	6755.8	7862.5	1949.9	30782.3
Norway	1979	442.5	1700.6	963.8	1693.0	260.5	5060.4
	1980	603.7	2264.4	1150.3	1424.5	314.1	5757.0
	1981	522.3	2103.4	1090.8	2160.9	300.3	6177.8
	1982	651.9	2075.7	999.9	2615.6	289.6	6632.7
	1983	571.4	1802.2	963.0	2197.2	262.5	5796.5
	1984	494.8	2077.4	1055.3	1983.7	281.0	5892.2
	1985	563.9	2305.1	1170.8	2059.5	328.4	6427.6
	1986	814.4	3449.9	1771.6	3007.5	459.9	9503.3
	1987	1024.0	3907.1	1926.1	3012.3	507.2	10376.7

Table continued/

Table IV.13 (continued)

Country		Group of products					
		Metal products	Non-electrical machinery	Electrical machinery	Transport equipment	Precision instruments	Total engineering products
Poland	1979	...	...	...	...	...	6013.6
	1980	196.4	2924.5	695.2	1622.9	289.5	6671.9
	1981	108.7	1995.7	527.2	1264.8	232.0	5097.6
	1982	102.8	1046.8	326.5	629.0	189.2	2793.3
	1983	120.4	1145.7	409.8	613.7	185.0	2978.6
	1984	123.7	1214.6	435.3	686.8	169.0	3216.9
	1985	140.1	1397.4	490.4	734.2	187.4	3637.6
	1986	170.7	1521.4	520.4	809.7	185.3	3972.2
	1987	...	...	...	...	...	...
Portugal	1979	86.2	824.5	322.9	481.5	101.9	1817.0
	1980	111.9	1172.8	467.7	670.4	150.3	2573.1
	1981	122.2	1259.5	498.3	896.2	152.9	2929.1
	1982	116.9	1230.4	528.1	804.4	155.6	2835.3
	1983	87.3	942.4	443.7	772.2	126.0	2371.6
	1984	80.2	731.8	422.8	531.2	97.3	1863.2
	1985	83.3	723.4	415.6	513.5	104.9	1840.7
	1986	134.4	1175.5	700.7	877.6	155.5	3043.6
	1987	221.3	1971.8	992.9	1524.3	232.2	4942.5
Spain	1979	314.6	2572.4	1203.6	1008.0	629.5	5728.1
	1980	404.9	3126.5	1528.5	1404.3	764.8	7229.1
	1981	329.7	2883.5	1363.0	1356.4	716.9	6649.4
	1982	366.1	3127.8	1494.0	1400.3	739.7	7127.9
	1983	282.9	2840.9	1429.3	1181.4	678.8	6413.3
	1984	307.7	3195.4	1295.2	1338.6	616.4	6753.3
	1985	308.8	3633.4	1339.1	1542.3	732.5	7556.1
	1986	521.4	5324.8	2384.6	2797.6	1080.9	12109.3
	1987	798.5	8179.6	3590.5	5303.3	1480.1	19352.0
Sweden	1979	771.9	3556.4	1958.4	2364.2	607.6	9258.5
	1980	960.3	4372.6	2328.8	2261.2	718.0	10640.8
	1981	801.5	3725.9	2037.7	2052.4	623.4	9240.8
	1982	708.0	3610.3	2039.2	2000.7	614.1	8972.4
	1983	662.3	3698.5	1982.4	1999.1	605.7	8947.9
	1984	697.3	3771.2	2270.2	2162.2	647.2	9548.1
	1985	749.8	4319.6	2553.1	2495.9	721.4	10839.8
	1986	993.4	5169.4	3031.9	3545.1	946.2	13686.0
	1987	1262.1	6723.8	3827.0	5045.3	1179.2	18037.4
Switzerland	1979	755.5	2830.3	1714.4	2657.4	816.8	8774.4
	1980	941.2	3660.9	2090.2	2917.9	1138.9	10749.1
	1981	829.1	3251.4	1820.5	2806.9	1038.6	9746.5
	1982	747.9	3039.6	1696.8	2788.1	878.4	9150.9
	1983	717.2	2973.1	1641.5	3066.3	846.1	9244.2
	1984	770.1	3106.1	1794.3	2448.6	898.0	9017.1
	1985	820.3	3538.3	1948.2	2485.5	930.0	9722.3
	1986	1184.0	5255.5	2876.1	4090.6	1356.5	14762.7
	1987	1515.1	6851.1	3664.9	5458.7	1591.7	19081.4
Turkey	1979	65.1	915.0	256.4	207.6	32.2	1476.3
	1980	80.4	865.3	281.4	195.8	54.5	1477.5
	1981	150.9	1254.8	346.2	323.5	74.1	2149.5
	1982	74.8	1407.3	386.4	494.9	90.9	2454.3
	1983	90.3	968.0	338.0	433.8	96.9	1927.1
	1984	94.5	1626.4	616.5	439.9	126.2	2903.5
	1985	102.7	1589.0	730.8	791.1	142.0	3355.6
	1986	150.2	2389.0	981.4	713.0	186.5	4420.1
	1987	199.4	2494.1	1021.2	539.0	239.4	4493.1
USSR */	1979	...	13800.0	...	3000.0	...	22800.0
	1980	...	16800.0	...	4660.0	...	25800.0
	1981	...	11900.0	...	4668.0	...	23100.0
	1982	...	14700.0	...	5328.0	...	27900.0
	1983	...	16100.0	...	6727.0	...	32100.0
	1984	...	15900.0	...	6530.0	...	30900.0
	1985	...	15400.0	...	6600.0	...	30500.0
	1986	...	...	...	...	...	...
	1987	...	...	...	...	...	...

Table continued/

Table IV.13 (concluded)

Country		Group of products					
		Metal products	Non-electrical machinery	Electrical machinery	Transport equipment	Precision instruments	Total engineering products
United Kingdom	1979	1637.4	10865.3	4912.7	11344.3	2218.0	30977.7
	1980	1998.0	12259.5	5676.8	12354.7	2668.1	34957.1
	1981	1636.4	11481.6	6345.1	9072.5	2653.4	31188.9
	1982	1661.7	12202.9	6823.1	9789.5	2626.5	33103.7
	1983	1679.8	12691.0	7372.4	10659.9	2720.3	35123.4
	1984	1857.8	14046.6	7882.1	9984.5	2929.3	36700.2
	1985	1942.3	15166.0	8509.2	11039.4	3088.4	39745.3
	1986	2410.6	18095.8	10345.2	13746.3	3685.0	48282.8
	1987	3100.1	23214.4	13298.4	17342.7	4368.8	61324.3
Yugoslavia	1979	278.3	3129.9	810.2	1039.9	310.0	5568.3
	1980	280.3	2632.1	758.5	766.2	246.1	4683.1
	1981	292.3	2604.5	744.2	943.9	259.0	4843.8
	1982	257.9	2274.0	628.6	949.5	221.5	4331.3
	1983	206.9	1723.5	569.1	576.7	183.6	3259.8
	1984	175.3	1539.3	548.4	519.0	183.1	2965.1
	1985	172.9	1703.9	562.0	661.8	212.2	3312.8
	1986	189.6	1840.3	679.0	695.2	252.7	3656.9
	1987	...	...	...	...	...	...
Canada	1979	1415.6	9890.1	3183.5	12920.5	1194.8	28604.4
	1980	1535.3	11371.1	3422.3	11931.6	1390.6	29651.0
	1981	1600.6	12836.0	4041.5	14128.6	1600.2	34206.9
	1982	1320.4	10735.2	3738.2	12029.0	1492.7	29315.4
	1983	1453.6	10791.5	4575.2	15694.3	1594.1	34108.6
	1984	1743.2	13446.4	5828.1	19882.1	1827.1	42726.8
	1985	1838.0	13808.7	5593.3	22710.4	1864.7	45815.1
	1986	1897.7	14904.2	6116.8	23822.5	1959.8	48701.2
	1987	2141.7	16531.3	6708.1	24702.4	2070.4	52153.8
United States	1979	3942.3	15727.0	13740.7	27092.6	3527.8	64030.4
	1980	4050.7	17540.7	15806.7	30489.5	3922.9	71810.5
	1981	4508.1	20382.7	19229.3	32857.7	4526.9	81504.7
	1982	4650.8	20343.7	20450.2	35310.0	4261.0	85015.7
	1983	4862.9	23453.9	25182.0	40399.4	4765.1	98663.2
	1984	6437.3	34706.2	36224.5	52288.8	6170.3	135827.1
	1985	7026.9	39504.6	38522.4	63809.6	6918.8	155782.2
	1986	7686.4	47260.9	43290.7	75873.9	8309.1	182421.0
	1987	8643.7	54811.7	47914.5	80342.8	9115.6	200828.2
Japan	1979	476.4	3269.2	1989.5	1778.9	1309.2	8823.2
	1980	547.7	3874.0	2261.2	2254.5	1459.3	10396.7
	1981	535.1	3656.9	2447.2	2710.8	1430.3	10780.4
	1982	570.7	3839.7	2437.8	1417.2	1422.6	9688.0
	1983	579.5	3741.7	2647.4	2517.9	1518.7	11005.3
	1984	613.4	4547.0	3402.7	2336.3	1788.3	12687.7
	1985	611.9	4792.0	3186.0	2596.9	1808.4	12995.3
	1986	751.0	5378.1	3823.7	3487.5	2034.3	15474.6
	1987	...	...	...	...	...	...

*/ Secretariat estimates.

Sources : United Nations Comtrade Data Base and Government replies to ECE questionnaires for Annual Reviews of Engineering Industries and Automation.

Definitions of items:

Total engineering products corresponds to SITC, Rev.2:69+7+87+881+884+885.
Metal products correspond to SITC, Rev.2:69.
Non-electrical machinery corresponds to SITC, Rev.2:71(excluding 716)+72+73+74+75.
Electrical machinery corresponds to SITC, Rev.2:716+76+77.
Transport equipment corresponds to SITC,Rev 2:78+79.
Precision instruments correspond to SITC, Rev.2:87+881+884+885.

Table IV.14 Exports of metal and engineering products of selected countries, by group of products, 1979-1987

(Millions of US dollars at current prices, f.o.b.)

Country		Group of products					
		Metal products	Non-electrical machinery	Electrical machinery	Transport equipment	Precision instruments	Total engineering products
Austria	1979	796.8	2168.7	1423.6	777.3	315.6	5482.1
	1980	921.8	2459.6	1547.3	835.5	359.2	6123.4
	1981	820.7	2185.3	1435.9	723.5	294.9	5460.3
	1982	831.6	2259.7	1565.8	804.8	285.2	5747.2
	1983	804.6	2498.5	1447.4	707.1	290.8	5748.4
	1984	745.4	2525.2	1453.4	727.3	298.2	5749.5
	1985	823.1	2906.7	1607.5	841.1	355.6	6533.9
	1986	1103.8	4107.7	2325.9	1046.4	481.5	9065.2
	1987	1303.0	4885.8	2968.3	1233.0	595.3	10985.3
Belgium	1979	1294.5	3621.8	2504.9	6802.9	316.8	14540.9
	1980	1461.9	4051.0	2877.6	6992.3	378.1	15761.0
	1981	1261.0	3498.9	2360.8	6231.3	357.1	13709.0
	1982	1170.3	3423.3	1994.4	6435.8	304.9	13329.1
	1983	1107.0	2998.7	1990.4	6659.5	341.9	13117.2
	1984	1116.5	3081.2	1879.9	5884.2	350.8	12319.0
	1985	1168.2	3721.4	2023.0	6681.8	360.6	13957.4
	1986	1463.0	4515.4	3152.0	10078.8	531.6	19778.8
	1987	1787.5	5740.8	3833.4	12866.0	656.7	24891.3
Bulgaria	1979	52.2	2217.4	619.3	695.8	143.4	3948.7
	1980	47.4	2546.8	748.2	959.7	153.5	4559.6
	1981	48.9	2634.2	694.4	1124.5	150.1	4720.3
	1982	50.0	2833.9	716.3	1317.3	118.2	5425.5
	1983	59.8	3365.6	774.4	1451.2	124.2	5911.5
	1984	66.0	4200.1	791.8	1044.0	132.0	6249.2
	1985	...	...	...	...	...	7206.0
	1986	...	...	...	...	...	7883.1
	1987	...	...	...	...	...	...
Cyprus	1979	3.1	8.9	10.2	8.9	8.8	40.0
	1980	6.1	13.3	18.5	8.7	12.2	58.8
	1981	8.8	16.6	15.5	6.7	7.9	55.4
	1982	5.2	17.7	27.4	2.2	9.1	61.7
	1983	6.3	15.2	30.9	2.5	15.6	70.5
	1984	5.7	18.1	24.9	6.2	19.2	74.2
	1985	5.7	20.5	26.9	4.3	12.1	69.5
	1986	5.3	19.1	40.0	5.2	8.0	77.5
	1987	...	...	...	...	...	...
Czechoslovakia	1979	123.5	4857.8	907.7	1541.4	130.6	7561.0
	1980	122.6	3924.7	775.7	2768.8	24.9	7616.7
	1981	95.2	4825.8	1060.3	1776.8	153.2	7911.3
	1982	212.1	5346.6	702.8	1810.6	197.9	8269.9
	1983	234.0	5773.7	820.0	1980.0	214.1	9017.9
	1984	230.3	6140.0	899.5	2061.5	216.7	9548.1
	1985	265.2	6254.7	927.8	2193.4	216.4	9847.4
	1986	300.8	7313.6	1060.0	2718.7	264.9	11658.1
	1987	...	...	...	...	...	...
Denmark	1979	316.6	2094.4	821.4	524.5	321.6	4078.4
	1980	407.9	2394.4	917.6	638.0	376.0	4733.9
	1981	366.8	2233.8	825.7	857.5	356.3	4640.1
	1982	365.1	2040.1	842.0	689.3	354.6	4291.0
	1983	348.9	1962.2	843.5	910.4	383.5	4448.4
	1984	351.4	1953.3	931.0	673.9	403.5	4313.2
	1985	366.0	2138.8	1091.6	795.5	433.2	4825.1
	1986	512.9	2871.0	1380.9	742.5	606.8	6114.2
	1987	631.3	3381.0	1593.1	999.9	693.1	7298.4
Finland	1979	240.0	954.2	414.7	847.9	94.6	2551.4
	1980	321.2	1183.9	598.4	724.4	110.6	2938.4
	1981	354.5	1288.5	561.0	987.2	99.1	3290.3
	1982	386.7	1421.0	584.6	1305.4	117.9	3815.6
	1983	298.8	1146.0	553.7	1384.3	130.4	3513.2
	1984	217.8	1206.8	560.1	1592.5	148.9	3726.1
	1985	257.2	1527.2	657.8	1238.7	158.5	3839.4
	1986	312.7	1911.1	999.3	1598.4	208.1	5029.6
	1987	431.8	2538.5	1325.8	1523.8	267.6	6087.5
France	1979	2905.0	12653.4	6410.7	16012.0	1938.3	39919.4
	1980	3453.1	13929.9	7074.3	15826.6	2214.0	42497.8
	1981	3195.5	12648.9	6369.8	14550.2	1955.1	38719.5
	1982	2924.1	11665.0	6105.2	13814.1	1842.7	36351.0
	1983	2700.4	11457.8	6291.4	13311.7	1886.5	35647.7
	1984	2582.8	11093.3	6396.2	13738.3	2031.4	35841.9
	1985	2452.3	12058.4	6929.3	13060.0	2351.0	36851.0
	1986	2939.1	15597.9	8760.3	17037.6	2682.0	47016.9
	1987	3316.4	18709.9	10809.8	21242.2	3355.1	57433.4

Table continued/

Table IV.14 (continued)

Country	Year	Metal products	Non-electrical machinery	Electrical machinery	Transport equipment	Precision instruments	Total engineering products
German Democratic Republic	1979	248.4	4646.5	1055.7	1787.0	699.2	8436.8
	1980	328.4	5248.5	1240.7	1925.8	802.6	9546.0
	1981	372.8	5602.5	1352.9	2254.7	915.5	10498.4
	1982	501.0	5971.4	1456.7	2505.6	783.7	11218.4
	1983	531.1	6576.8	1571.0	2587.0	855.1	12121.0
	1984	520.0	6768.7	1574.7	2655.0	903.7	12422.1
	1985	431.8	6630.5	1598.7	2746.9	1000.9	12408.8
	1986	527.4	7379.6	1905.2	2863.1	1187.8	13863.1
	1987	...	...	...	...	...	...
Germany, Federal Republic of	1979	5854.0	34175.7	14677.7	28159.7	4684.7	87551.9
	1980	6640.9	36688.8	15832.9	31153.3	5239.5	97294.2
	1981	5774.3	32439.0	13728.9	30878.8	4794.9	89286.6
	1982	5816.0	32037.2	13907.3	34468.4	4828.8	92940.8
	1983	5539.5	30392.7	13268.1	31525.4	4743.4	87257.8
	1984	5250.1	30469.9	13520.0	31327.5	4909.5	87025.4
	1985	5629.0	34456.2	14903.2	33889.6	5685.1	95883.6
	1986	7800.9	48342.9	21053.9	45108.0	7831.2	131914.2
	1987	9255.9	57602.2	25743.4	57168.6	9368.8	161167.4
Greece	1979	66.5	22.5	88.5	21.1	1.3	199.9
	1980	110.6	27.0	95.0	34.8	6.3	273.6
	1981	116.8	39.4	110.1	52.7	5.8	324.8
	1982	81.7	55.5	90.0	51.7	6.1	285.0
	1983	95.9	37.5	71.6	43.6	4.3	252.9
	1984	80.2	38.0	62.2	23.2	4.7	208.2
	1985	72.2	27.6	72.8	31.8	5.1	209.7
	1986	74.9	36.5	94.7	34.4	7.5	247.9
	1987	...	...	...	...	...	...
Hungary	1979	109.3	1083.7	723.4	890.1	170.7	2977.2
	1980	135.5	1118.5	726.1	939.6	186.6	3106.3
	1981	139.3	1087.5	735.0	866.0	210.0	3037.7
	1982	137.5	1123.5	728.8	923.2	218.4	3131.3
	1983	113.6	1085.1	698.3	867.1	192.7	2956.8
	1984	100.9	996.4	682.5	898.0	203.0	2880.9
	1985	99.3	1115.9	757.7	989.4	250.1	3212.5
	1986	130.2	1209.9	881.4	1119.4	289.8	3630.7
	1987	...	...	...	...	...	...
Ireland	1979	162.9	665.6	309.7	175.6	217.5	1531.3
	1980	205.4	896.2	463.9	205.4	247.3	2018.2
	1981	170.0	1065.7	461.1	159.4	266.5	2122.6
	1982	186.2	1248.5	517.7	210.0	301.7	2464.1
	1983	188.6	1536.9	508.7	206.5	375.7	2816.4
	1984	197.6	2003.0	594.7	162.0	425.1	3382.4
	1985	221.3	2365.9	616.8	101.8	442.2	3748.0
	1986	267.8	2997.7	752.6	90.5	516.6	4625.3
	1987	311.6	3955.5	932.1	123.7	624.6	5947.5
Italy	1979	3050.4	10965.6	4214.6	6809.8	842.1	25882.6
	1980	3563.7	13862.2	4367.9	7040.4	1008.6	29842.8
	1981	3492.2	12835.3	4486.4	6870.5	941.0	28625.4
	1982	3237.0	12282.3	4588.1	6003.7	921.4	27032.5
	1983	3171.6	12469.9	4601.9	6231.7	977.2	27452.3
	1984	2928.2	12173.4	4599.2	6022.8	1038.9	26762.5
	1985	2919.4	13838.8	4795.6	6194.0	1175.2	28923.0
	1986	3852.2	18513.1	6266.5	8115.4	1578.8	38325.9
	1987	4540.4	22561.5	7737.5	10295.6	1940.6	47075.6
Netherlands	1979	1186.0	4444.3	4220.7	2460.1	804.8	13115.9
	1980	1395.4	4997.0	4521.5	3055.6	874.8	14844.2
	1981	1274.8	4484.1	3696.4	2831.5	833.9	13120.7
	1982	1421.5	4650.4	3472.4	2652.7	859.4	13056.2
	1983	1260.2	5023.1	3199.0	2530.3	818.1	12830.8
	1984	1166.9	5353.2	3263.9	2224.4	871.8	12880.2
	1985	1283.1	5670.5	3558.4	2269.2	963.7	13744.8
	1986	1738.1	7738.4	4890.2	3315.6	1259.2	18941.6
	1987	2208.0	9334.8	5687.8	4405.3	1490.9	23126.8
Norway	1979	188.0	638.6	330.8	1311.5	86.9	2555.9
	1980	238.4	820.7	374.1	1057.5	125.1	2615.8
	1981	210.2	849.5	357.1	1089.8	136.5	2643.1
	1982	211.9	829.3	362.3	1421.7	124.8	2950.0
	1983	190.7	738.9	374.6	1276.2	112.4	2692.7
	1984	188.1	767.2	335.5	980.6	103.3	2374.7
	1985	171.5	758.9	302.2	1489.9	122.3	2844.8
	1986	224.7	937.3	433.7	1835.8	159.2	3590.6
	1987	293.0	1134.4	533.0	1998.4	195.8	4154.6

Table continued/

Table IV.14 (continued)

Country		Group of products					
		Metal products	Non-electrical machinery	Electrical machinery	Transport equipment	Precision instruments	Total engineering products
Poland	1979	372.7	...	...	...	200.5	7584.7
	1980	377.7	3160.1	1039.0	1969.4	199.4	7923.5
	1981	307.2	2821.0	897.9	1440.9	179.8	6753.5
	1982	204.1	2202.3	636.1	1135.7	138.7	5273.6
	1983	222.0	2062.9	676.9	1097.1	140.1	5178.9
	1984	243.1	2144.2	537.3	1012.4	138.4	4884.2
	1985	250.6	2153.8	552.6	972.4	142.4	4923.1
	1986	334.7	1597.5	630.7	1079.7	124.6	4654.2
	1987	...	...	...	...	...	...
Portugal	1979	83.6	98.7	186.6	128.3	33.4	530.6
	1980	116.0	131.9	268.9	215.2	49.9	781.9
	1981	127.3	155.6	236.8	137.3	48.9	705.9
	1982	114.9	183.7	250.9	148.4	42.4	740.4
	1983	96.2	247.3	284.9	175.7	33.1	837.3
	1984	119.4	295.7	360.0	243.5	31.4	1050.1
	1985	135.2	291.8	378.9	213.8	34.2	1054.0
	1986	154.2	327.3	452.3	345.6	51.1	1330.6
	1987	189.4	428.9	611.2	467.6	54.9	1752.0
Spain	1979	819.4	1677.8	695.6	2417.4	91.8	5701.9
	1980	873.6	1886.7	856.1	2718.0	116.7	6451.1
	1981	806.1	1930.9	783.0	2428.3	113.5	6061.8
	1982	835.5	1907.6	774.2	2846.5	114.2	6477.8
	1983	727.1	1359.8	750.7	3052.8	100.6	5991.1
	1984	689.6	1779.2	812.2	3666.2	91.9	7039.1
	1985	720.3	2094.5	836.2	3675.3	131.7	7458.0
	1986	794.2	2515.7	1187.4	4707.6	163.4	9368.4
	1987	915.1	3215.6	1472.8	5955.8	216.7	11776.1
Sweden	1979	1031.0	4840.0	2221.8	4255.8	403.0	12751.6
	1980	1195.8	5653.6	2414.5	4204.9	499.4	13968.1
	1981	1101.0	5171.4	2252.3	4396.0	486.6	13407.3
	1982	966.7	4685.5	2294.9	4269.4	449.2	12665.7
	1983	963.9	4654.5	2143.9	4400.1	482.2	12644.5
	1984	1031.3	4937.9	2286.4	4677.8	524.1	13457.5
	1985	1085.3	5347.0	2558.3	4831.0	621.9	14443.5
	1986	1360.6	6789.7	3070.9	6445.7	766.5	18433.5
	1987	1574.3	8033.8	3573.6	7630.4	954.8	21766.8
Switzerland	1979	1011.8	5759.5	2294.5	364.7	3123.5	12554.0
	1980	1102.3	6678.5	2407.8	378.5	3526.3	14093.3
	1981	983.2	5898.6	2260.7	442.4	3365.1	12950.0
	1982	960.4	5810.3	2222.1	420.1	3066.0	12478.9
	1983	942.7	5362.9	2277.4	416.4	2941.1	11940.4
	1984	977.8	5213.2	2097.0	442.5	2948.9	11679.4
	1985	1041.0	5941.7	2189.9	412.9	3254.0	12839.4
	1986	1426.5	8401.5	3221.4	605.6	4459.9	18115.0
	1987	1733.9	10168.4	4045.6	829.8	5312.9	22090.7
Turkey	1979	8.0	8.7	9.7	25.2	0.0	51.6
	1980	14.3	14.5	18.5	49.8	0.1	97.2
	1981	53.8	47.3	44.8	116.2	0.9	263.0
	1982	84.6	121.5	94.1	85.6	1.8	387.5
	1983	82.6	132.1	64.9	66.6	0.6	346.7
	1984	66.6	121.1	111.4	120.7	0.6	420.4
	1985	183.6	403.9	123.3	117.7	18.6	847.1
	1986	122.5	196.3	136.3	82.0	12.6	549.8
	1987	186.7	677.3	299.3	108.8	24.8	1296.9
USSR */	1979	69.0	5277.0	812.0	3607.0	195.0	11770.0
	1980	73.0	6484.0	852.0	3954.0	346.0	12720.0
	1981	80.0	4200.0	850.0	3600.0	350.0	10960.0
	1982	100.0	4538.0	871.0	3814.0	206.0	11330.0
	1983	100.0	4767.0	1024.0	2751.0	250.0	11610.0
	1984	70.0	4900.0	1030.0	3400.0	200.0	11380.0
	1985	50.0	4700.0	1050.0	3400.0	200.0	11350.0
	1986	...	...	...	...	...	...
	1987	...	...	...	...	...	...

Table continued/

Table IV.14 (concluded)

Country		Group of products					
		Metal products	Non-electrical machinery	Electrical machinery	Transport equipment	Precision instruments	Total engineering products
United Kingdom	1979	2535.2	15592.7	5252.4	10528.2	2091.7	36000.2
	1980	3042.2	20141.6	6521.0	12879.9	2669.6	45254.3
	1981	2571.0	17950.4	6016.5	9872.5	2638.6	39048.9
	1982	2437.3	16586.3	6044.6	8981.1	2537.9	36587.3
	1983	2007.6	13869.9	5660.0	8232.8	2527.3	32297.5
	1984	1963.0	14949.6	5746.0	8158.4	2684.8	33501.7
	1985	2095.1	16773.9	6536.5	8600.2	3089.7	37095.4
	1986	2147.2	19165.3	7676.3	10311.3	3707.5	43007.5
	1987	2532.6	23668.8	9468.6	14189.5	4315.4	54174.9
Yugoslavia	1979	293.0	720.0	606.0	688.3	48.8	2356.1
	1980	315.7	843.9	807.4	890.4	66.0	2923.4
	1981	438.1	1101.5	1009.8	999.1	83.8	3632.2
	1982	484.5	1203.7	1043.3	1091.5	91.4	3914.3
	1983	475.8	1147.4	970.1	947.9	81.4	3622.7
	1984	382.0	1145.5	933.0	1074.3	86.2	3621.0
	1985	380.3	1134.4	970.6	1386.9	77.1	3949.4
	1986	421.0	1309.4	1022.3	1143.1	85.4	3981.1
	1987	...	...	...	...	...	...
Canada	1979	813.6	4499.7	1179.9	10717.1	418.6	17628.9
	1980	855.6	4640.1	1456.0	10578.4	517.9	18048.1
	1981	980.2	5667.7	1761.3	12238.4	579.9	21227.6
	1982	842.3	5295.2	1746.5	14779.1	642.2	23305.2
	1983	835.1	5752.9	1945.6	17592.5	734.8	26860.8
	1984	1089.4	6857.8	2734.7	22696.1	781.2	34159.1
	1985	1076.4	6571.8	2816.5	24753.2	722.2	35940.1
	1986	1193.7	6723.8	2656.2	26013.4	845.3	37432.5
	1987	1355.6	7704.6	2923.4	25296.9	949.5	38229.9
United States	1979	3264.6	32514.3	12361.6	25542.4	5227.4	78910.3
	1980	4017.3	41035.6	14892.3	28583.8	6279.3	94808.2
	1981	4542.5	46480.7	16570.0	32478.2	7035.7	107107.1
	1982	3829.0	43239.9	16221.4	27492.5	7048.1	97830.8
	1983	3344.0	37886.5	16495.3	28034.8	6695.8	92456.4
	1984	3423.8	41786.2	18571.8	29496.6	7014.0	100292.4
	1985	2861.2	42348.8	17328.5	34625.4	7304.3	104468.2
	1986	2673.3	41791.1	18745.6	34885.1	7697.2	105792.3
	1987	3138.3	47359.0	22378.3	40482.2	8467.8	121825.6
Japan	1979	3217.7	15909.6	16844.8	25720.7	4700.9	66393.7
	1980	4088.8	19405.4	22095.1	34369.5	5603.4	85562.2
	1981	4472.5	24485.9	27954.1	41272.4	6443.8	104628.7
	1982	4422.4	22048.0	25071.8	38037.5	5351.6	94931.3
	1983	4038.9	25199.2	29344.6	39147.4	5863.7	103593.8
	1984	3907.1	30479.8	37213.8	45558.5	6547.0	123706.1
	1985	3564.2	32611.3	37436.1	49151.5	7231.4	129994.5
	1986	4027.7	42101.0	44837.1	59427.8	8647.7	159041.3
	1987	...	...	...	...	...	...

*/ Secretariat estimates.

Sources : United Nations Comtrade Data Base and Government replies to ECE questionnaires for Annual Reviews of Engineering Industries and Automation.

Definitions of items:

Total engineering products corresponds to SITC, Rev.2:69+7+87+881+884+885.
Metal products correspond to SITC, Rev.2:69.
Non-electrical machinery corresponds to SITC, Rev.2:71(excluding 716)+72+73+74+75.
Electrical machinery corresponds to SITC, Rev.2:716+76+77.
Transport equipment corresponds to SITC,Rev 2:78+79.
Precision instruments correspond to SITC, Rev.2:87+881+884+885.

IV.3 DEVELOPMENTS IN THE EXPORT PRICES OF ENGINEERING PRODUCTS

Export price indices of selected countries for which the relevant data were available for the main groups of engineering products (compiled on the basis of national currencies) are shown in table IV.15. Although the development of export prices differed for individual product groups in countries listed, some common features are apparent. [IV.29]

The lowest price increase over the period 1979-1986 for most countries and most years was recorded for exports of electrical machinery. In Japan, the export price of electrical machinery showed the most dramatic drop in 1983-1986. On the other hand, the category of transport equipment has shown a steady rise in prices since 1979.

Aside from Japan, the Netherlands and the Federal Republic of Germany showed the lowest overall export price increases for the 1979-1986 period. For the Netherlands, precision instruments followed electrical machinery in posting the lowest gains in export prices, while for Hungary and the United Kingdom it was the metal goods industry.

The most dramatic gains of the countries listed for the entire engineering sector in 1979-1986 were shown by Sweden, with more than a 60 point jump for the entire sector and a 100 point increase in the case of precision instruments. Swedish export prices also showed the highest increase of all countries listed for all other engineering categories. For the entire sector, Sweden was followed by the United Kingdom with a 50 point increase and by Finland with 40 points. As in Japan, the single category reflecting the highest export price increases was transport equipment and the second highest was non-electrical machinery in Hungary and the Netherlands. In Finland, non-electrical machinery was first, followed by transport equipment.

While table IV.15 shows export price trends for individual engineering subsectors in individual countries, table IV.16 reflects the development of export prices in the principal developed market economies for selected machinery and equipment groups. As seen in table IV.16, major changes in the export price structure took place in 1981 and 1986. Figures for these years reflect in part the effects of the changing relationship between different currencies which contributed to the distortion of the traditional pattern of international trade in general and in engineering industries in particular.

Another characteristic through 1981 was the prevalence of high interest rates in the market economies, and the length of time these had been in effect. Some domestic markets became so attractive to foreign exporters, whose products were comparatively cheap - often cheaper than those made by domestic manufacturers - that the latter saw their home markets contracting.

To counteract such influence, engineering manufacturers were led to accept lower profit margins, or even export at a loss. Against a background of falling investment and consumption, increasing inflation and intense competition in export markets, the onset of a recession and increasing unemployment, the engineering industries were forced to reappraise their strategies. It may not be too much to say that 1980 marked the beginning of a new era in engineering wherein the various branches came to appreciate that fundamental changes were necessary to ensure their future viability. Both the market economies and the centrally planned economies came to the same conclusion: that there was need for rationalization and automation and careful reappraisal of investments.

By the end of 1981, the situation began to stabilize. The aggregated index of machinery and transport equipment remained unchanged for the whole period 1982-1985. Export prices of non-electrical machinery decreased slightly in 1983 and 1984, and returned to the 1983 level in 1985. The development in 1983 and 1984 followed the pattern of capital investment in those years, of which non-electrical machinery constitutes an important part.

Similarly, export prices of electrical machinery declined in 1982-1984, and remained at the 1984 level in 1985. This was the result of a dramatic price erosion in telecommunication apparatus and domestic electrical equipment, and the reverse development in prices of electric power machinery and switchgear, which followed the trend of non-electrical investment goods. The export prices of transport equipment, and road motor vehicles in particular, remained virtually unchanged between 1982 and 1985, and increased in 1986, mainly as a consequence of the depreciation of the United States dollar. With the further decline of the dollar through 1986, this trend continued. [IV.30]

Table IV.15

Export price indices in major groups of engineering industries in selected ECE member countries and Japan, 1979-1986

(1980 = 100)

Country	ISIC Rev.2 Code [a]	1979	1980	1981	1982	1983	1984	1985	1986
Finland	38	92.8	100.0	106.8	113.4	122.9	129.0	133.2	136.4
	381	93.1	100.0	105.7	110.3	119.6	124.8	128.1	130.5
	382	93.7	100.0	108.5	118.1	127.7	135.0	141.4	144.2
	383	95.1	100.0	102.3	105.7	117.3	119.4	119.4	119.9
	384	91.5	100.0	107.7	113.9	122.3	130.9	137.1	144.1
	385	96.8	100.0	110.6	116.3	122.6	125.1	122.3	122.0
France [b]	38	...	...	228.7	263.5	290.0	317.6	...	...
	381	...	...	250.2	292.7	320.8	360.0	...	...
	382	...	...	203.9	241.8	260.4	286.4	...	...
	383	...	...	222.4	239.4	264.7	294.7	...	...
	384	...	...	252.4	291.6	328.8	352.8	...	...
	385	...	...	198.7	232.6	265.7	293.0	...	...
Germany, Federal Republic of	38 [c]	95.6	100.0	104.6	109.9	112.9	116.0	119.6	122.2
	381	94.4	100.0	104.4	110.8	113.5	117.1	120.3	123.0
	382	95.5	110.0	104.5	110.1	113.1	116.2	120.2	123.2
	383	97.5	110.0	103.9	107.7	110.6	110.3	112.3	113.1
	384 [c]	94.7	100.0	105.3	111.2	114.5	118.1	122.0	1 24.9
	385	96.0	100.0	103.5	107.3	109.6	120.0	124.9	128.3
Greece	38	...	100.0	...	...	183.3	218.3	264.1	304.0
	381	...	100.0	...	...	191.7	250.7	306.9	323.4
	382	...	100.0	...	...	...	...	...	...
	383	...	100.0	...	...	177.2	194.9	233.3	297.5
	384	...	100.0	...	...	...	...	...	...
	385	...	100.0	...	...	...	...	...	...
Hungary	38	102.9	100.0	104.4	107.1	114.1	120.5	128.0	...
	381	97.8	100.0	105.6	108.3	111.8	117.8	124.0	...
	382	105.7	100.0	104.9	107.8	116.0	122.5	131.6	...
	383	101.8	100.0	102.6	103.6	109.6	114.8	119.1	...
	384	103.0	100.0	105.6	109.9	118.5	126.0	134.3	...
	385	104.2	100.0	103.9	105.9	110.0	117.0	129.3	...
Netherlands	38 [d]	...	100.0	105.0	110.0	113.0	115.0	118.0	120.0
	381	94.0	100.0	103.0	110.0	113.0	116.0	119.0	121.0
	382	94.0	100.0	106.0	113.0	116.0	119.0	121.0	123.0
	383	97.0	100.0	105.0	109.0	111.0	112.0	115.0	115.0
	384 [d]	94.0	100.0	104.0	111.0	115.0	119.0	122.0	125.0
	385	...	100.0	105.0	108.0	113.0	115.0	118.0	120.0
Portugal	38	82.0	100.0	119.9	154.2	...	...	...	...
	381	87.2	100.0	123.5	156.7	...	...	...	...
	382	82.4	100.0	129.3	195.8	...	...	...	...
	383	75.9	100.0	112.6	142.8	...	...	...	...
	384	86.1	100.0	118.9	139.0	...	...	...	...
	385	90.3	100.0	121.8	139.5	...	...	...	...
Sweden	38	92.3	100.0	109.9	123.3	139.4	146.7	156.0	162.4
	381	91.7	100.0	105.9	118.9	134.4	143.8	154.4	160.9
	382	91.5	100.0	109.5	120.8	133.6	140.9	148.6	157.3
	383	95.6	100.0	111.7	129.9	142.3	142.3	145.4	149.8
	384	91.2	100.0	109.7	122.9	143.5	153.4	166.3	170.7
	385	97.1	100.0	120.8	137.3	165.7	177.8	195.5	205.7
United Kingdom	38	...	100.0	...	118.0	127.0	136.0	145.0	153.0
	381	...	100.0	...	111.0	116.0	124.0	132.0	138.0
	382	...	100.0	...	118.0	126.0	135.0	146.0	153.0
	383	...	100.0	...	120.0	123.0	132.0	144.0	156.0
	384	...	100.0	...	116.0	130.0	141.0	143.0	151.0
	385	...	100.0	...	124.0	137.0	142.0	150.0	153.0
Japan	38	95.0	100.0	101.4	105.1	98.6	99.4	97.8	85.3
	381	89.0	100.0	105.6	109.3	94.0	96.2	93.7	72.4
	382	97.7	100.0	99.2	101.6	96.9	96.3	95.7	88.8
	383	100.0	100.0	96.5	98.7	91.3	90.1	86.0	73.9
	384	93.2	100.0	104.1	110.0	109.3	111.7	112.0	101.0
	385	97.7	100.0	99.2	101.6	96.9	96.3	...	...

Source: Government replies to ECE questionnaires for *Annual Reviews of Engineering Industries and Automation.*

a ISIC, Rev.2 code:

Division 38:	Engineering products (including metal products)
Major Group 381:	Metal products
Major Group 382:	Non-electrical machinery
Major Group 383:	Electrical machinery
Major Group 384:	Transport equipment
Major Group 385:	Precision instruments

b 1970 = 100.

c Excluding shipbuilding, aircraft and aerospace industries.

d Excluding shipbuilding and aircraft industries.

Table IV.16 Export price index of machinery and transport equipment for selected countries, 1979-1986

1980=100

	1979	1981	1982	1983	1984	1985	1986
Power generating machinery other than electric							
Japan	98	108	100	101	99	99	137
Germany, Fed. Rep. of	94	87	86	85	79	79	110
United States	90	114	123	125	132	136	139
Agricultural machinery and implements							
Japan	103	103	90	96	99	98	122
Germany, Fed. Rep. of	94	86	84	82	75	75	103
United States	88	111	115	121	124	127	130
Metalworking machinery							
Japan	101	107	107	108	107	110	144
Germany, Fed. Rep. of	93	85	84	82	76	77	108
United States	87	110	118	120	123	126	130
Textile and leather machinery							
Japan	98	104	96	100	101	103	140
Germany, Fed. Rep. of	94	84	83	82	76	76	106
United States	87	104	111	114	116	117	123
Machines for special industries							
Japan	101	104	102	98	103	107	121
Germany, Fed. Rep. of	94	84	82	80	74	74	103
United States	88	114	123	124	125	126	126
Electric power machinery and switchgear							
Japan	97	99	90	93	92	90	108
Germany, Fed. Rep. of	94	84	82	80	73	73	100
United States	86	107	115	117	119	123	125
Telecommunication apparatus							
Japan	107	100	91	88	87	84	106
Germany, Fed. Rep. of	96	81	77	73	66	63	86
United States	93	106	110	113	115	115	119
Domestic electrical equipment							
Japan	99	104	95	92	89	84	106
Germany, Fed. Rep. of	96	84	81	79	71	70	96
United States	90	110	115	118	120	122	123
Road motor vehicles							
Japan	96	108	102	104	106	107	136
Germany, Fed. Rep. of	94	85	83	81	75	75	103
United States	89	110	119	124	128	130	134

Source : Monthly Bulletin of Statistics, United Nations, May 1988.

General note : These indices are of the Laspeyres type (base weighted) shifted when necessary so that 1980=100. They are converted from the national currency by its weighted average exchange rate in the base period. It should be noted that these indexes, being expressed in terms of US dollars, reflect not only changes in prices but also changes in the parity between national currencies and the dollar. It should be pointed out that the categories of goods encompass a broad variety of products; hence different trends in country indexes occur within the same heading. For example, since Japan's exports of transport equipment include large values for ships, the trend of Japan's index for that particular category depends to a much greater extent upon the price of these products as compared with other countries.

IV.4. INTERNATIONAL SPECIALIZATION AND CO-OPERATION IN THE ENGINEERING INDUSTRIES

In the field of international specialization and co-operation in research and development and in production, engineering industries play a key role both in the market economies of western Europe, North America and Japan and in the centrally planned economies of eastern Europe. [IV.31] For western Europe, the primary objective of such co-operation is to reduce the technology gap with the United States and Japan. In the United States, a major impetus for sharing technology and production with trade partners comes either from strategic considerations or from the need of those partners buying military hardware to insist on offset and co-production agreements. Japan's strategy focuses on offshore integrated process and production licensing. Similar considerations apply to the Soviet Union and the other centrally planned economies of eastern Europe, with the added consideration of viewing such specialization and co-operation endeavours as integral to the cohesiveness and effectiveness of the Council for Mutual Economic Assistance (CMEA).

Western Europe : from ESPRIT to EUREKA

The members of the European Economic Community (EEC) have undertaken a plethora of specialized R and D co-operation projects financed and carried out jointly with varying degrees of success, such as JET, RACE, BRITE and ESPRIT. In 1985, the EEC allocated an ECU 1.1 billion budget for such programmes, primarily in advanced technology sectors and industry related projects, most of them affecting engineering sectors. However, R and D efforts at the Community level will remain minor in comparison with efforts at the national level, private sector R and D and non-EEC multinational co-operative projects.

Of the projects sponsored by EEC itself, ESPRIT has been the most ambitious research programme. It entailed an initial year-long ECU 11.5 million pilot phase followed by a five-year ECU 1.5 billion programme devoted to five research areas: advanced microelectronics, advanced information processing, software technology, office automation and computer-integrated manufacturing (CIM).

During the 1979-1985 period, EEC launched several other R and D programmes in advanced technology, including an ECU 50 million grant for a biotechnology programme in genetic engineering for the years 1981-1985 and an ECU 40 million grant for microelectronics research. In 1983, the EEC Commission also introduced an ECU 175 million research programme, known as BRITE, to run through 1987, in industrial technologies to be applied in a broad range of sectors, with priority for high-technology industries, including electrical engineering. Other programmes include Ariane, the joint space project, and RACE, a joint project in telecommunications.

Moreover, in 1985, EEC liberalized its anticollusion rules for agreements in R and D, joint production and marketing co-operation, which should facilitate the commercialization of the results of programmes such as ESPRIT. Exemptions extend to both reciprocal licensing arrangements and joint ventures, with exemptions for joint production valid only for those products that have been decisively improved by joint R and D.

The end of the period under review, i.e. 1985, saw the genesis of EUREKA, the latest and potentially most promising industrial co-operation effort in western Europe, outside the institutional confines of EEC. The programme, which was spearheaded by France, has now become a 19-nation co-operative venture covering some $5 billion worth of high-technology industrial projects involving from the smallest to the largest firms in Europe.

According to Daniel Greenberg, editor of the Washington-based newsletter *Science and Government Report,* by 1987 EUREKA had already become Europe's biggest transnational undertaking, exceeding even the Ariane rocket programme. Membership is limited to west European countries and their companies. A project to be funded must originate with an industrial firm and must also involve at least one other firm in another west European country. Universities and government laboratories can participate, but industry is the initiator and the home base. The Governments of the industrial partners contribute roughly half the costs, usually on the approval of a single reviewer.

Among some 100 projects initiated, those directly involving engineering industries include:

- Advanced fishing boats. Participating countries: France, Spain, Norway; cost: $56 million over five years;

- Fast mobile robots to deal with disasters and terrorism. Participating countries: France, Italy, Spain; cost: $100 million over six years;

- Cranes and other devices for dangerous or difficult construction jobs. Participating countries: France and United Kingdom; cost: 22 million over five years;

- Non-polluting coal-fired power station. Participating countries: Federal Republic of Germany, France; cost: $11 million over five years;

- Traffic-control systems. Participating countries: France, United Kingdom, Italy, Sweden, Federal Republic of Germany; cost: $15.5 million in first of eight years;

- Software factories that can produce advanced computer programs. Participating countries: Denmark, Finland, France, Italy, United Kingdom; cost: $141 million over six years; and

- Highly-automated hospital rooms and intensive-care units. Participating countries: Federal Republic of Germany, Netherlands, United Kingdom; cost: $13 million over five years.

Eastern Europe: the CMEA Comprehensive Programme

One of the long-standing objectives of the Council of Mutual Economic Assistance (CMEA) is to promote rationalization of industrial production within the CMEA member countries through specialization, with each country producing certain types of machinery, components and other products necessary to one or more of the others. Such production reportedly accounted for 21.1% of mutual exports of industrial products within CMEA in 1984.

This specialization is an extremely complex undertaking which can only be summarized but not described in detail in the present study. According to the CMEA Secretariat, there are 93 basic agreements on specialization and co-operation in the machine-building sector, covering 17,000 individual types of products and sub-products. The machine-building sector accounts for 83% of specialized production. Multilateral agreements also cover 18 types of synthetic rubber, five types of chemical and biochemical fodder additives, 23 types of pharmaceuticals, a large number of plant-protection agents and some 3,000 subcategories of various small-tonnage chemicals. The chemical sector accounts for slightly less than 12% of specialized production, with other sectors accounting for about 5%.

The key purchaser of most of these specialized products is the USSR. According to information provided by the CMEA Secretariat, it bought 78.9% of Bulgaria's, 74.9% of Hungary's, 76.2% of the German Democratic Republic's, 81.8% of Poland's, 60.5% of Romania's and 73.2% of Czechoslovakia's specialized production in 1984.

Transport equipment, the largest single category, accounted for 29% of specialized production in the machine-building sector. Metalcutting machine tools accounted for 8.3% of the machinery involved, while energy and electrotechnical equipment comprised 5.1%, mining and metallurgy and oil-processing equipment 4.2% and other industrial equipment 27%.

The German Democratic Republic and Czechoslovakia played a leading role in sales of specialized metalcutting machine tools in 1984 (52 and 22% of the total, respectively). In 1984, leaders in the export of specialized electrotechnical equipment were Czechoslovakia (25%), the German Democratic Republic (23%) and Bulgaria (22%). Specialized equipment in the material-handling sector was produced mainly by Bulgaria (64%) and the German Democratic Republic (20%). Under the 1984 programme, tractors and agricultural machinery came mainly from the German Democratic Republic (47%), the USSR (28%) and Bulgaria (12%). The German Democratic Republic produced over 47% of the mining, metallurgical and oil-processing equipment, and Poland another 36%. Transport equipment under the specialization agreements was provided by the German Democratic Republic (28%) and Hungary (12%).

In December 1985, all the CMEA countries signed the "Comprehensive Programme for the Scientific and Technical Progress of the CMEA Member Countries up to the Year 2000", hereafter called the Comprehensive Programme. The agreement is intended to establish priorities for technological development and identify specific sectors for co-operation between CMEA member countries. Although 2000 has been set as the target year for full implementation, it has been agreed that over half the programme's 93 specific tasks should be ready for industrial application by 1989. The Comprehensive Programme focuses on five high-technology areas, in the order of priority listed by the document, as follows:

- The use of electronics in the national economy;
- Comprehensive automation;
- Accelerated development of the nuclear power industry;
- New materials and technologies for their production and processing; and
- Accelerated development of bioengineering.

The specific projects relating to the engineering industries mentioned by the document include the following:

- Integrated control systems;
- Industrial robots and manipulators, including robots with artificial vision and capacity for reacting to spoken commands;
- Automated process for production of ultra-precise equipment and instruments;
- Standardized sub-assemblies for automated machinery, including mechanical, hydraulic, electrotechnical and pneumatic components;
- A standardized series of technical facilities for automation of lifting and transport equipment, with emphasis on robotics and special monitoring equipment;
- Fast-breeder reactors;
- Improvement of continuous-casting technology and secondary steelmaking for high-quality steels; and
- Development of industrial lasers with applications in cutting, welding, thermal processing and pattern-cutting operations.

The United States aerospace offset/shared production programmes

The United States aerospace industry is continually increasing the number and scope of foreign shared production programmes, often referred to as offset programmes. This has resulted in the transfer of manufacturing rights and technology for military aerospace products to foreign countries and their corporations, in part in the interest of improving mutual defence, and partly in answer to requests from trade partner countries to help ease the cost burden of such military purchases and to improve the local defence-industry infrastructure or gain technological expertise. Such shared production or offset agreements often fall into the category of broadly defined countertrade.

On the civilian side, start-up costs in excess of $1 billion for both large transport aircraft and engines, along with the heavy financial risks associated with to-day's markets, have provided an impetus for United States manufacturers to seek foreign partners. Thus, foreign customers may succeed in gaining shared production or offset agreements through the purchase of United States civil and military aerospace products, thereby easing their foreign exchange burdens and at the same time increasing employment opportunities and manufacturing capabilities in their country.

Recent examples of such international division of labour include the Boeing 767 large transport plane with a combined Japanese and Italian content of 15%, and joint efforts involving Fairchild Industries and Saab-Scania of Sweden, with 50% participation each, in the production of 34-passenger SF-340s.

Both United States large aircraft engine producers - General Electric and Pratt & Whitney - participate in shared production programmes. General Electric and Snecma of France jointly produce both the CFM 56-2 for use in retrofitting late model McDonnell Douglas DC-8s and Boeing military KC 135s and CFM 56-3s for the new Boeing 737-300 and possibly early models of a proposed new 150-seat aircraft. Pratt & Whitney, together with a Japanese consortium, Rolls Royce of the United Kingdom, and manufacturers from the Federal Republic of Germany and Italy plan to co-produce a new engine designed specifically for the 150-seat passenger aircraft.

Moreover, co-operation with other countries in the field of automated production also takes place between corporations through licensing of technologies to existing foreign companies or the establishment of subsidiary manufacturing plants overseas.

Joint ventures

The establishment of joint ventures is a common form of international co-operation in engineering industries of the market economies, but it is a relatively new one in east-west economic relations. At the end of 1987, there were a total of 166 east-west joint ventures which were either active or about to be put into operation, [IV.32] and of which approximately one third were in the engineering industries. For example, Bulgarian partners established a joint venture in 1981 with Fujitsu Fanuc (Japan) for maintenance, service and development of machine tools; Czechoslovak Tesla with N.V. Philips (Netherlands) for the manufacture and marketing of video tape recorders; Polish Technocabel CIE IMPEXMETAL with FLT Metaux (Belgium) for the manufacture and marketing of cutting tools incorporating natural and synthetic diamonds; Romanian Industrial Centre for Electronics, Technology and Computers with Control Data Corporation (United States) for the manufacture and marketing of peripheral equipment for computers. Hungary and the Soviet Union are in the lead, both in the number of joint ventures with western partners and in the value of contracts. The quickening pace of business contacts is clear evidence that this form of co-operation serves the interests of all partners involved.

IV.5. SPECIAL FORMS OF TRADE IN ENGINEERING PRODUCTS: EXPORT OF TURNKEY PLANTS, COUNTERTRADE AND BUYBACKS

While the years 1984 and 1985 were characterized by the beginning of general recovery from the recession of the preceding years, this recovery did not concern the entire ECE region and did not affect most of the developing countries. Continued liquidity constraints in many countries, particularly in the developing world, but also in some of the centrally planned economies of eastern Europe, combined with mandated or *de facto* import restrictions, have given a boost to the search for alternative ways of maintaining trade flows, while limiting the outflow of foreign exchange. One such controversial alternative, which encompasses a variety of trading practices ranging from non-monetary barter to the sale of sophisticated turnkey plant with buyback commitments, is countertrade. Most broadly, countertrade can be defined as any conditional trade transaction which imposes trade-related performance requirements.

In the most classic form of countertrade, an exporter agrees to take or buy products in return or to provide some other concession aimed at equalizing, or at least partially balancing, the importer's foreign-exchange expenditure. Although these concessions usually take the form of buying something the importer has to sell, they may be as varied as providing related or unrelated technology to diversify the buyer's export base, improving the importer's international marketing skills or creating new jobs in an unrelated sector of the economy. In the case of military sales, one form of such countertrade is offsets or shared production, as discussed in subchapter IV.4, in the context of the United States aerospace industry.

Countertrade is not a new phenomenon. What is new is the dramatically increasing volume of such trade, the growing number of countries that require or conduct countertrade and the growing sophistication both on the part of those requiring countertrade and those willing and able to offer it. The impact of countertrade on the engineering exports of countries in the ECE region is inversely proportional to the import priority of the given product. Thus, for example, luxury goods would be most subject to countertrade demands, while such essentials as food and pharmaceuticals the least.

Although it is difficult to generalize owing to the specific nature of engineering products that hold different import priorities in various countries, a review of actual transactions indicates that engineering goods that play a significant role in straightforward counterpurchase transactions tend to be mainly in the transportation sector and agricultural equipment. In more sophisticated forms of countertrade, such as the sale of turnkey plants with buyback commitments or in the area of offsets, the engineering industries obviously play a dominant role.

Another marked change in recent years is the transformation of countertrade from being primarily a phenomenon of east-west trade to include many North-South transactions, as well as trade between developing countries. In the 1970s, the main purpose of countertrade was to find outlets for east European exports and to ensure an inflow of western products and technology. In the 1980s, many developing countries have begun to use countertrade successfully, not only to reduce their trade deficits and help service their external debts, but also to help overcome protectionist barriers in their exports to the industrialized countries. There is also an evident shift away from strictly product-for-product countertrade towards greater emphasis on obtaining longer-term access to new markets and on industrial co-operation involving the transfer of capital and technology.

Strict barter - the exchange of product for product, in which no money changes hands and the obligations are not transferable to third parties, such as trading houses - is limited primarily to intergovernmental trade agreements and plays no significant role in countertrade as a whole and in the engineering sector specifically. By far the most prevalent form of countertrade involves commodity-based counterpurchase transactions - in which an exporter sells a product and is paid for it, but at the same time commits itself to purchase a specified amount of goods or services from the original buyer, usually at a later time and against commercial payments; in the engineering sectors this would typically involve the sale of equipment and subsequent purchase of commodities.

More and more countries are seeking the advantages offered by buybacks and other product-compensation agreements. In their simplest form, buybacks involve the sale of plant, equipment and technology, usually but not always on a turnkey basis, and the subsequent purchase of resultant product. In more sophisticated forms, buybacks use the buyer's buying leverage to extract additional concessions (beyond such considerations as price, quality and delivery terms). These concessions can be in the form of capital investments or the provision of capital goods; they can and usually do involve technology transfers; they almost always include supplier financing, at least for the interim period between the time of the import of the plant and equipment and the time the facility goes on stream; and, finally, by definition, buybacks include the supplier's assistance in international marketing.

Estimates of the current volume of countertrade as a share of world trade vary tremendously - ranging from as low as 1% by the International Monetary Fund to as high as 40% by some countertrade proponents and press reports. Estimates of the General Agreement on Tariffs and Trade (GATT) Consultative Group of Eighteen place countertrade at 8% of world merchandise trade, and the OECD's Working Party of the Trade Commission on North South Trade notes that the volume of countertrade, both in absolute terms and as a share of world trade, is significant and, more importantly, is growing rapidly.

The centrally planned economies of eastern Europe still account for the largest share of global countertrade, but that ratio is declining rapidly since that of other geographic areas - particularly South-East Asia, the Middle East and Latin America - has grown at a fast pace in the last three years. The fact remains that no exact data are available on total and regional countertrade volumes. This is due to a number of factors, key among them being the degree of secrecy surrounding such deals, particularly in the private sector, as well as the fact that such deals often entail a degree of price distortion and definitional confusion which complicate statistical compilations. A fragmentary analysis of countertrade in the engineering sectors would require a compilation of reported transactions.

Reasons for special forms of trade in the engineering industries

While the debt crisis and the ensuing balance-of-payments problems, as well as a growing trend towards protectionism, are the key factors that spur most countries to consider countertrade, a number of other factors also play an important role. These considerations apply primarily at the government, policy-making level, but are valid regardless of whether countertrade is required, encouraged or merely permitted, and whether the party seeking such countertrade is the Government, its State trading organizations or private-sector traders, for the sake of convenience, all three of them being referred to here as the countersellers.

For the primary exporter, the company most likely to be selling engineering products in a countertrade transaction, countertrade most often represents an added cost, greater complication, and increased risk - all discussed separately below. However, many companies - not least of them in the more troubled segments of the engineering industry - realize that countertrade can offer a competitive advantage, or at least be used as a positive marketing tool to keep or expand existing market shares and penetrate new markets.

The following list highlights the rationale used by countersellers in considering countertrade as an answer to specific problems:

- In the light of mushrooming external debts and liquidity constraints, it allows a counterseller to continue trading when foreign exchange is not available;

- It links the import of commodities and equipment to an increase in exports. It may also ease the import of necessary spare parts, components and materials that it would otherwise be impossible to import, given currency and import constraints;

- It does not increase overall indebtedness and requires only interim financing for the period between import and export. In addition, interim financing in countertrade is frequently the suppliers' rather than the buyers' problem. Similarly, countertrade does not affect a country's foreign-exchange reserves, which may be earmarked for debt servicing. Under these circumstances, countertrade offers the allure of allowing trade to continue without affecting the balance of payments;

- It enables the counterseller to dispose of his surplus goods either at lower, undisclosed prices or at nominal prices that conform to an international agreement, but nevertheless involve an effective discount in terms of the goods received in exchange. This makes it possible to overcome certain constraints of regulated commodity agreements or to overcome tariff and quota restrictions;

- It makes it possible to provide some preferential terms such as subsidies or discounts, which the countersellers concerned do not wish other trading partners to know about. This, combined with the fact that most often a third party is responsible for marketing the products and sales channels are more complicated, may make it possible to overcome certain other non-tariff barriers;

- It is compatible with the increasingly prevalent ideology of "bilateralism" that characterizes many countersellers' trade policies, which call for balanced trade on a country-by-country basis;

- A number of countersellers view countertrade - particularly buyback of manufactured goods under licence - as a means of ensuring that they receive up-to-date technology, since the seller has a stake in the licensed product. Countertrade can also help a Government's industrialization plans through longer-term buyback arrangements. Under such agreements, the seller of plant and equipment is obliged to take the end product; thus the counterseller licensee is likely to receive the most up-to-date technology, and any innovations and improvements in the process will be passed on during the licence period. The rationale is that the licensor, bound to take back resultant products, will make sure that the production capacity is as efficient and modern as possible;

- It is viewed by some countersellers as an instrument of industrial policy to favour particular industries or the export sector in general. In some countries, offset agreements have been entered into specifically to expand labour-intensive industries and to reduce nominal unemployment;

- Countertrade products may penetrate non-traditional markets previously untapped by the counterseller, thus expanding or diversifying its export markets. After fulfilling a countertrade obligation, the counterseller may be able to sell directly to these markets. In some cases, this may involve new buyers in the same geographic market, allowing for expansion or preservation of market share;

- By allowing access to the exporter's international marketing network, it provides the counterseller with other advantages, such as reduced marketing costs without loss of established buyers; new buyers; insight into the exporter's selection of products and into market receptiveness to newly introduced products; elimination or reduction of financing costs; very close contacts with the supplier because of the bilateral nature of the deals;

and the ability to move products that, all variables being equal, could not be moved any other way, because of inferior quality, delivery terms or excessive supply on world markets; and

- The existence of a countertrade policy can provide leverage for the counterseller, particularly in the case of State trading organizations or other major government procurement transactions, to obtain either more competitive offers or offers that, in addition to product or technology, may provide other tangible or intangible benefits to the local economy.

Countertrade can be advantageous to a counterseller in a country where pricing distortions between official exchange rates and free-market rates prevail, and where the domestic price structure at the official rate is inconsistent with world market prices. A domestic producer can acquire more product by countertrade than by accounting in inflated prices. Many countries see countertrade as a way to carry out trade when the market-price system no longer allows realistic pricing of goods and when the assignment of shadow prices is consequently necessary. The shadow price may be a new actual price, specified in the countertrade arrangement, or it may appear in terms of the volume of goods.

In a sense, countertrade in this situation provides a way of overcoming problems arising from overvaluation of a currency. It becomes the equivalent of an export subsidy plus an import tax. Unlike a devaluation that operates across the board for all trade transactions, such countertrade operations produce selective, case-by-case devaluations. An exporter in the country with the overvalued currency accepts a discount on the price of the item at the official exchange rate. Each transaction carries with it a distinct price; thus, when several countertrade transactions have been negotiated within a single country, the effect may be similar to that of a multiple exchange rate.

Constraints on countertrade

Perhaps the most tangible and direct indictment of disadvantage associated with countertrade is the fact that it is a more complex, risky - and consequently costly - way of doing business than the more conventional forms of trade. While the costs are theoretically shared by all parties in a countertrade transaction, in fact, they more often than not are borne by the counterseller.

These higher costs result from a number of factors. The need for countertrade presupposes that there is not a ready, cost-effective export market for the goods in question. Their disposal thus requires a price discount to the end-buyer plus, often, additional transportation, handling and storage charges, as well as a commission for any intermediaries, e.g. trading houses.

Whereas some trading houses say that their countertrade profits (as opposed to the commissions and fees they charge their clients, or the discounts, i.e. price rebates, they must give to dispose of the countertrade product) average 1-2%, increasing to 4% in high-risk situations, several companies interviewed contend that margins are much higher. They base their views on the widely varying quotations that they sometimes receive on the same deal from different trading houses.

Less obviously, the prices of countertrade transactions must also incorporate a share of the partners' administrative, legal and technical costs - expended not only on the one particular deal, but also on the identification and negotiations of other countertrade transactions that may never have come to fruition; this cost can be considerable, given the fact that the success ratio of deals explored to deals completed is very low.

Still another category of costs involves compensation for increased risks inherent in a countertrade transaction. These risks include, but are not limited to, uncertainties about future deliveries, in terms of quality and time (particularly relevant for perishable goods), about whether the necessary approvals for linking imports and exports as well as other regulatory requirements will be fulfilled and, finally, as regards the possible negative impact on the market place stemming from the final disposal of the countertraded goods.

The price distorting effects of countertrade may by the same token have a beneficial effect on a specific transaction or, for that matter, for the economy as a whole. For example, countertrade can help overcome internal accounting obstacles. Some countertrade deals are structured to provide a central bank loan to the relevant production units, rather than a convertible-currency expenditure, with the loan being secured by eventual countertrade proceeds.

On the negative side, this may result in differential export incentives to some sectors of the economy at the cost of those sectors that are more dependent on the higher-cost imports. It may also give undue support to inefficient production and trading structures, in so far as it provides a form of local protection to producers of high-cost or otherwise uncompetitive goods.

The drawbacks and supplementary costs of countertrade must be weighed against the expenses an indebted counterseller would incur if it tried to follow traditional multilateral trading practices and the availability of conventional trade financing. Depending on the premium it carries (i.e. the total of the above-mentioned costs), a countertrade deal could prove to be more profitable than financing a direct import. For some countries, at certain times, additional financing may simply not be available at any cost.

Moreover, to the extent that countertrade does distort price structures, it may negatively affect the internal allocation of domestic resources. Higher factor prices in a particular sector could be interpreted as an opportunity for further investment in that sector or, alternatively, they could lead to incorrect evaluation of projects that utilize inputs or capital goods purchased at prices that diverge significantly from those on the open market.

While the issue of net effective cost of countertrade is difficult enough to address, the more fundamental concern about countertrade has to do with its macroeconomic impact on aggregate exports, i.e. the issue of "additionality", as well as the extent to which it may cause distortions in the effective allocations of resources in the economy. To date, no effective mechanisms have been developed to quantify such effects, much less to control them.

A reasonable argument can be made - supported by considerable circumstantial evidence - that a mandated countertrade policy does not result in additional exports, but rather in the shifting of the structure of a country's exports, in all likelihood to its economic disadvantage. Some countries have sought to curtail this "shifting" effect by making commodities available in countertrade only on condition they are distributed in new markets.

A related potential disadvantage of mandated countertrade is that, rather than increase the competitiveness of offers, it may indeed make them less attractive. For example, bidding, especially for major government procurement contracts, may become less intense if countertrade or offsets are required. This automatically reduces the competitiveness of the operation and potentially increases the cost to the counterseller.

IV.6. INTERNATIONAL TRADE IN PATENTS, LICENCES AND KNOW-HOW CONCERNING ENGINEERING PRODUCTS

Patents

The World Intellectual Property Organization (WIPO) publishes annual statistics on the number of patents issued and patents applied for by country. The number of patents issued by those ECE member countries for which data are available and by Japan in the years 1980 and 1985 is shown in table IV.17. Although a sectoral breakdown was not available, it is a fact that the engineering industries account for a very large share of the total. Thus the geographic trends that can be discerned from these statistics should apply to the engineering industries under review in the present study.

The total number of patents issued by the countries reviewed was 27 percentage points higher in 1985 than in 1980. The increase of patents issued to foreign entities, i.e. non-residents, grew more than twice as fast, or by 61 percentage points, with the number of patents issued to non-residents accounting for 41% of the total in 1980 and 52% in 1985.

On a regional basis, western Europe showed the most significant increase, with the total number of patents issued rising by 83 percentage points in 1985 over the 1980 level. Patents issued to non-residents increased even faster, or by 97 percentage points. Non-resident patents accounted for 73% of the total in 1980, and 79% in 1985.

The largest individual country gains in this region were posted by Italy, whose 47,924 patents issued in 1985 represented an increase of some 500 percentage points over 1980 and were the highest number of patents issued by any country in the region. Malta and the Netherlands each posted a 260 percentage point increase, followed by Sweden (170). In terms of total number of patents issued in 1985, France was the second largest, with 37,530 (up 34 percentage points), followed by the United Kingdom with 33,480 (up 41), and the Federal Republic of Germany with 33,377 (up 65).

On the down side, Portugal posted the largest decline, down 58 percentage points, followed by Denmark (down 36), Cyprus (down 35), Greece (down 27), Turkey (down 20) and Spain (down 12). These were the only countries in the region to show a lower number of patents issued in 1985 than in 1980.

The centrally planned economies of eastern Europe as a whole showed a 10 percentage point decline in the total number of patents issued, but the number of patents issued to foreigners increased by 16 percentage points. However, these represented only a small share of the total, with 4.8% in 1980 and 6% in 1985. This decline was accounted for by only two countries - the Soviet Union, whose 74,745 patents issued in 1985, by far the largest number in the region, represented a 20 percentage point drop over 1980; and Poland, whose number dropped by 42 percentage points, to 4,467. The largest increase was posted by the German Democratic Republic, with 12,705 patents, representing a 118 percentage point increase.

North America as a whole posted a 5 percentage point gain, with the increase in patents issued by the United States going up by 16 percentage points to 71,669. Of these, patents issued to foreigners accounted for 45%, up 30 percentage points from the 24,675 patents issued to foreigners in 1980, which accounted for some 40% of the total.

Japan posted a 9 percentage point increase in the total number of patents issued, but the number of patents issued to foreigners declined by almost 4 percentage points, as did their share in the total, from 18% in 1980 to 16% in 1985. Japan's 50,100 patents issued in 1985 was the third largest number of all countries reviewed, following the Soviet Union's lead of 74,745 and the United States' 71,669.

Licences

Comprehensive comparative data on trade in licences and know-how for all the countries in the ECE region is to the best knowledge not being compiled at present, either in aggregate figures or in sector-specific breakdowns. Individual national statistics do provide a significant amount of information, which could serve as a starting point for such a comprehensive survey. Such an undertaking, however, exceeds the resources provided for the present study.

For example, in the market economies, United States statistics provide aggregate figures on royalties and licence fees earned by United States companies and their foreign subsidiaries, as well as revenues from licence fees and royalties, notably in the United States Bureau of Economic Analysis surveys, International Transactions in Royalties, Licensing Fees, Film Rentals, Management Fees etc., with Unaffiliated Foreign Residents (BE-93), as well as in a number of benchmark foreign investment surveys. The estimates are disaggregated by country and linked to direct investment benchmark survey data. There are plans to add details on the type of intangible assets involved, beginning with the survey covering 1977.

Available data show for example that total receipts from royalties and licence fees by United States residents amounted to $6 billion in 1985, up from $5.7 billion in 1984. Of this, in 1985, $4.3 billion came from payments by United States overseas subsidiaries and only $1.5 billion from foreign unaffiliated companies. Payments by United States companies for licence fees and royalties in 1985 amounted to some $850 million, of which $470 million was to United States company affiliates overseas.

Of the centrally planned economies of eastern Europe, for example, Poland provides extensive data both on the number of licences imported and exported and the seller and buyer country involved, as well as the costs and revenues of the licence transactions and the foreign-exchange costs related to licensed production. As an illustration, in the period 1971-1975, Poland imported 316 licences, most of them from the Federal

Republic of Germany (76), followed by France (50) and the United States (33). In the period 1976-1980, the number dropped to less than half of that of the preceding period, with a total of 136 imported licences, of which the Federal Republic of Germany provided 39, France 23 and the United States and Italy 13 each. In the period 1981-1984 (latest data available), Poland bought only 4 licences - 1 from Austria in 1984 and, in 1983, 2 from the USSR and 1 from the United States.

On the export side, in the period 1971-1975, Poland sold 42 licences, of which 10 went to the Federal Republic of Germany and 6 to the German Democratic Republic. In 1976-1980, the number of licences sold rose to 61, of which 6 went to Czechoslovakia and 6 each to the Federal Republic of Germany and the German Democratic Republic. In 1980-1984, the number dropped to 31 licences, 16 of which were sold in 1984 alone, including 3 each to France, Japan, the Federal Republic of Germany and the German Democratic Republic.

In 1980, Poland paid $66.5 million in licence fees and royalties, against revenues from its own technology sales of $4.8 million. In 1984, licence fees amounted to $14.2 million and revenues from licence sales to $780,000. [IV.33]

Table IV.17

Patents issued to local and foreign entities, by country of issue, 1980 and 1985

	1980			*1985*		
	Residents	*Non-residents*	*Total*	*Residents*	*Non-residents*	*Total*
Western Europe						
Austria	1 227	4 745	5 972	1 073	7 669	8 743
Belgium	837	5 081	5 918	786	9 289	10 075
Cyprus	1	65	66	0	43	43
Denmark	192	1 453	1 645	200	854	1 054
Finland	439	1 467	1 906	563	1 597	2 161
France	8 438	19 622	28 060	9 835	27 695	37 530
Germany, Fed. Rep. of	9 826	10 362	20 188	13 215	20 162	33 377
Greece	1 114	942	2 056	1 127	2 098	3 225
Ireland	24	1 407	1 431	20	1 022	1 042
Italy	1 810	6 190	8 000	10 065	37 859	47 924
Luxembourg	73	1 025	1 098	55	3 878	3 933
Malta	1	11	12	1	19	20
Netherlands	417	2 907	3 324	744	11 223	11 967
Norway	276	1 843	2 119	229	1 936	2 165
Portugal	95	2 200	2 295	42	918	960
Spain	1 485	7 739	9 224	1 498	7 617	9 115
Sweden	1 394	3 604	4 998	1 650	11 870	13 520
Switzerland	1 475	4 486	5 961	2 634	11 906	14 540
Turkey	32	452	484	61	324	385
United Kingdom	5 158	18 646	23 804	6 087	28 393	33 480
Yugoslavia	77	578	655	207	846	1 053
Subtotal	34 391	94 825	129 216	50 092	187 218	236 312
Eastern Europe						
Bulgaria	1 271	102	1 373	1 567	372	1 939
Czechoslovakia	6 763	307	7 070	6 420	1 425	7 845
German Democratic Republic	4 455	1 371	5 826	11 487	1 218	12 705
Hungary	760	1 018	1 778	1 197	898	2 095
Poland	5 736	1 962	7 698	3 894	573	4 467
Romania	1 194	814	2 008	2 156	632	2 788
USSR	92 897	113	93 010	73 275	1 470	74 745
Subtotal	113 076	5 687	118 763	99 996	6 588	106 584
North America						
Canada	1 503	22 392	23 895	1 355	17 342	18 697
United States	37 152	24 675	61 827	39 554	32 107	71 669
Subtotal	38 655	47 067	85 722	40 909	49 449	90 366
Japan	38 032	8 074	46 106	42 323	7 777	50 100
Total	224 154	155 653	379 807	233 320	251 032	483 362

Source: Industrial Property Statistics, 1980 and 1985, World Intellectual Property Organization (WIPO), Geneva, 1982, 1987.

IV.7. STANDARDIZATION IN ENGINEERING INDUSTRIES

Economic development in general, and industrial development in particular, would be unthinkable without international trade. In order to permit manufacturing companies to import materials and components and to export their products, international standards must be agreed upon. [IV.34] They serve as important references when dealing with an international market with varying production and trading practices, languages and culture. Therefore, standardization is a prerequisite for any international exchange of goods, for interchangeability to raise the competitivity of goods in the world market, for quality assurance and for the international division of labour. Specialization and technological progress in particular call for international co-operation.

Barriers to international trade are, however, frequently set up owing to the differences in national standards which may be due to many factors peculiar to a country such as the level of technological development, environmental conditions and national policies concerned with the protection of home industries, and the building up of a competitive strategy. Even if financial and political barriers to trade are lowered or eliminated, there still remain many technical impediments. The creation of incompatible national standards, and regulations based upon them, can hinder or even prevent the passage of goods across national frontiers. [IV.35] The concept 'reference to standards' has been used for many years in various technologies. In most cases its use has been restricted to the national level. In recent years, however, the tendency had been to use harmonized standards, i.e. international standards, and to legislate by reference to these standards.

The international harmonization of national technical specifications is the task of specialized international organizations, the most important being the International Organization for Standardization (ISO) and the International Electrotechnical Commission (IEC). [IV.36]

World standards assuring interchangeability and mutual compatibility of equipment discourage, on a world-wide basis, the manufacture of an unnecessary diversity of common items to serve the same purpose. This contributes to economy of manufacture and, far from inhibiting, it actually facilitates the design of equipment embodying such components. Such standards also enable equipment made in one country to work effectively in co-ordination with equipment made in another. [IV.37]

Standards represent international agreements to facilitate the multilateral flow of goods and expertise. Standardization becomes important at that stage in the development of a technology when the technology needs the stabilizing effect of uniformity in certain operations. When a technology is new or is changing rapidly, standardization is not always a good idea, and the developers of such technologies are hardly ever eager to take the time to agree on standard solutions for their technical problems. However, as a technology matures, the lack of standard solutions becomes inconvenient, and may even impede the further development of the technology. At that point, standardization work is seen to have a pay-off because, in providing a stabilizing effect for certain aspects of the technology, it allows more time and energy to be spent on those aspects of the technology where innovation remains productive. [IV.38]

While it seemed, a few years ago, that the standardization process was nearly completed for a wide range of traditional engineering products, recent economic and environmental demands accelerated the process of technological innovation. The need to save energy and material and the call for environmentally acceptable technological solutions, on the one hand, and the challenges of information technology, on the other, stimulated the advent of new generations of "intelligent" measuring and controlling instruments of much higher precision and reliability, with complex control functions incorporated and new generations of data processing and telecommunication equipment. In a broad variety of engineering products, mechanical controls and mechanisms were replaced by electronics.

These technological developments called for the revision of existing and the creation of new standards. The standardization process had to keep pace with the actual developments and the subsequent pressing needs of manufacturers and users for a more rapid development of standards. [IV.39] As a consequence, the process of international standardization in engineering products accelerated in the period under review.

International Standards specifications lay down the requirement that serves as the basic quality criterion for inspection. The standard test methods provide a basis for the determination of quality characteristics. Thus international trade in engineering products is not dependent solely on standardization of the products themselves. Much work is carried out internationally to facilitate trade procedures as such. Recently, an agreement was reached within the Working Party on Facilitation of International Trade Procedures, a subsidiary body of the UN/ECE Committee on the Development of Trade, on a standard which will greatly facilitate computer interchange of trade data.

International trade transactions generate a massive transfer of information along the trade and transport chain: to move one piece of a consignment, up to 50 parties in different countries may have to create, transmit, receive, process, check, correct and file more than 50 separate documents with an average of over 360 copies per shipment. Electronic data processing and teletransmission are more and more used in international trade to reduce costs and speed up information transfer. [IV.40]

Today's manufacturing systems involve a number of associated machines - machine tools, robots, telecommunication equipment, computers, transport and handling equipment, individual work-stations and many related processes and higher-level activities, such as production planning, inventory control, office automation, etc. It is therefore a costly and technically difficult task to interface the different pieces of equipment (from various manufacturers) into a compatible system, not to mention the technical barriers to trade which may hinder international exchange of such equipment and services. Standardization should overcome these problems and permit integration of the various elements in automated systems within the manufacturing environment.

The objective of ISO is to promote the world-wide development of standardization and related activities with a view to facilitating international exchange of goods and services. ISO brings together the interests of producers, users, Governments and the scientific community. Its work is carried out through more than 2,000 technical bodies. More than 20,000 experts from all parts of the world participate each year in the ISO technical work which, to date, has resulted in the publication of some 6,400 standards. In accordance with a formal agreement, stating the relationship between the two organizations, ISO and IEC complement each other, forming a system for international standardization. To ensure the necessary technical co-ordination, they have established a joint ISO/IEC Technical Programming Committee. In addition it has been decided to establish a new joint committee for information technology standardization. In this connection ISO and IEC have set up the Joint Technical Committee 1 - Information technology (JTC 1) to assume the leadership in what is already a race between standardization and new technological developments.[IV.41]

Table IV.18 provides basic information on the standardization activities of the ISO technical committees primarily concerned with basic mechanical engineering, industrial automation systems (ISO/TC 184) and information processing systems (former ISO/TC 97). ISO/TC 22 - "Road vehicles", the last item in table IV.18, is introduced to give an indication of a product- or application-oriented ISO Technical Committee and its work-load.

Although it would appear that the standardization process, especially in the traditional mechanical and manufacturing sectors, is well established and relatively complete, a large proportion of the international standards in these sectors were issued only during the period under review (1980-1986). This could perhaps be due to the growing demand for internationally agreed standards, to facilitate trade and interchangeability in this rapidly innovating area of manufacturing technologies, reflecting the use of computer controlled machines and the related interfaces needed for the processing of complicated metal parts at high speeds and with high reliability. The components are linked by electronic controls that dictate what will happen at each stage of the manufacturing sequence - thus creating a demand for standardization in the field of information technology. Information related technologies have therefore become an even more important area in international standardization. Standards on interface conditions to ensure compatibility and interchangeability for an integrated system covering both the network and the equipment have become urgent and important.

Such technological developments have resulted in the revision of many ISO standards already formulated and have also created an international demand to formulate ISO standards in new areas to meet the pressing needs of industry and trade. As a consequence, it would appear that the process of international standardization in traditional mechanical and manufacturing engineering has also accelerated during the period under review, as illustrated by the percentage increase for 1980-1986 indicated in table IV.18.

Machine tools play a primary role in the field of non-electrical engineering and are the most essential part of the manufacturing technologies. Machine tools, with their high resolution and integration capabilities, have become necessary in the manufacture of highly technically intensive products. Over 80 International Standards in the field of machine tools have been formulated since the establishment of ISO/TC 39. ISO/TC 39 - "Machine tools", was established in 1947 to formulate standards in the field of all machine tools for the working of metal, wood and plastics, operating by removal of material or by pressure. Numerous standards such as the ones presented in table IV.19 issued by ISO in connection with acceptance conditions, testing for accuracy and test conditions for various types of machine tools are very important in this respect.

Although the present growth in machine-tool production is not significant, considerable changes have taken place in the composition of that production and in the mode of operation of the machine tools. In general, NC-machines constitute an increasingly important share of total machine-tool production. ISO standards in the field of numerical control of machine tools are listed in table IV.20.[IV.42]

ISO Technical Committee TC 29 "Small tools", established in 1947, has formulated some 200 International Standards, mainly for cutting tools for use in manufacturing. With the rapid innovation in the field of manufacturing technologies, the tool industry is looking for new technologies for tooling and tooling control. Apart from the development of new geometries and materials for cutting tools, considerable research is in progress to develop tool condition sensors for the detection of worn and broken cutting tools in operation - to minimize or even to eliminate the downtime in unmanned operation of machine tools and automated systems. Other areas where standardization may become necessary in the near future are tool holding, tool changing and tool storage mechanisms and coding of tools.

In 1983, a specialized Technical Committee, ISO/TC 184 "Industrial automation systems", was established to deal with standards in the field of industrial automation systems encompassing the application of multiple technologies, i.e. information systems, machines and equipment and telecommunications.

This Technical Committee has established five subcommittees, namely: SC 1 - Numerical control systems; SC 2 - Robots for manufacturing environment; SC 3 - Manufacturing application languages; SC 4 -

External representation of product definition data; and SC 5 - System integration and communication.

The programme of work of ISO/TC 184 includes the preparation of standards in the application area, such as basic reference models for automated and computer-aided manufacturing (CAM) systems, manufacturing message specifications (MMS), electrical and mechanical interface characteristics and line-control protocol. The ECE Working Party on Engineering Industries and Automation maintains working contacts with ISO/TC 184. [IV.43]

Robots have become an important component of modern manufacturing systems, primarily in view of the high machine utilization and repetitive functioning. Being a new field, significant technical advances are being made in the design and capabilities of these systems. In extending their capabilities and the possibilities of their integration into computer-based manufacturing systems, software is expected to play an increasingly important role. There is also a continuing need for increasing the intelligence of robots, including interfaces with other types of equipment. ISO identified the requirements of the industry at an early stage of development and established sub-committee ISO/TC 184/SC 2 "Robots for manufacturing environment", dealing with terminology, symbols, performance, safety, requirements for programming methods, data communication, mechancial interfaces and safety. International standards are under preparation in these areas. Programming languages require the human operator to specify manipulator positions and trajectories. While new high-level word-modelling languages automatically generate manipulator positions and trajectories, such languages can be used only with robots controlled by a general-purpose computer and only a limited number of currently installed robots have this capacity.

Regarding the activities of the former ISO/TC 97 "Information processing systems", special mention should be made of the preparation of standards in the fields of telecommunications and information exchange between systems; software development and system documentation; interconnection of equipment; information retrieval, transfer and management for Open Systems Interconnection (OSI); and languages.

Recognizing the need for compatibility between distributed systems and networks by means of common standards, ISO has developed a seven-layer model for network architecture (ISO 7498:1984) that offers a standard approach to the design and operation of computers and communication systems. This is done by integrating each system into a unified communication network which ISO has defined. In this system, which is known as the Open Systems Interconnection (OSI), a real system would contain one or more computers' associated software, peripherals, terminals, human operators, physical processes etc. that would form an autonomous whole capable of performing information processing. The seven layers of the OSI architecture fall into two distinct categories: layers 1-4 deal with local area network (LAN) component and data transfer; layers 5-7 are closer to the user's view of a particular application and so are largely software-based. [IV.44]

Some of the ISO Standards and draft International Standards (DIS) and draft proposals which have direct relevance to the Open Systems Interconnection (OSI) are listed in table IV.21.

OSI, as mentioned above, has a layer structure and all standardization work is layer-oriented. Therefore, to implement a user function or application one needs a "pile" of standards. This is in fact a cross-section through all the layers. But just identifying a set of standards for a certain user function, although necessary, is not sufficient to permit the construction of interworking products. Each OSI layer service definition, together with its corresponding layer protocol specificiations, defines the total functionally specified for that layer in the Reference Model. A "functional" standard goes beyond this general definition of a "layer" standard and emphasizes aspects of the standard relevant to the product design and to procurement processes. In functional standardization, the emphasis is on the function to be provided to the product user. The above-mentioned OSI base standards, by means of which interworking between products can be achieved, are then important only in support of the required function. When a number of ways exist in combining protocols from the adjacent OSI layers to perform a particular function, it is certain that each valid one, as well as some of the invalid ones, may be chosen by at least one user, unless some constraints on this freedom can be agreed upon and maintained. Functional standardization is therefore a convergence process, to minimize the "degrees of freedom" that are not of user interest. The importance of "functional standards" is felt in both ISO/TC 184 and ISO/IEC JTC 1 and relevant work has commenced within these technical bodies.

In the field of ISO/TC 44, "Welding", over 60 International Standards have already been produced. To cater for the requirements of robotic welding with special reference to multi-spot/multi-head welding, TC 44 has commenced preliminary work on items such as: cylinders for multi-spot welding, electrodes and taper fits for spot-welding equipment, straight electrodes with taper male, and threaded connection for fixing electrodes. Some of the ISO standards have been revised to keep abreast with the rapidly changing technology and the strong dependence on electronics and information technology. This trend implies the use of powerful computing systems extending from small computers to a hierarchized computer network. Table IV.22 lists some of the basic standards in this field which should be of interest.

The above brief review of the work of some of the ISO technical committees formulating standards in the field of manufacturing shows that areas such as machine tools, numerical control, robots, mechanical and electrical devices, mechanisms, interfaces, the CAD/CAM systems and the controlling and communication systems are constantly undergoing very significant changes with the development of automated manufacturing. These technologies developed individually as specialized disciplines. However, with the present day requirements of manufacturing industry, they had to be gradually integrated to meet specific requirements. The development of mechatronics (mechanics-electronics), computers and the various types of information processing equipment and systems have contributed considerably to the integration of these disciplines.

It could be damaging to the industry if the technology from various parts of the world (and mainly from the ECE region) were introduced without adequate co-ordination and co-operation, thus generating problems of incompatibility, diversity of design, variety and possible obstacles to international exchange of goods and services. ISO has taken up this challenge very successfully with the co-operation of international organizations such as IEC (IEC/TC 44 - Electrical equipment of industrial machines' electrical interfaces: IEC/TC 65 - Industrial process measurement and control - programmable controllers, systems considerations), the International Telegraph and Telephone Consultative Committee of the International Telecommunication Union (CCITT), the Institute of Electrical and Electronics Engineers (IEEE) and the industry as a whole and has initiated some of the necessary ISO standards which are already having a tremendous impact in ensuring compatibility of the various subsystems, machines, mechanisms and interfaces.

Table IV.23 illustrates the vast contribution of ECE member countries to the standardization process. Of the 164 secretariats of ISO technical committees, 90.2% are located in the ECE region. When one takes into account the secretariats of sub-committees and working groups, where the detailed technical work is largely carried out, the situation is even more dramatic: with 94.3% of the present 2,209 secretariats located in the region. This is further proof of the unique position of ECE member countries in world industrial development.

Table IV.18

Selected ISO Technical Committees active in the field of engineering industries

Technical Committee	*Title*	*Year of establishment*	*Total number of standards valid by December 1986*	*Number of standards issued during 1980-1986*	*Percentage increase 1980-1986*	*Number of new standards (DIS, DP, WI) currently in preparation*
TC 2	Fasteners	1947	110	81	74	51
TC 4	Rolling bearings	1947	44	23	52	18
TC 10	Technical drawings	1947	53	38	72	56
TC 14	Shafts for machinery and accessories	1947	10	2	20	15
TC 29	Small tools	1947	205	95	46	108
TC 39	Machine tools	1947	81	45	56	83
TC 41	Pulleys and belts (including veebelts)	1947	45	25	56	39
TC 44	Welding and allied processes	1947	69	38	55	88
TC 57	Metrology and properties of surfaces	1947	12	9	75	10
TC 60	Gears	1947	12	3	25	26
TC 97	Information processing systems [a]	1960	180	115	64	378
TC 118	Compressors, pneumatic tools and pneumatic machines	1965	12	6	50	15
TC 123	Plain bearings	1967	25	21	84	17
TC 131	Fluid power systems	1969	73	51	70	82
TC 176	Quality assurance	1979	1	1	100	6
TC 184	Industrial automation systems	1983	9	1	11	35
TC 22	Road vehicles	1947	230	158	69	220

Source: ISO Catalogue 1987, ISSN 0303-3309, Geneva 1987.

a As from 1987, ISO/IEC Joint Technical Committee (JTC) No. 1 "Information Technology".

Table IV.19

Selected ISO standards on test and acceptance conditions for machine tools

Reference	*Title*
ISO 1701:1974	Test conditions for milling machines with table of variable height, with horizontal or vertical spindle - Testing of accuracy
ISO 1708:1983	Acceptance conditions for general-purpose parallel lathes - Testing of accuracy
ISO 1984:1982	Acceptance conditions for milling machines with table of fixed height with horizontal or vertical spindle - Testing of accuracy
ISO 1985:1985	Acceptance conditions for surface grinding machines with vertical grinding wheel spindle and reciprocating table - Testing of accuracy
ISO 1986:1985	Acceptance conditions for surface grinding machines with horizontal grinding wheel spindle and reciprocating table - Testing of accuracy
ISO 2407:1984	Acceptance conditions for internal cylindrical grinding machines with horizontal spindle - Testing of accuracy
ISO 2423:1982	Acceptance conditions for radial drilling machines with adjustable arm height - Testing of accuracy
ISO 2433:1984	Acceptance conditions for external cylindrical grinding machines with movable table - Testing of accuracy
ISO 3070-1:1975	Test conditions for boring and milling machines with horizontal spindle - Testing of accuracy - Part 1: Table-type machines.
ISO 3070-3:1982	Acceptance conditions for boring and milling machines - Testing of accuracy - Part 3: Planer-type machines with movable column
ISO 3089:1974	Self-centring, manually operated chucks for machine tools - Normal accuracy - Acceptance test specifications (geometrical tests)
ISO 3190:1975	Test conditions for turret and single spindle co-ordinate drilling machines with vertical spindle - Testing of accuracy
ISO 3655:1986	Acceptance conditions for vertical turning and boring lathes with one or two columns and a single fixed or movable table - General introduction and testing of accuracy
ISO 3686:1976	Test conditions for turret and single spindle co-ordinate drilling and boring machines with table of fixed height with vertical spindle - High-accuracy machines - Testing of accuracy
ISO 3875:1980	Conditions of acceptance for external cylindrical centreless grinding machines - Testing of accuracy
ISO 4703:1984	Acceptance conditions for surface grinding machines with two columns - Machines for grinding slideways - Testing of accuracy
ISO 6155-1:1981	Acceptance conditions for horizontal spindle capstan, turret and single automatic lathes - Testing of accuracy - Part 1: Machinable bar diameters greater than 25 mm
ISO 6480:1983	Conditions of acceptance for horizontal internal broaching machines - Testing of accuracy
ISO 6481:1981	Acceptance conditions for vertical surface-type broaching machines - Testing of accuracy
ISO 6779:1981	Acceptance conditions for broaching machines of vertical internal type - Testing of accuracy
ISO 6898:1984	Open-front mechanical power presses - Capacity ratings and dimensions
ISO 6899:1984	Acceptance conditions for open-front mechanical power presses - Testing of accuracy
ISO 7008:1983	Woodworking machines - Single-blade circular saw benches with or without travelling table - Nomenclature and acceptance conditions
ISO 7009:1983	Woodworking machines - Single-spindle moulding machines - Nomenclature and acceptance conditions
ISO 7572:1984	Conditions of acceptance and installation for work-holding fixed tables of machine tools
ISO 7947:1985	Woodworking machines - Two-, three- and four-side moulding machines - Nomenclature and acceptance conditions
ISO 7987:1985	Woodworking machines - Turning lathes - Nomenclature and acceptance conditions
ISO 8956:1986	Acceptance conditions for copying attachments, integral or otherwise, for lathes - Testing of accuracy

Source: As for table IV.18.

Table IV.20

ISO standards in the field of numerical control of machines

Reference	*Title*
ISO 841:1974	Numerical control of machines - Axis and motion nomenclature
ISO 2806:1980	Numerical control of machines - Vocabulary bilingual edition
ISO 2972:1979	Numerical control of machines - Symbols
ISO 3592:1978	Numerical control of machines - NC-processor output - Logical structure (and major words)
ISO 4336:1981	Numerical control of machines - Specification of interface signals between the numerical control unit and the electrical equipment of an NC-machine
ISO 4342:1985	Numerical control of machines - NC-processor input - Basic part programme reference language
ISO 4343:1978	Numerical control of machines - NC-processor output - Minor elements of 2000-type records (Post-processor commands)
ISO/TR 6132:1981	Numerical control of machines - Operational command and data format
ISO 6983-1:1983	Numerical control of machines - Program format and definition of address words - Part 1: Data format for positioning, line motion and contouring control systems
ISO 7846:1985	Industrial real-time FORTRAN - Application for the control of industrial processes

Source: As for table IV.18.

Table IV.21

ISO standards (issued and in preparation) in the field of Open Systems Interconnection (OSI)

Reference	*Title*
Application layer (OSI):	
ISO/DP 8571-1-4	Open Systems Interconnection: File transfer, and management Part 1:1986 General description Part 2:1986 The virtual file store definition Part 3:1986 The file service definition Part 4:1986 The file protocol specification.
ISO/DIS 8649-2	Open Systems Interconnection: Definition of common application service elements: association control
Session layer (OSI)	
ISO/DIS 8326	Open Systems Interconnection: Basic connection-oriented session protocol specification
ISO/DIS 8327	Open Systems Interconnection: Symmetric synchronization protocol
Transport layer (OSI)	
ISO 8072:1986	Open Systems Interconnection: Transport service definition
ISO 8073:1986	Open Systems Interconnection: Connection-oriented transport protocol specification
Network layer (OSI)	
ISO/DIS 8473	Open Systems Interconnection: Data communication protocol for providing the connectionless-mode network service
Data Link layer (OSI)	
ISO/DP 8802-2	Local Area Networks - Part 2: Logical link control
Physical layer (OSI)	
ISO/DIS 8802-4	Local Area Network - Token-passing bus access method and physical layer specification

Source: ISO Central Secretariat, Geneva

Table IV.22

ISO standards in the field of welding

Reference	*Title*
ISO 7284:1984	Resistance welding equipment - Particular specifications applicable to transformers with two separate secondary windings for multi-spot welding, as used in the automobile industry
ISO 7286:1986	Graphical symbols for resistance welding equipment - Bilingual edition
ISO 7931:1985	Insulation caps and bushes for resistance welding equipment
ISO 6848:1984	Tungsten electrodes for inert gas shielded arc welding, and for plasma cutting and welding - Codification
ISO 5826:1983	Transformers for resistance welding machines - General specifications applicable to all transformers
ISO 5827:1983	Spot welding - Electrode back-ups and clamps
ISO 1027:1983	Radiographic image quality indicators for non-destructive testing - Principles and identification
ISO 700:1982	Power sources for manual metal arc welding with covered electrodes and for the TIG process
ISO 669:1981	Rating of resistance welding equipment
ISO 6947:1980	Fundamental welding positions - Definitions and values of angles of slope and rotation for straight welds for these positions
ISO 3777:1976	Radiographic inspection of resistance spot welds for aluminium and its alloys - Recommended practice

Source: As for table IV.18.

Table IV.23

Distribution of ISO technical secretariats (1987) [IV.45]

Member body	*Host country of the secretariat*	*Technical Committees*	*Sub-Committees*	*Working Groups*	*Total*
ABNT	Brazil (non ECE country)	-	2	-	2
AFNOR	France	27	110	238	375
ANSI	United States	16	66	212	294
BDS	Bulgaria	-	1		1
BSI	United Kingdom	23	94	287	404
CSBS	China (non ECE)	-	1	1	2
CSN	Czechoslovakia	1	3	6	10
DIN	Germany, Fed. Rep. of	26	107	277	410
DS	Denmark	3	3	24	30
GOST	USSR	10	32	13	55
IBN	Belgium	4	9	21	34
ICONTEC	Colombia (non ECE)	-	-	1	1
IPQ	Portugal	2	-	4	6
IRANOR	Spain	-	3	4	7
IRS	Romania	2	5	-	7
ISI	India (non ECE)	5	16	5	26
JISC	Japan (non ECE)	3	11	31	45
MSZH	Hungary	1	1	3	5
NNI	Netherlands	5	14	50	69
NSF	Norway	2	9	10	21
ON	Austria	2	4	10	16
PKNMiJ	Poland	1	4	7	12
PNGS	Papua New Guinea (non ECE)	-	-	1	1
SAA	Australia (non ECE)	4	4	26	34
SABS	South Africa, Rep. of (non ECE)	3	2	5	10
SCC	Canada	5	30	66	101
SFS	Finland	-	3	9	12
SII	Israel (non ECE)	1	1	2	4
SIRIM	Malaysia (non ECE)	-	-	2	2
SIS	Sweden	12	23	72	107
SNV	Switzerland	4	19	17	40
TSE	Turkey	-	2	-	2
UNI	Italy	2	22	40	64
Total		164	601	1 444	2 209

Source: *ISO Memento 1987*, ISSN 0536-2067, Geneva 1987.

REFERENCES

IV.1 *Bulletins of Statistics on World Trade in Engineering Products.* UN/ECE annual publication (issues covering 1979-1986); and *Monthly Bulletins of Statistics;* United Nations publication (for the same years).

IV.2 See also. [IV.1] Sales numbers of individual issues of the UN/ECE *Bulletins of Statistics on World Trade in Engineering Products* from 1979:

1979	-	E/F/R.81.II.E.13
1980	-	E/F/R.82.II.E.5
1981	-	E/F/R.83.II.E.8
1982	-	E/F/R.84.II.E.5
1983	-	E/F/R.85.II.E.11
1984	-	E/F/R.86.II.E.10
1985	-	E/F/R.87.II.E.10
1986	-	E/F/R.88.II.E.14

The Working Party on Engineering Industries and Automation also holds annual meetings on Questions of Statistics Concerning Engineering Industries and Automation. The tenth meeting (second Joint Meeting with the Conference of European Statisticians) was held in October 1988.

IV.3 Subchapter IV.1 of the present study contains a shortened version of chapter 3 of Recent Changes in Europe's Trade, as prepared by GEAD for publication in the *Economic Bulletin for Europe,* vol. 39, No. 4, Pergamon Press 1987.

IV.4 "Eastern imports of machinery and equipment: 1960-1985", *Economic Bulletin for Europe,* vol. 38, No.4.

IV.5 A comparative analysis of east European and selected newly industrialized countries' (Argentina, Brazil, Mexico) inflows of technology, including machinery and equipment from the west, was issued recently. It concludes that the combined imports of western technology of these three NICs were considerably larger than those of eastern Europe and that this gap did not appear to be diminishing. Argentina, Brazil and Mexico had to cut their imports in the early 1980s. For more details see K. Poznanski, "Patterns of Technology Imports, Inter-regional Comparisons", *World Development,* vol. 14, 1986, No. 6, pp. 743-756.

IV.6 See United Nations *Monthly Bulletin of Statistics,* section on External Trade. It should be noted that the United Nations classification has not always been strictly adhered to in this subchapter, but the differences are minor.

IV.7 The source of these price indices is the United Nations *Monthly Bulletin of Statistics,* various issues, Special Table C.

IV.8 For most east European countries, data on exports of engineering goods by country of destination are not available.

IV.9 See "Aspects of capital formation in manufacturing industry" in *Economic Survey of Europe in 1986-1987,* UN/ECE publication, Sales No. E.87.II.E.1. For example, in the United Kingdom, recent revisions of capital stock figures reflect the fact that service lives have been reduced by installments from just over 30 years for assets installed before 1950 to about 23 years for assets installed from 1970 onwards (p.86). Also, in the United States, a uniform 15 years for all manufacturing machines had been assumed for the calculation of capital stock, whereas the new revised estimates show lifetimes of 20 years or more for almost one third of the branches. However, service lives are on average shorter in the Federal Republic of Germany, which suggests that there is likely to be considerable variation between countries.

IV.10 Even if this assumption leads to an overestimate of the stock of imported capital, the error is likely to be small since, on average, only 7% of the cumulated engineering goods are more than 20 years old. Extending the assumed lifetime reduces the share of capital accumulated in the recent past.

IV.11 Retirement rates tend to be very low in eastern Europe and the Soviet Union, generally varying around 1-2%. See section 4.4 "Investment" in *Economic Survey of Europe in 1984-1985, p.138* UN/ECE publication, Sales No. E.85.II.E.1.

IV.12 Given the size of Japan's economy and share of world trade, its share in world imports of engineering goods seems to be exceptionally low. In comparison, France's, the Federal Republic of Germany's and the United Kingdom's volumes of engineering goods imports during 1981-1985 were each roughly three times the size of Japan's.

IV.13 The cyclical downturn in the developed market economies was concentrated first on stocks and then on fixed investment, the latter curtailing the demand for imported machinery and equipment.

IV.14 During 1976-1980, eastern Europe's imports of engineering goods from the *non-socialist* countries fell at an average annual rate of 6%. In 1981-1985, the average rate of decline slowed to 2%. This latter figure conceals an upturn in import volumes in 1984-1985. A more detailed analysis of these developments is presented in "Eastern imports of machinery and equipment" in *Economic Bulletin for Europe,* vol. 38, No.4.

IV.15 For recent studies of these developments in the west, see "Aspects of capital formation..." and for the east "Investment", section 3.4, pp 163-165, both in *Economic Survey of Europe in 1986-1987.* UN/ECE publication, Sales No. E.86.II.E.1.

IV.16 See GATT, *International Trade,* various issues.

IV.17 See e.g." Exports of manufactures from eastern Europe and the Soviet Union to the developed market economies" in *Economic Bulletin for Europe,* vol. 35, No. 4.

IV.18 The weakness in these imports was documented in an ECE secretariat study, "Eastern imports of machinery and equipment" in *Economic Bulletin for Europe* vol.38, No.4.

IV.19 This assumes a two-year time lag for the absorption of capital goods.

IV.20 See *Economic Survey of Europe in 1986-1987,* section 5.2. UN/ECE publication, Sales No. E.87.II.E.1.

IV.21 Engineering goods imports accounted for over one half of total incremental imports in 1986. *Ibid.*, pp. 302-303.

IV.22 While in preceding subchapter IV.1 the development of trade in engineering technologies and products (engineering imports) was analysed from the global (world-wide) viewpoint - on the basis of a methodological approach used in the ECE General Economic Analysis Division (GEAD) - the present subchapter IV.2 summarizes the geographic and product pattern of the ECE regional trade as presented in individual *Annual Reviews of Engineering Industries and Automation*, covering the period 1979-1986.

IV.23 Where available, tables IV.13 (imports) and IV.14 (exports) include figures for 1987.

IV.24 1987 data on engineering exports for Japan and some other countries were not available by the time of final editing of the present study.

IV.25 Opening statement by A. Papandreu, the Prime Minister of Greece, to the ECE Symposium on East-West Business Opportunities and Trade Prospects, held at Thessaloniki (Greece) from 8-11 September 1986 (report of the Symposium, document TRADE/SEM.8/2, paras.7-10).

IV.26 "Recent Changes in Europe's Trade" in *Economic Bulletin for Europe,* vol. 39, No.4.

IV.27 See also subchapter IV.1 of the present study and papers presented to the ECE Symposium at Thessaloniki [IV.25] e.g. East-West Trade in the Engineering Industry: an overview of developments and prospects (TRADE/SEM.8/R.18); Czechoslovakia and International Trade in Engineering Goods (TRADE/SEM.8/R.33) and Community Trade in Engineering Products with East European countries, 1977-1984 (Conference paper presented by the EEC secretariat).

IV.28 See report of the ECE Symposium at Thessaloniki [IV.25] (TRADE/SEM.8/2, para. 32 - Workshop III on Engineering Industries).

IV.29 Table IV.15 illustrates the development of export prices in ten reporting countries only. Regarding producer price indices, Government replies to ECE questionnaires have been received from 20 countries: see table 6 in *Annual Review of Engineering Industries and Automation 1986,* vol. I. UN/ECE publication, Sales No. E.88.II.E.16.

IV.30 Table IV.16 shows the development of export prices by various types of machinery and equipment in three leading exporting countries (see also subchapter IV.2). In other countries, depending on the structure of the engineering industries (e.g. with no automotive industry), the trends in export (and producer) prices might be different.

IV.31 See also subchapter II.3 of the present study, on gross capital formation and R and D expenditure in the engineering industries.

IV.32 For detailed information, see *East-West Joint Ventures - Economic, Business, Financial and and Legal Aspects.* UN/ECE publication, Sales No. E.88.II.E.18.

IV.33 The Senior Advisers to ECE Governments on Science and Technology (SAST), in order to promote international co-operation in the field of technology transfer, regularly update the *Manual on Licensing Procedures in Member Countries of the United Nations Economic Commission for Europe.* The Manual is published by and available from Clark Boardman Co., Ltd., 435 Hudson Street, New York, NY 10014.

IV.34 O. Sturen "Responding to the challenge of the GATT Standards Code". ISO Secretariat, Geneva 1983.

IV.35 C.J. Stanford "The Impact of World Standards on International Trade". IEC Secretariat, Geneva 1983.

IV.36 The ECE Working Party on Engineering Industries and Automation maintains close working contacts with both the ISO and the IEC secretariats, as well as with their technical committees active in the field of engineering standardization. The present subchapter IV.7 is largely based on material and information received from ISO. Relevant IEC activities are described in *Trends in the Electrical and Electronics Industries,* pp.100-111, document ECE/ENG.AUT/31.

IV.37 Same source as [IV.35]

IV.38 O. Sturen: "International Technology as Codified in Standards". ISO Secretariat, Geneva 1985.

IV.39 International Electrotechnical Commission *Annual Report 1984.* IEC Secretariat, Geneva 1985.

IV.40 For a more detailed description of ECE activities in this field, see e.g. draft chapter IV of the present study, as issued in 1986 (document ENG.AUT/AC.10/R.2/Add.15).

IV.41 The fact that the first ISO/IEC Joint Technical Committee (JTC 1) was created in 1987 in the field of information technologies is proof of the need to improve co-ordination of standardization of automated (computer-based) systems. In this respect, see also the brief evaluation of the ISO/TC 184 activities in the present chapter or the description of the Manufacturing Automation Protocol (MAP) communication standardization design, as introduced by General Motors in 1980-1982 on the basis of IEEE standards and the OSI model, see e.g. *Software for Industrial Automation* pp.97-102. UN/ECE publication, Sales No. E.87.II.E.19.

IV.42 Table IV.20 includes only standards issued before 1985. Further activities (since 1986) were undertaken within the scope of ISO/TC 184, Subcommittee 1 "Numerical Control Systems".

IV.43 The ECE Working Party on Engineering Industries and Automation maintains close working contacts with ISO and IEC in many fields of common interest, including active ISO and IEC participation in techno-economic studies and seminars undertaken by the Working Party. The ECE Working Party has had direct liaison with ISO/TC 184 since its creation in 1983.

IV.44 See also reference [IV.41]. Recent trends in selected telecommunication equipment and systems are described in *The Telecommunication Industry: Growth and Structural Change.* UN/ECE publication, Sales No. E.87.II.E.35.

IV.45 Table IV.23 is based on 1986 figures. In 1987, the total number of secretariats increased from 2209 (1986) to 2359 (i.e. 164 for TC, 644 for SC and 1551 for WG).

CHAPTER V NATIONAL CONTRIBUTIONS TO THE PERFORMANCE OF THE ENGINEERING INDUSTRIES

V.1. INTRODUCTION

The present chapter reviews the national statements provided by individual countries during the preparation of the study. Four countries (Austria, Czechoslovakia, the Ukrainian SSR and the USSR) provided their contribution in time for the first draft of the study. [V.1] Six more countries (Belgium, Bulgaria, the German Democratic Republic, Italy, Poland and Sweden) prepared their statements for inclusion in the second draft. [V.2] As the study was being finalized, additional information was received from Hungary and the United Kingdom. Background material provided by the United States was used throughout the study and incorporated in individual chapters. [V.3]

Most of the national statements which follow were shortened during the final editing phase in order to minimize possible repetition of information already given in preceding chapters and to avoid discrepancies in data presented. This chapter thus has more of a complementary character and reflects some selected views on specific aspects of the performance of engineering industries within the national economies. [V.4]

V.2. AUSTRIA

In the period from 1979 to 1985, [V.5] the value of production in the Austrian engineering industries, in both nominal and real terms, increased at a rate above the average of west European countries. As it is common for engineering industries to increase their productivity every year, it has become imperative for the Austrian engineering industries to improve their efficiency in order for them to become more competitive with respect to regions such as the Pacific area and the Far East; this of course relates also to increasing the degree of automation.

The annual increase of production has, however, not been enough to keep the level of employment in the engineering industries. According to statistics, the number of persons employed in these industries has gone down by 3% a year. This development illustrates the post-industrial society. Employment is guaranteed only if the service sector is able to absorb those employees who have become redundant in the industrial sector. Thus the connotation of the term "industry" must change, particularly because industrial products without services are becoming scarcer. The services connected with production may rightly be considered a new developing sector.

The statistics show that producer prices in Austria have not risen as much as in all the other western industrialized countries. This gives rise to the suspicion that the Austrian economy has maximized production as well as employment - rather than income. Profit as the main factor in the theory of entrepreneurial action, which has also formed the economic structures, will, therefore, have to play a major role in future studies on restructuring.

Structure analysis and policy

In 1983, the Austrian Institute for Economic Research (WIFO) launched an economically relevant research programme on structuring - to be carried out periodically - by order of the Austrian Federal Ministry of Finance. A preliminary report mentions various phenomena, some of which go beyond official statistics and the usual comparative statistical approach. Pressured by obvious structural problems characteristic of the 1980s (engineering industries being included), new efforts must be made to remedy the situation.

At the same time, economists have rediscovered the microlevel of structural changes and the importance of motivations and decisions made within companies, thus reviving Schumpeter [V.6] and his image of the entrepreneur. Economists have started to replace the theory of foreign trade, which had been expanding for centuries, by a bold concept of a general theory of international production. They have recognized the life cycle of both manufacturing processes and products and, finally, they have questioned the way of using structural instruments and started investigating them.

Promotion of non-material investments

Investments are essential for implementing the latest technological developments in the manufacturing process. The conversion of traditional to new manufacturing processes and new products requires considerable investments in the transition phase. Yet the share of industrial investments in accrued value has been diminishing, and it will continue doing so in the long run whilst a strategy to maintain the share of engineering products in GNP is applied. As the productivity of capital in the industrial sector is increasing, it will be necessary in part to shift from the so-called material to non-material investments. Non-material investments are investments in staff education, in research and development, in making a product marketable and in marketing as such. In terms of volume, these three categories of non-material investments may well have already caught up with material investments. As non-material investments are gaining in importance for the development and marketing of new products, it would make sense to consider non-material investments as being an economic aim to be followed up.

Corporate organizations and the change of economic structures

Owing to changes in the overall world economy, economic insecurity has grown and corporate strategies as well as organizational questions have gained in importance. The Structure Report of the Austrian Institute for Economic Research (Austrian Structure Report 1984) arrives at the following conclusions, relevant to the engineering industries:

- It seems that an enterprise that pursues active strategies, such as conquering new markets, manufacturing new products, raising the degree of specialization and complying with customers' specifications and requests, has better chances than a company that aims at a lowering of costs while keeping the product unchanged.
- It seems that divisionally structured corporations and corporations which have more decentralized decision-making structures (such as profit centres, allocation of achievements and profits, information systems) are more successful than functionally structured corporations. This thesis may not be valid for all types of corporations and markets, namely for those requiring a matrix organization.
- It seems to be generally accepted that mergers have not as a rule brought the expected success; on the other hand, no time-tested alternative method is currently available which might reduce capacity better than concentration when considered from an economic point of view that includes labour-market policy.

A company which actively plans and regularly checks its strategies and organization almost certainly has an advantage over others which remain passive and merely react to external circumstances.

V.3. BELGIUM

Table V.1 sets out the main results achieved by the engineering industries in Belgium in the years 1983-1986.[V.7]

The volume of orders of the metalworking industries rose by 15% between 1980 and 1986, deliveries rose by 16%, and employment fell by 17%.

Product prices in the metalworking sector had risen only slightly, by 2.2% in 1986.

The reduction in the number of jobs was slightly up in 1986 relative to the two preceding years. This decline primarily concerns metal structures and marine engineering, railway equipment, agricultural machinery and electrical goods.

Exports increased by 8% in 1986 over 1985. Exports to other countries of the European Economic Communities rose by 12%. The trend for exports outside Europe (Asian countries, OPEC countries, the United States) was more of a downward one.

Imports were 10% in 1986.

V.4. BULGARIA

Capital formation

In the period 1979-1985, Bulgaria [V.8] continued to give priority by comparison with the other branches of industry to the development of the engineering industries and to pursue the rapid development of their material and technical base.

Total capital investment in the mechanical engineering and metalworking industry in 1979-1985 amounted to 5,339.4 million leva, with annual averages of: 470.8 million leva in the period 1976-1980: 574.5 million leva in 1979-1980 and 838.1 million leva in 1981-1985. As a proportion of average annual capital formation in industry, average annual capital spending in mechanical engineering and metalworking amounted to 18.3% in 1976-1980, 22.7% in 1979-1980 and 23.3% in 1981-1985.

Capital spending in the electrical engineering and electronics industry totalled 1,657.2 million leva in the period 1979-1985. Average annual rates of such investment were 121.2 million leva in 1976-1980, and 279.6 million leva in 1981-1985. Capital spending in the electrical engineering and electronics industry represented lower proportions of total capital spending in industry and the economy as a whole than did capital spending in the mechanical engineering and metalworking industry. As a proportion of average annual industrial investment, average annual capital spending in the electrical engineering and electronics industry amounted to 4.7% in 1976-1980 and 7.8 per% in 1981-1985.

In the engineering industries, the implementation of the investment programme during the period 1979-1985 brought with it the intensive development and modernization of production facilities. This is shown by the figures for the development of the proportion of investment devoted to modernization and rebuilding in the engineering industries (tables V.2 and V.3).

The output of the engineering sector during the period 1981-1985 rose by some 52%, i.e., twice as fast as the output of industry as a whole. This had a favourable impact on the entire process of investment. There was a rise in the importance of engineering for the technical and technological modernization of all sectors of the economy and, in particular, for the introduction of automated technologies, in order to increase the country's export potential.

Bulgaria's investment policy in the engineering sectors has been aimed at pursuing the fundamentally new course of intensive development of the national economy through technological innovation and restructuring in all sectors. This being so, priority has been given to the development of production capacity, namely in the following areas:

- Electronics, including the manufacture of computers and communication equipment;
- Electrical engineering (including electrical drives for industrial robots and NC-machine tools);
- Industrial automation equipment including flexible manufacturing systems;
- Building of hoisting and transport machinery;
- Heavy engineering (the manufacturing base of which has been intensively developed in the past years); and
- Shipbuilding using leading-edge technologies.

Employment

The rapid development of the engineering technologies, and of electronics in particular, requires the orientation of a greater proportion of the growth in the number of skilled industrial workers. Employment in engineering rose from 358,289 in 1979 to 364,994 in 1980 and 399,887 in 1985. In other words, a total of 41,598 employees entered the engineering industries during the period under review (66.7% of the total industrial work-force increase).

The great majority of new skilled industrial workers have, in accordance with the priority given to its development, entered the field of electronics. During the period under review, the number of persons employed in State and co-operative electronics and electrical engineering enterprises increased by 19.5%; the corresponding increase in the mechanical engineering and metalworking industries was 7.5% and that in industry as a whole 4.75%.

Owing to the rapid development in the engineering industries, there was also an increase in the share of the engineering work-force within the country's total work-force. Table V.4 shows the changes in engineering employment in relation to employment in industry and the national economy as a whole. It also shows the proportion of the able-bodied population which is employed in the engineering industries.

During the period 1981-1985, the gross industrial output per production worker of State and co-operative enterprises grew by 56.5% in the electronics and electrical engineering industries, by 27.7% in the mechanical engineering and metalworking industries and by 20.6% in industry as a whole. The relation of growth rates in the labour productivity during this period was similar: hourly productivity increased by 55.9% in the electronic and electrical engineering industries, by 28.7% in the metalworking and mechanical engineering, and by 21.8% in industry as a whole.

The improvements in the technological and technical base of engineering and the increasing sophistication of engineering products are leading to substantial changes in the structure of the work-force. While the total number of persons employed in engineering is rising, some decline is occurring in the proportion of manual workers relative to engineers and technicians. There are differing influences on the number of manual workers: on the one hand, the commissioning of new capacity requires an increase in the manual work-force, while on the other the introduction of high-output machinery and advanced technologies both frees manual workers for other tasks and requires them to have higher levels of general education and occupational training than before.

Table V.5 compares the changes in the structure of the work-force in mechanical engineering, electrical engineering and industry as a whole during the period 1980-1985. (The figures relate only to State industrial enterprises.)

It must be emphasized that the proportion of engineering and technical staff is substantially higher in engineering and manufacturing than in industry in general. The new scientific and technological trends entail higher requirements as regards the skill patterns and levels of education of the work-force in all sectors of the economy, particularly engineering. In order to meet the increased and constantly changing demand for highly skilled workers in the future, Bulgaria commenced in the period 1980-1985 the revamping of its secondary and higher education system.

Technological change

Bulgaria's technology policy as regards the engineering industries in 1979-1985 was aimed at meeting the need to increase the range and the quality of engineering products and to implement progressive structural changes in the sector. Increases were achieved in the output of products of strategic and structural importance for the economy and manufacture commenced of, amongst others, new types of machine tools, electric and other trolleys, computers, ships. There were rapid developments in the field of robotics and significant advances in microelectronics (e.g. the manufacture of solid-state components).

The main paths of technological development in the Bulgarian engineering industries in 1979-1985 were the following:

- Application of electronics: the introduction of advanced computer technology and automated assembly systems, the building of equipment and systems for active control and diagnosis, etc.;
- Development of equipment for biological engineering and biotechnologies;
- Development of communication equipment, digital processing technologies, building of computerized information systems etc.;
- Development of a number of new engineering materials (metallurgy, metal casting, plastic materials etc.);
- Development of new types of numerically controlled metalworking machine tools;

- Design and implementation of flexible manufacturing systems (FMS) and computer integrated manufacturing (CIM) systems;
- Comprehensive automation of manufacturing processes through increases in the proportion of fully automated production cells and centres, automated production lines, industrial robots and manipulators etc; and
- Development of computer-aided design and engineering based on the use of microcomputer systems and of specialized software.

By comparison with 1979, the number of new and improved products in 1985 rose by a factor of 1.6. Similarly the use of new raw and semi-finished materials increased by a factor of 3.4 and the number of new and improved technologies employed by a factor of almost 19. The most important efforts for the technological modernization of industrial production were those to develop the automated technologies; the increase in this respect was by a factor of 2.2. This shows the important role of the engineering sector and, in particular, of the production of computers and other automation-oriented equipment in the technological modernization of the Bulgarian economy.

In the period up to 1990, the material and technical base of the economy as a whole will be subject to large-scale technical innovation. The draft State Plan for the period 1986-1990 embraces the development of more than 200 technological systems covering 850 principal and associated technologies. This will sharply reduce inputs of time, energy and raw and semi-processed materials and increase the proportion of "science-intensive" products and processes.

International specialization and co-operation

In the period 1979-1985, Bulgaria made active and profitable use of the advantages of international co-operation and economic integration among the CMEA countries in order to develop its engineering industry and, thereby, to achieve its main social and economic objectives. During the period in question, Bulgaria expanded its activity and, at the same time, consolidated its participation in the work of CMEA sectoral organs and international economic organizations (the CMEA Committee on Co-operation in Engineering Industries, the Inter-governmental Commission on Computers, the CMEA Commission on the Radio and Electronics Industries, INTERATOMENERGO, INTERTEKSTILMASH, INTERELEKTRO etc.).

In recent years, the processes of socialist economic integration have been pursued and have gradually been taken down to lower levels of economic activity. This pattern has found expression in the development of direct links and the merging of scientific and industrial co-operation within the framework of joint scientific and manufacturing associations and enterprises. International specialization and co-production (ISCP) has already confirmed itself as the principal form of the international socialist division of labour and has acquired new breadth and depth. ISCP now involves all the main subsectors and groups of engineering activity.

During the period 1976-1980, Bulgaria was a participant in some 50 multilateral agreements on ISCP; under them, it specialized in some 700 products and was the sole manufacturer of 230 of these. Consequently, it gave priority to developing the manufacture of: electric and internal-combustion-engined fork-lift trucks; machine tools, including numerically controlled; food-processing machinery; communication equipment; hydraulic and pneumatic equipment; electrical engineering equipment; plastics-processing equipment; production lines for the manufacture of building materials; machinery sets for wine growing, horticulture, vegetable growing and tobacco farming; and various capital goods for the electric power, iron and steel, chemical, and mining and mineral-dressing industries. Particular mention must be made of the role of specialization and co-operation in the development of the manufacture of computer equipment. During the period under review, arrangements were made for the large-scale manufacture in Bulgaria of computer peripherals, including storage devices, of which Bulgaria is the main exporter among the CMEA member countries.

Over 100 contracts and agreements on ISCP within the framework of CMEA were in force during the period 1981-1985; of these, more than 60 concerned general engineering, while some 40 concerned electronic and electrical engineering. Bulgaria was a party to 62 of these agreements and therefore specialized in some 1,500 products. The contribution of ISCP to the improvement of the structure of the engineering industry was greatly increased, in accordance with the scientific and technological progress in two areas:

(a) The creation and development of advanced technologies, process lines and machine-tool-based systems in those areas in which Bulgaria had gained prominence as a result of specialization prior to 1980, e.g. warehousing systems, transport and loading systems, automated process lines and modules, flexible manufacturing systems, metalcutting machine tools, including numerical controls, automated manufacturing lines for the building industry and ceramics, midi- and mini-computers, electronic telephone exchanges, ships and vessels, monitoring and control systems in agricultural machinery, raw-material- processing systems for fruit and vegetable conserves, automated guided vehicles (AGV) etc.; and

(b) The establishment of the foundations for the creation and development of new, advanced subsectors and types of activity, e.g. biotechnology equipment, robotics, fibre optics, laser equipment, laboratory and industrial nuclear equipment, advanced medical apparatus for diagnosis and therapy, software products and systems etc.

A striking indicator of the Bulgarian engineering industry's participation in ISCP is the growth in the share which machines and equipment manufactured under specialization and co-production agreements in accordance with CMEA policy represent of engineering exports: from barely 24% in 1970, this share rose to 45.4% in 1975, 45.7% in 1980 and 49.3% in 1985.

There were also significant changes in the structure of Bulgarian exports of specialized machines and equipment by subsector and by type (see table V.6). During the period under review, Bulgaria's contribution to the CMEA countries' total exports of machines and equipment covered by ISCP increased both in absolute and in relative terms. Bulgaria's substantial participation in ISCP is, indeed, an outcome of the fact that the products in which it specializes are intended primarily for export to CMEA countries. The products in which Bulgaria specializes within the framework of CMEA are also exported to many of the developing countries of Asia, Africa and Latin America, as well as to a number of developed market economies (equipment for the food and tobacco industries; electric and internal-combustion-engined fork-lift trucks; electric monorail hoists; equipment for the chemical, mining and mineral-enrichment industries; automatic telephone exchanges; agricultural machinery; bearings; electric motors; electrical instruments; various types of ships etc.).

Bulgaria engages in various forms of trade and industrial co-operation in the engineering field with the majority of its trading partners throughout the world: industrial co-production, joint companies, co-operation in the purchase and granting of licences and know-how, exchanges of specialists, joint construction of engineering plants and provision of technical assistance. This co-operation and the expansion of the abilities of the Bulgarian engineering industry have led both to an increase in the proportion which complete facilities (entire process lines, machining systems, installations or factories, including turnkey arrangements) represent of total engineering exports and to proportional growth in the supply of components, assemblies and machines. From 6% in 1975, the share of complete facilities in Bulgarian engineering exports rose to 10% in 1980 and to over 14% in 1985.

During the period under review, co-operation between the CMEA countries was improved and enriched over the so-called "research - development - production - application" chain. In this respect, there has recently been considerable growth of direct links between engineering enterprises in various countries, with a view to reinforcing co-operation in research, technology and manufacturing. Bulgaria and the Soviet Union have set up joint science-and-manufacturing associations having wide powers and, as regards the field of their joint activity, a common management, a common science and technology policy, a common capital investment policy and common manufacturing. The principal contribution of these science-and-manufacturing associations concerns the development of bilateral and multilateral specialization.

Main developments in 1979-1985

During the period 1979-1985, the Bulgarian engineering industry proved itself to be a dynamic sector growing faster than industry in general and other sectors of the economy. The share of gross output of engineering industries in total industrial production rose from 20.4% in 1981 to 25.1% in 1985. This is indicative of the fact that the average annual growth in engineering output was some 9%.

As regards manufacturing resources, engineering is one of the leading branches of industry. During the period 1979-1985, it came second in both absolute and proportional terms (13.1%) in the country's capital spending; only the housing, communal and consumer services sector did better (20.2%). There was also a positive trend in the participation of engineering in capital formation. The share of machines and facilities of indigenous origin rose from 59.6% in 1979 to 62.5% in 1985. In the latter year, the machines and equipment produced by the Bulgarian engineering industry for capital formation projects represented 17.2% of its gross output.

The place of engineering in the national economy is also determined by its growing exports: whilst engineering accounted for 44.5% of total exports and 47.4% of industrial exports in 1979, the figures had risen to 54.4 and 57.6% respectively by 1985.

Electronics was the fastest developing of the branches of engineering. While mechanical engineering and metalworking industry production recorded an average annual growth rate of 6.45%, the electrical engineering and electronics industry achieved an average annual rise of 13.3%. There was also steady growth in the electrical engineering and electronics industry's share of total engineering output. The subsector's export capacity is also growing. Thus, in 1979-1985, the exports of the electrical engineering and electronics industry increased by a factor of almost 2.5, while the exports of the mechanical engineering and metalworking industry increased by a factor of a little over 2 and those of engineering as a whole by a factor of 2.2.

During the period 1979-1985, there was also a rapid development of non-electrical engineering with a view to modernizing the material and equipment base of engineering and of other sectors of the economy by the supply of, amongst others, high-output automated machines, automated production lines and systems.

Chemical engineering grew particularly fast. Bulgaria produces a variety of machines and equipment and growth in the output of these is exceptionally rapid. Taking 1970 as a base, gross output had risen by a factor of 14 by 1980 and by a factor of 31 by 1985. The supply to the chemical industry of indigenous modern, complete process lines and plants and individual apparatus suitable for the use of new raw material, like the application of processes involving higher than usual parameters (temperature, pressure etc.), is of great importance for Bulgaria. In the first place, this supply was sufficient to meet a significant proportion of home demand, with the result that conditions were created in the chemical industry for the development of raw material into high value-added products and for growth in the plastics, microbiology and pharmaceutical sectors and in household and speciality chemicals etc. In addition, Bulgaria was able to export plants and complete installations for the manufacture of sulphuric acid, soda ash, calcium carbide, formalin etc. The latest technology was introduced for the manufacture of air-cooled heat exchangers as individual items and in complete installations.

There was also rapid growth in the manufacture of machinery and equipment for the metallurgical industry. During the period 1970-1985, gross output of such products rose by a factor of 16. Manufacture began of assemblies for continuous-casting plants, rolling-mills, machines for the blocking of the iron flow in blast furnaces, large-capacity crushers and grinders, etc.

Bulgaria has substantial experience and a long history of manufacturing metalcutting machine tools and forging and pressing equipment. Taking 1975 as a basis, Bulgarian output of such machines was higher by a factor of 3.7 in 1980 and by a factor of 5.7 in 1985. The salient feature of the manufacture of metalworking machines in the period 1979-1985 was the steady growth in the proportion of highly productive NC-machines tools.

Transport engineering in Bulgaria is developing in two main areas: hoisting and transport machinery for the mechanization of material-handling in works and warehouses (trucks, monorail hoists, warehouse equipment) and equipment for long-distance transport (motor vehicles, railway rolling-stock). During the period 1979-1980, transport engineering was of structurally decisive importance. As a result of the international specialization and co-production with CMEA countries and of the expansion and consolidation of co-operation with foreign firms, the manufacture of hoisting and transport machinery developed particularly fast: by a factor of 4.7 in the period 1970-1985.

Together with traditional equipment (battery trucks and electric monorail hoists), Bulgaria mastered and began the production of a number of new versions of higher quality products with improved technical and economic parameters (e.g. internal-combustion-engined trucks). In a progressive development, the manufacture of battery-driven trucks is declining each year in favour of that of motor trucks. However, the number of battery-driven trucks produced during the period under review remained high: 902,202. For the mechanization of building and loading and unloading operations, a total of 11,229 cranes was produced in 1979-1985. Manufacture commenced of high-capacity cranes with increased lift, and of overhead travelling cranes (single- and double-beam) with lifting capacities of from 5 to 250 tonnes, and of gantry cranes etc. During the period 1979-1985, the share of hoisting and transport machinery in total exports of industrial machines and equipment averaged 19.9%.

The other main branch of transport engineering is the motor vehicle industry. Thanks to extensive co-operation with the USSR and Czechoslovakia, Bulgaria is developing its own manufacture of commercial vehicles. Bus building is also carried out in co-operation with Czechoslovakia, Hungary and Romania. There was a moderate rise in the production of buses during the period under review: the total number of units produced was 15,934, and the figure for 1985 was 1.4 times that for 1979. Bulgaria also makes a wide range of vehicle trailers (drop-sided and self-emptying) for the building industry and agriculture, as well as semi-trailers with three different carrying capacities. Output of trailers grew moderately during the period to total 63,701 units; in 1985, it was 1.3 times higher than in 1979. During the period 1970-1985, the gross output of the motor vehicle industry increased fivefold.

For its part, the gross output of the shipbuilding industry increased by a factor of 2.3 between 1970 and 1985. During the period 1979-1985, Bulgaria built multi-purpose and specialized vessels. It introduced new, leading-edge technologies and began to use an automated system for the designing of ships and their equipment.

Manufacture of food-processing machinery grew moderately: taking 1970 as a base, output had increased by a factor of 2.6 by 1980 and by a factor of 3.7 by 1985. Growth was, however, faster in some areas of this activity than in others. Thus, during the period 1980-1985, manufacture of complete installations for the making of bread and bread products increased by a factor of 6, while the manufacture of complete installations and individual machines for tobacco processing centres rose by a factor of 5. The manufacture of biotechnology equipment increased 14% in a single year (1984-1985) - such equipment is of great importance for the automation of processes in the food industry.

The main tasks facing the agricultural machinery industry during the period under review were to re-equip agriculture and gradually to transfer it to an industrial footing and also to produce up-to-date, competitive machinery. In close co-operation with CMEA countries, Bulgaria has developed, amongst others, high-efficiency grain combines and attachments to them for the harvesting of sunflowers, soybeans and maize, as well as rotary cultivators, machines for wheat-growing and irrigation machines and equipment. To meet the needs of its own agriculture and for export, Bulgaria produced during the period 1979-1985 a total of 42,555 wheeled or crawler tractors, 225,714 self-propelled silage combines, 11,334 ploughs, 5,851 cultivators, 3,036 centrifugal pneumatic fertilizer distributors, 4,556 tractor spray units etc.

A particular feature of the development of electrical engineering in the period 1980-1985 was the switch to the manufacture of complete installations and systems for the generation, transformation, transmission and distribution of electricity for the supply of industrial facilities, agriculture etc., for the domestic market and for export. The electrical engineering industry enriched its product range. It commenced manufacture of: extensive series of complete electric drives for NC-machine tools and robots and of high-torque, thyristor-controlled motors for NC-machine tools;

electrical and pneumatic instruments etc. Total production included: 16,286 electric generators; 9,133,800 electric motors; 50,204 power transformers; and 1,895 lifts. Production of household appliances included 2,866 heating furnaces; 1,187,000 cookers; 736,000 washing-machines; 672,000 electric boilers.

Hence, during the period under review, engineering confirmed its position as a leading sector of the Bulgarian economy. This is borne out by the increase in the sector's share of the country's total production volume and the consolidation of engineering's role as the principal capital-forming sector, a branch of decisive importance for the structure of industry in general and an efficient contributor to the country's international trade.

V.5. CZECHOSLOVAKIA

In Czechoslovakia, the engineering industries form the base of industrial production and have considerable impact on the national economy as a whole through their deliveries of machinery and equipment to all other branches.[V.9]

During the years 1979-1986, the Czechoslovak engineering industries registered a more dynamic development than other industrial branches. Thanks to its growth rate (over 28%), it was the leading sector in terms of share of total industrial output.

One of the main factors contributing to the strong growth in engineering was the increase in labour productivity. It was underpinned by a rise in the capital equipment/worker ratio that was achieved in large measure through the introduction of highly-automated production equipment. There was widespread introduction of industrial robots, manipulators, flexible manufacturing systems and integrated production lines. The quantity of modern capital equipment rose by 36.8% and the capital equipment/worker ratio improved by one third.

Engineering output was adapted to the needs of domestic and foreign customers by changes in the selection of goods produced and improvements in their technical and economic parameters and quality. Robots and manipulators, electronic components and assemblies, automated process-control systems and hydraulic components and assemblies all accounted for substantially increased shares of total output. Manufacture of equipment for nuclear and thermal power stations stabilized, as did that of special equipment for the chemical and pharmaceutical industry. Building commenced of a new generation of heavy trucks with a wide range of special equipment, as well as of buses, construction machinery, agricultural machinery and tractors. The makers of equipment for the textile, printing and leather industries responded to their customers' needs by multiplying the automatic functions and improving other parameters of their machines.

Structural changes have been taking place in the engineering industries aimed at intensive development of the national economy. These structural changes have been linked to co-operation within the framework of the socialist economic integration of the CMEA member countries. Czechoslovakia takes an active part in the comprehensive programme of scientific and technical progress of the CMEA member countries in implementing 45 projects in the field of industrial automation and application of microelectronics.

So-called "state target programmes" related to engineering production are influential in ensuring the required structural changes. Within the framework of these target programmes, production of particular engineering industry product groups grew more rapidly in the period 1980-1985 in comparison with other industrial branches. For example, production of industrial robots and manipulators increased 13 times, electronic components and parts by 96%, hydraulic elements by more than 50% and systems for automated process-control by nearly 84%. State target programmes for 1986-1990 for the engineering industries include items 04 "Hydraulic elements and aggregates", 05 "Robotization of technological processes", 06 "Efficient semiconductor converters and devices" and 07 "Development of material and technical bases for electronization of the national economy". In 1986, goods worth 13,070.70 million koruny were produced under these target programmes.[V.10]

Czechoslovakia is among those industrially-developed countries that have managed to keep the volume of their exports on an upward path. The proportion of machinery and equipment in the overall volume of exports has risen to over 55%, i.e., to a level comparable with that attained in the world's most industrially advanced countries. Imports of machinery and equipment account for roughly one third of Czechoslovakia's total imports.[V.11]

Czechoslovak engineering exports continued to go primarily to socialist countries. Such countries now take some 85% of these exports, the leader being the USSR, which accounts for over 60%.

Czechoslovakia and the CMEA countries are intensifying their system of specialization and co-operation in the manufacture of machinery and equipment. Trade in specialized products was three times higher in 1985 than in 1980. In 1985, products manufactured under specialization arrangements accounted for 27% of the exports and almost 24% of the imports of machinery and equipment.

Some 15% of Czechoslovak engineering exports go to market economies, including developing countries. Czechoslovakia is endeavouring to create a new basis and new forms for its trade in engineering products with developing countries.

Centrally planned economies predominate in Czechoslovak imports of machinery and equipment. Approximately 20% of such imports come from industrially developed market economies.

The trend in engineering production price indices since 1979 reflects the changes in the prices of raw materials, energy and other materials. These changes have also influenced the trend in export price indices.

Historically, Czechoslovak engineering production, exports and imports have borne the stamp of mechanical engineering; this imprint has become still more marked during the 1980s. Transport equipment and non-electrical machinery together account for 85% of Czechoslovakia's engineering exports, which have traditionally included: agricultural machinery and tractors; goods vehicles and cars; metalcutting machine tools and machine tools for mechanical working; electric locomotives; and machinery and equipment for the textile, leather, printing, building, metallurgical, chemical and food industries.

A similar situation obtains with regard to Czechoslovak imports of machinery and equipment, of which non-electrical goods and transport equipment now account for a much increased and a much reduced proportion, respectively, and electrical engineering products represent only a small fraction.

V.6. GERMAN DEMOCRATIC REPUBLIC

The German Democratic Republic's mechanical engineering industries [V.12] comprise the following branches of industry: heavy mechanical engineering; construction of machine tools and processing machinery; general manufacturing of machines; agricultural machinery; and transport equipment. A total of 1,193 enterprises belong to the sector of mechanical engineering. In 1985 they provided a total industrial output amounting to 99.2 billion mark, i.e. 18.9% of the industrial output as a whole. The index of gross output in the field of mechanical engineering increased from 100 in 1980 to 126 in 1985 (see table V.7).

Besides electrical engineering and electronics, mechanical engineering exerts an essential influence on the structure of the German Democratic Republic's industry. Its task consists in paving the way for a broader implementation of new key technologies. In this way mechanical engineering influences to a high degree the qualitative changes in the material and technical basis of the national economy, and, above all, the development and the overall application of such key technologies as flexible manufacturing systems, technologies and equipment for production and application of new materials, equipment for biotechnology and plant for acquiring and preparing raw materials, as well as installations for generating energy, recycling used materials and protecting the environment. These tasks can only be fulfilled by the rapid development of manufacturing technologies through the comprehensive introduction of microelectronics, new materials and automated production processes and information technology in general.

In this respect, it has become necessary also to develop other branches of the national economy determining scientific and technological progress even more significantly than mechanical engineering itself. As a result, the share of mechanical engineering in total industrial production decreased from 22.8% in 1981 to 18.9% in 1985 (see table V.8), although the gross output of mechanical engineering increased steadily during the same period (compare tables V.7 and V.8).

Despite the comprehensive rationalization measures adopted in mechanical engineering and in the German Democratic Republic's economy as a whole, there has been an increase in the number of workers employed. In 1981, there were about 963,500 people working in mechanical engineering; by 1985 this number had risen to more than 973,000. These figures illustrate the role of mechanical engineering as a key industry (see table V.9).

At present, rapid progress is being made in training methods and programmes and further education of employees in this sector. The varying educational aims are determined by the increasing automation of the entire production process based on a wide application of microelectronics and information technology at enterprise level. The relevant national authorities are now endeavouring to develop training and education schemes for engineers and economists according to scientific and technological needs, as expected by 1990.

The ongoing technological change has an influence on mechanical engineering itself as well as, through its products, on other branches. In this way, e.g., the intensive application of microelectronics, both in mechanical engineering technologies and in engineering products, causes a considerable increase in the quality and reliability of devices and equipment, increased labour savings, decreasing material and energy consumption and production costs in general in all branches of the national economy. At the same time, there is a trend towards the automation and increasing flexibility of manufacturing processes utilizing new or rarely applied production methods, such as surface finishing of materials, beam technology (laser-, ion- and electron-beam technology), ultra-precision technology and others, as well as the necessary information processing throughout the factory (e.g. computer integrated manufacturing). The production of machine tools and textile and printing machinery is an essential line of the German Democratic Republic's mechanical engineering. During the period 1986-1990, the production of machine tools will be increased about 1.5 times and the manufacture of textile and printing machines about 1.4 times. The development of products and machine systems for both agriculture and the food industry is another goal in the country's mechanical engineering.

The role of supplier to other industrial sectors is now and will in future continue to be another substantial structural aspect of mechanical engineering. Its products, such as gears and couplings, pumps and compressors, cast products, armatures, hydraulic and pneumatic units, tools, antifriction bearings and standardized parts, are needed by the manufacturers of final products.

In particular, the development and manufacture of equipment and plant for obtaining, processing and finishing energy carriers, raw materials and waste products, power-station equipment, air-conditioning, dust-removing and water-purification plant, and products for the building industry and transport will have an increasing effect on the technology provided by mechanical engineering and on its products. Medical and laboratory technology will become more and more important, as well as manufacturing capacities for biotechnology devices and equipment, for both research and production.

The German Democratic Republic's mechanical engineering industry has the goal of producing 40-50% of its total production volume in the form of comprehensive, integrated manufacturing systems by 1990. Development trends of selected products in mechanical engineering between 1981 and 1985 are indicated in table V.10.

The special importance of mechanical engineering exports becomes obvious when the share of goods production of mechanical engineering is compared with its share in the export of industry as a whole. Although in 1985 the share of mechanical engineering amounted to only 18.9% of total industrial production, it reached a 46.6% share in total exports. Its share in total imports, however, was only 26.8% in the same period. The shares of mechanical engineering export and import in total industry between 1981 and 1985 are shown in tables V.11 and V.12. In future, too, mechanical engineering will increasingly govern the growth rate of export.

International specialization and co-operation in mechanical engineering is based on agreements between CMEA member countries, with special emphasis on the co-ordinated development and manufacture of new products, machines, equipment and systems having a high degree of finishing. The interrelations between the partners in the division of labour will adapt more and more to new processes in science, technology and manufacturing, and will call for flexible response to desires and changes in demand.

V.7. HUNGARY

In 1985, the output of the Hungarian [V.13] engineering industries recorded an increase in terms of volume of 18.4% over 1980 (and of 3.7% over the preceding year). A total of 24.7% of the 1985 gross output value of industry was accounted for by the engineering industries and 32% of those employed in industry worked in the engineering sector. In 1980, 23.0% and in 1985, 23.9% of total industrial goods sales was accounted for by the engineering industries and even higher shares were achieved in engineering exports (41.5% and 43.8%, respectively) (see table V.13).

The branch pattern of the engineering industries reflected the changes in the production dynamics of individual subsectors. The largest growth shares were registered in the output of telecommunication equipment and vacuum engineering, while a declining share was recorded in the metal products industry. The development of the manufacturing branches of electronic engineering, agricultural and food processing machinery, road vehicles and subassemblies was above average (see table V.14).

Table V.15 illustrates the development in employment by individual subsectors (branches) of the engineering industries (1984-1985).

V.8. ITALY

Italian industry, which has consolidated its position in recent years, is in fact based on the country's engineering industries.[V.14] There have been three different chronological phases to this consolidation:

- 1974-1978, a period distinguished by a readjustment phase;
- 1979-1981, a period of expansion; and
- 1981 onward, a phase described as selective evolution.

Readjustment

The readjustment phase, which followed the first oil crisis, in 1974, marked the end of a positive trend. The steep rise in oil prices had a serious effect on many Italian industries at grips with, *inter alia*, considerable problems over productivity and labour costs that were too high in relation to the international competition. This readjustment phase was not one of great technological advances, because investment had been appreciably reduced. In order to overcome their difficulties, enterprises applied themselves primarily to reducing their indebtedness.

Expansion

The development phase that began towards the end of the 1970s was noteworthy for a policy of heavy decentralization applied to the production of the major industries. This policy proved to be effective for two main reasons:

(i) In the first place, the switching of production quotas from the end-producers to the intermediate suppliers made production capacity more flexible, with a higher potential output, and more capable of supplying a range of products that were improved as regards quality and functional innovation; and

(ii) This trend enabled Italian industry to make the transition from a situation of almost total resistance to progress, and products that were not sufficiently diversified to be competitive, to a structure typified by a larger range of competitive products, resulting in greater profitability for enterprises and a very wide range of choice for the consumer by price and by quality.

Selective evolution

The third phase, referred to as selective evolution, began towards the end of 1980, and was typified, at least initially, by profound change in the process of development. This was a period when the role of small enterprises, which had been of key importance in the expansion phase, underwent decisive change, above all in relation to investment. In addition, the relationship between large and small enterprises became far more advantageous for the former, having regard also to other indicators such as turnover, the value added for each hour of labour (seen as a criterion for the assessment of labour productivity) and the return on capital.

This selective evolution phase even witnessed a resumption of investment, which was no longer connected with the increasing of production capacity; this put an end to the previously sacrosanct relationship between investment and production technology. The result was a slowing down in the rate at which production workers were laid off, the decline being offset by the replacement of unskilled labour by specialist labour.

It follows that all technological and organizational changes with a positive effect on both the process and the product may be regarded as basic to the development of Italian industry during the first half of the 1980s.

The positive results achieved by the Italian economy during the past few years, thanks, *inter alia*, to the favourable international situation following the fall in the price of oil and other raw materials, could never have happened without the weighty contribution of the engineering industries. This by no means uniform phenomenon did not affect all subsectors of the electrical and electronics industries. The differences in the development of the key sectors of the engineering industries in 1984, relative to 1980, may be seen with clarity from table V.16, drawn up by the Bank of Italy on the basis of ISTAT data.

Whereas the metal industry and the agricultural and industrial machinery industry did not succeed in equalling their 1980 results in 1984, and have continued with their downward trend, the situation is completely different for the industries producing transport and electrical equipment, and above all those producing office equipment and precision instruments. In the context of Italian industrial organization, the growth of production in these three subsectors offsets the somewhat disappointing results recorded by the other subsectors. Even so, it must be pointed out here that the performances of some subsectors at the start of the 1980s are difficult to repeat, having regard to the changes in the international economic situation; consequently, the protection of some production quotas may be justifiably regarded as the main objective of Italian industry, which has to face extremely aggressive competition on the international level, more especially from the newly industrialized countries (NIC).

V.9. POLAND

Overall economic developments

In 1983-1985, a three-year plan was implemented in Poland with the aim of surmounting the economic slow-down of 1979-1982, whose combined effects had reduced the amount of national income generated by almost one quarter. [V.15.]

Positive results were achieved in 1983-1985. The decline in national income dating back to 1982 was not only halted but reversed. During 1983-1985, the amount of national income generated increased by 16%, industrial output rose by 16.3% and gross agricultural production by 9%. In industry, despite the fact that the value of output in 1985 was still lower than in 1978, the output of certain industrial goods already exceeded that level.

The three-year plan was also fulfilled in terms of the rate of export growth, and was overfulfilled by almost 5% as regards exports to the centrally planned economies. The planned increase in exports of manufactured goods, however, was not attained. Despite the undoubted achievements of the three-year plan, per capita national income generated in 1985 was 15% lower than in 1978, industrial output was 6% lower and agricultural production 8% lower.

Engineering industries - the key sector for technical progress

The way for the Polish economy to overcome the slow-down of the early 1980s is through development in the context of profound structural change. That requirement was determined during the past three years and will in the near future define the direction of development in the engineering industries which, because of the wide application of the goods produced by those industries, is a pre-condition for technical progress and structural change in the national economy as a whole.

At the present stage of development, the main objectives of the engineering industries are:

(i) To create the conditions for the growth of labour productivity in the economy as a whole so as to ensure economic development during stabilization and even periodic falls in employment. This calls for more rapid technical progress towards improving the efficient use of machinery and equipment and especially progress in the automation of industrial processes, as well as further development in the use of information technologies;

(ii) To reduce the specific consumption of energy and materials in production and operation, both in the manufacturing industry itself and amongst the end-users of engineering products;

(iii) To intensify agricultural production through better harmonization of the supply of modern means of production with the changing pattern of agriculture; and

(iv) To increase exports of processed goods in the face of limited supplies of imported raw and other materials.

The relative performance of the engineering industries in the economy as a whole

In 1983-1985, industrial output recovered to the level achieved before the slow-down of the early 1980s. During those years, the rate of growth of engineering industries output considerably outstripped the rate of growth of industrial output as a whole. While the average annual rate of growth of industrial output at fixed prices amounted to 5.2%, the engineering industries achieved output growth of 7.2%. As a result, engineering output in 1985 exceeded the 1980 level by 5.5%, making it possible to re-establish the long-term trend, halted in 1981-1982, towards a higher share of engineering industries output in industrial output as a whole. That share (at fixed prices) increased from 17.2% in 1970 to 23.2% in 1980 and 24.7% in 1985.

During 1983-1985, total exports of engineering products (in 1982 prices) increased on average by 15.8%, with exports to the centrally planned economies rising by 33.4% and those to other countries declining by 26.2%. The decline occurred mainly in 1985. The growth in exports of engineering products was smaller than the growth in exports as a whole and, as a result, their share in total Polish exports decreased in 1983-1985 from 43.8% to 39.4%.

Research and development, fixed capital formation and employment

After a sharp decline in 1979-1982, appropriations for the development of the engineering industries during the following three years were gradually increased. Capital investment in the engineering industries in 1985 was 35% greater (at fixed prices) than in 1982, although it represented only 54% of the 1978 level and 80% of the 1980 level. As a result, capital investment during the five-year period 1981-1985 amounted to 53% of investment for the previous five-year period (at fixed prices). The engineering industries accounted for roughly 21% of total capital investment in industry.

With a low rate of replacement, the gross value of capital stock in engineering industries increased in 1983-1985 by 9.1%, but the rate of renewal of fixed assets in 1981-1985 amounted on average to approximately 4%, as compared with 10% in the previous five-year period. The decline in the rate of renewal of machinery and equipment in the engineering industries accounts for the decline in the share of machinery and equipment in the fixed capital of engineering industries from 54.1% in 1980 to 45.2% in 1985.

Appropriations for R and D in the engineering industries represented approximately 60% of those in industry as a whole, a much larger share than in the case of capital investment. Approximately 75% of those expenditures went for the development and introduction of new machinery, equipment and products, and some 22% for the automation of production and for the development of modern production methods and technologies.

The engineering industries, like the economy as a whole, have experienced a significant decline in employment in the 1980s because of the retirement of a considerable proportion of the work-force and the departure of many women to raise children, as well as a smaller increase in the number of new workers due to the demographic situation. Employment in industry in 1981-1985 fell by almost 245,000, i.e., by more than 4.5%. Engineering industries accounted for 211,000, or 86% of that decline. Employment in this sector fell from 1,625,000 in 1980 to 1,488,000 in 1982 and 1,414,000 in 1985, a decline of 13% over the five-year period. Falling employment mostly affected manual workers (down by 173,000, or 15%, during the five-year period), and to a lesser extent other groups.

The structure of the engineering industries

The growth pattern and sectoral structure of the Polish engineering industries in 1982-1985 are illustrated in table V.17. [V.16]

In this respect, engineering industries may be broken down into three main groups:

(i) The rapidly developing electronics industry and production of precision instruments, with average annual rates of output growth over the three-year period of 14.2% and 9.4%, respectively;

(ii) Subsectors whose growth rates are close to the average for the engineering industries as a whole (7.1%) - machinery and equipment (6.9%) and the electrical equipment industry (8%); and

(iii) Subsectors whose share in total engineering industries output is decreasing: the metal products industry (average annual growth rate 5.1%) and the transport equipment industry (5.7%).

A positive trend which became apparent especially in 1985 has been the intensive development of instrument-making, in which output growth that year amounted to 13.2% as against an average of 7.7% in 1983-1984, including the manufacture of measuring apparatus and medical equipment, which had hitherto shown slow growth.

Technological changes and their impact on the engineering industries

In the second half of the 1980s, as part of the general drive towards improved cost-effectiveness, scientific and technical development work in the iron and steel and engineering industries is to be concentrated on such objectives as:

- Achieving progress in the rational use of energy, raw and other materials through the development of technologies and designs which are non-material-intensive in production and operation, the substitution or fuller utilization of raw materials, and the improved use of waste products and secondary raw materials;

- Improving the use of existing capital stock by encouraging modernization and the introduction of modern equipment, technologies and manufacturing arrangements;
- Increasing output through improved labour productivity which in turn is based on better organization of working time, the awarding of bonuses for the automation of production processes, and the introduction of modern machinery and equipment;
- Developing the engineering industries and the use of electronics as the basis for technical progress;
- Improving the quality of output and the efficiency, durability and reliability of products and installing more up-to-date machinery, equipment and apparatus;
- Producing new materials and products for export which have appropriate technical performance characteristics and comply with quality and aesthetic requirements;
- Producing new products for the domestic market which meet high specifications of efficiency, quality and design;
- Improving labour organization and production processes, with particular emphasis on the rational use of working time and of skilled, and in particular, highly skilled staff; and
- Expanding the technological base and increasing the application of information technology.

In view of the constraints in the area of employment, great attention is being given to the automation and robotization of production processes. Research and development work in this area will concentrate mainly on the following technologies: metalcutting and metal forming, welding and pressure-welding, mould assembly and inspection, heat treatment, corrosion-resistant coatings and electroplating, metal casting, and technologies eliminating work which is heavy, monotonous or hazardous to health.

The electronics industry

The main objectives in the development of electronics are:

- To improve the technology for manufacturing semiconductor components for VLSI microcircuits;
- To develop the technology and production facilities for electronic components suitable for surface wiring;
- To develop production facilities for new electronics materials; and
- To expand the production of manufacturing equipment, especially automated assembly lines for integrated circuits.

Systematic production growth as a result of the introduction of new technologies will make it possible to expand the application of electronics in the economy, better supply the needs of the domestic market and increase exports.

The main application areas for electronics are:

- High-performance household electronic equipment using LSI microcircuits and new assembly technologies;
- Radar and radio-navigational equipment for maritime use, communications, and air and land transport;
- New types of computers, including the R-47 computer, minicomputers with 32-bit processors and microcomputers for various purposes; peripheral equipment such as thermal and laser printers, graphic colour displays, Winchester discs, plotters, etc.; and
- Equipment for industrial automation such as miniaturized pressure cells, thin-layer temperature-sensitive elements and the new generation of electrical servomotors.

Structural developments in the metal products industry

In the metal products industry, the extent of structural changes was limited by the moderate rates of growth in the sector as a whole (16% at fixed prices in 1983-1985). The pattern of growth within the subsector in 1983-1985 was as follows (at fixed prices):

Castings	6%
Metal structures	7%
Metal tools	25%
Metal goods for industrial use	24%
Metal goods for household use	26%

The manufacture of new types of special rolling-contact bearings started in small batches and production of barrel-shaped roller bearings began under licence from the Koyo Company (Japan). Mass production of taps and dies with an inclined angle of attack, bimetallic bandsaws and variable-inductance digital pick-ups was introduced in the tool-making industry.

In the area of consumer goods, production began of components for mechanical and automated household appliances, and a number of products were improved using electronic components and parts.

Structural developments in the manufacture of non-electrical machinery

With the fairly high growth in the manufacture of non-electrical machinery, amounting to 18% at fixed prices in 1983-1985, there was considerable differentiation in the pace of growth in various subsectors. The pattern of growth (at fixed prices) within the subsector was as follows during the period under review:

Power equipment	25%
Mining machinery	24%
Metallurgical and foundry equipment	-6%
Machine tools and machining equipment	37%
Chemical industry machinery	-14%
Mineral industry machinery	58%
Light industry machinery	38%
Food-processing machinery	-9%
Construction machinery	19%
Agricultural machinery	8%
Loading and conveying equipment	27%

The main structural developments in terms of the product range were towards an increased share of high-efficiency automation and electronic goods. In the machine-tool and construction machinery industries, the quality and functioning of machinery were improved through the introduction into production of modern machinery components, such as clutch and braking systems, electronic control systems and safety devices.

In the machine-tool industry, the pattern of output of metalcutting machine tools showed a trend towards a higher share of NC-machine tools, automatic lathes, heavy and super-heavy machine tools equipped with a wide range of special accessories, as well as specialized machine tools and fully automated machining lines. In the area of machine tools for mechanical working, a decline was recorded in the production of general-purpose crank-operated and friction-drive presses, while the output of highly efficient rolling mills, automatic presses, drop-forging hammers and forging presses grew. A number of NC-machine tools and automated manufacturing systems for machining and precision stamping were brought into production. [V.17]

Structural developments in the electrical and electronics industries

In the electrical and electronics industries in 1983-1985, the pattern of growth in the electronics sectors (49% at fixed prices) was higher than for the electrical equipment industry (26%). As a result, the share of electronics in the output of the electrical engineering industry rose from 9.1% in 1982 to 11.1% in 1985.

For individual groups of products, the pattern of production growth (at fixed prices) in 1983-1985 was as follows:

Electric power equipment	13%
Cables and conductors	39%
Other electrotechnical equipment	29%
Professional and household electronic and telecommunication equipment	49%

The most rapid development in 1983-1985 occurred in the manufacture of electronic components and household electronic equipment, whose share in the output of the industry under review rose as a whole by 5% over the three years.

One of the main developments in this area was the use of thyristors in electrical equipment, which involved expanding the manufacture and application of electric-powered devices containing electronic components to control and process the flow of electric energy. Devices of this kind introduced in Poland before 1985 made it possible to reduce the total annual consumption of electric power by 1.5%.

In the area of electronics, the following products were developed and brought into mass production:

(i) Materials for electronic components

- Single-crystal silicon;
- LSI and VLSI microcircuits, including microprocessor systems;
- Optoelectronic components for fibre-optic technology; and
- Miniaturized passive elements with enhanced technical performance.

(ii) Household electronic equipment

- Low-energy colour television sets; and
- High-quality stereophonic radio receivers and electrical sound systems.

(iii) Electronic communication equipment

- Various telephone exchange systems;
- Telegraph and telephone exchanges for rural areas; and
- A multiplex telephone communication system.

Structural developments in the manufacture of transport equipment

The transport equipment industry was, together with the metal products industry and non-electrical machinery the slowest engineering sector to develop. Only in two areas - automobiles and aircraft - was production growth slightly higher than the average for the engineering industries.

The following rates of production growth (at fixed prices) were observed in 1983-1985:

Rolling stock	21%
Automobiles	24%
Tractors	21%
Bicycles and motorcycles	6%
River and sea transport vessels	19%
Aircraft	23%

As regards the technical development of the motor vehicle industry, work was carried out to meet the deficiencies in the product range and improve performance characteristics, in particular by reducing fuel consumption. The following new designs were developed:

- Light diesel-powered delivery vehicles with a load capacity of 1.5 tonnes and a mass of 3.5 tonnes;

- Cross-country and four-wheel-drive vehicles to replace obsolete designs;
- Vehicles with a load capacity of 3.5 tonnes to fill the gap between light delivery vehicles and vehicles with medium load capacity;
- Trolleybuses; and
- Large-capacity buses, in co-operation with the Hungarian company Ikarus.

All models of vehicles now in production are gradually being modernized and new versions are being developed, including:

- Small cars - so-called "face lifting";
- Polonez passenger cars;
- Lorries of medium and large carrying capacity (with a partly removable cab); and
- Buses with medium and large seating capacity.

Manufacture of a new range of C 90-type diesel engines began, and engines for medium-capacity and large-capacity lorries were modernized. New technologies were used in the manufacture of tractors under licence. Production processes were automated and flexible manufacturing systems introduced.

In the area of railway rolling stock, design and research was aimed chiefly at meeting the requirements of the main customer, the Polish State Railways, and also at bringing products into line with the requirements of the International Union of Railways. In the field of production technologies, work focused on ways of reducing material and energy inputs, as well as on improving product durability and quality.

Structural developments in the manufacture of automation means and precision instruments

The manufacture of automation means and precision instruments, together with electronics, was the subsector which showed the most dynamic growth. All the areas of this industry expanded rapidly, and the largest growth was achieved by information technology and office automation facilities, whose share in the output of the engineering industries rose over three years by more than 3%.

The pattern of production growth for the subsectors (at fixed prices) in 1983-1985 was as follows:

Automation means	21%
Information processing	37%
Office automation	84%
Measuring apparatus	27%
Optical instruments	29%
Medical equipment	18%

The technical development of electronic computing equipment in 1981-1985 was concentrated on four basic groups of products:

- Medium-size computers, minicomputers and microcomputers;
- Line and matrix printers;
- Disc, tape and cartridge storages; and
- Terminal stations.

Computing capacity was increased and the functions and configurations of computers expanded to create more powerful computer systems. Several types of microcomputers were brought into production and the output of computers and minicomputers increased. Production started of several types of peripheral equipment, including displays and printers using microprocessors.

Developments in computer technology made it possible to build systems such as:

- An automated system for designing and testing microprocessors;
- A modular terminal system for designing and building flexible manufacturing configurations; and
- A microcomputer-based system for calibrating and checking electricity meters.

Technical advances in automation facilities and measuring apparatus consisted mainly in developing systems and recent applications of automation using microprocessor control, and in eliminating the electromechanical and mechanical components used hitherto. Production of a third-generation analogue control system began, and a pneumatic control system was updated for use with computer systems. A standard range of new controllers, electronic recorders and digital panel indicators was introduced. A pollution measuring and warning system was developed. The utilization of microelectronics in electronic measuring and controlling instruments was increased.

V.10. SWEDEN

The engineering industries traditionally constitute the core of Swedish industry. According to the ECE data bank, in the period under review, their share in total industry represented 32-35% in terms of gross output, 37-39% in terms of value added and 42-43% of employment in total industry. [V.18]

Figure V.1.

World market share in metalworking and mechanical engineering of Sweden, 1980-1986

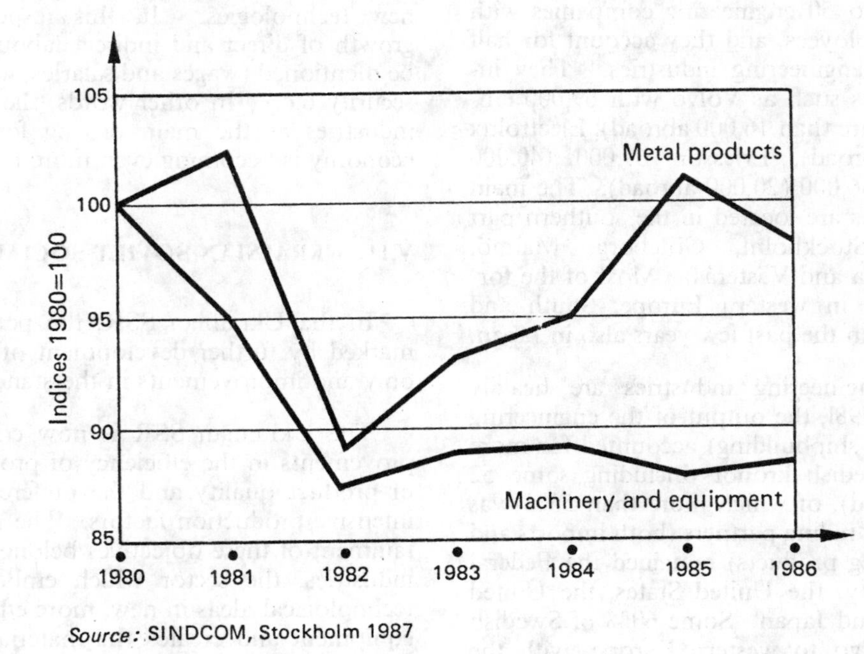

Source: SINDCOM, Stockholm 1987

Figure V.2.

World market share in knowledge-intensive engineering products of Sweden, 1980-1986

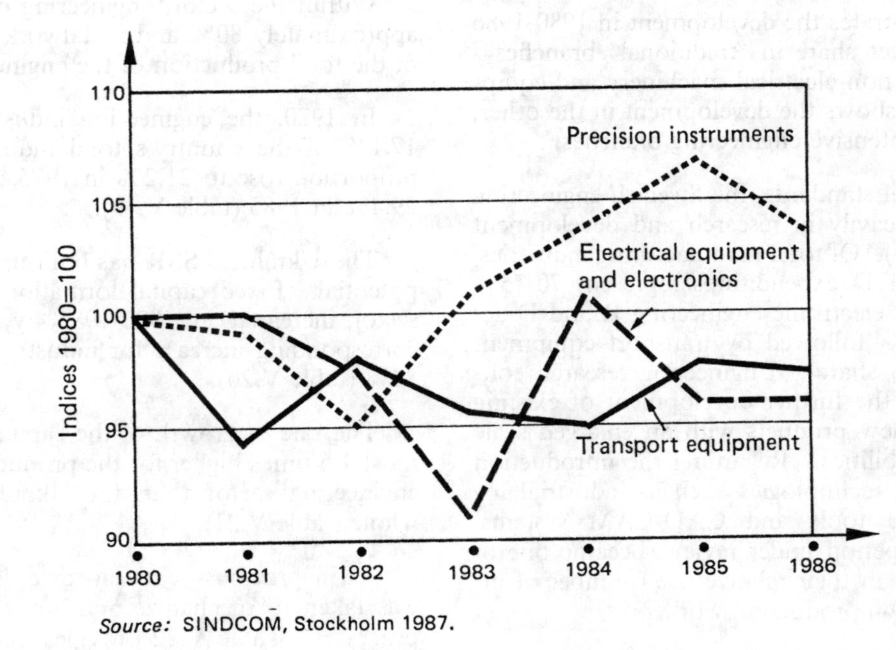

Source: SINDCOM, Stockholm 1987.

Whilst in the past decade the total number of employees in industry has decreased, the number of people working in engineering industries has grown. Approximately one tenth of the total working population (some 4.2 million in 1982), i.e. some 400,000-450,000 people - almost half the employees in manufacturing industries - worked for an engineering enterprise. This included more than 80,000 women.

There are close to 50 engineering companies with more than 1,000 employees, and they account for half the employment in engineering industries. They include large companies such as Volvo with 69,000 employees (of which more than 16,000 abroad), Electrolux 102,000 (62,000 abroad), Ericsson 69,000 (40,000 abroad) and ASEA 56,000 (20,000 abroad). The main manufacturing centres are located in the southern part of the country (Stockholm, Göteborg, Malmö, Linköping, Eksilstuna and Västerås). Most of the foreign subsidiaries are in western Europe, South and North America, and in the past few years also in Japan.

The Swedish engineering industries are heavily export-oriented. In 1981, the output of the engineering industries (excluding shipbuilding) accounted for more than 120 billion Swedish kronor (including some 52 billion of value added), of which more than 53% was exported. The main trading partners (both imports and exports of engineering products) remained the Federal Republic of Germany, the United States, the United Kingdom, Norway and Japan. Some 60% of Swedish engineering exports go to western Europe, with the Nordic market accounting for approximately 20%. More than two thirds of exports are realized in traditional fields of mechanical engineering and transport equipment, with an increasing share being assumed by innovated machinery and equipment (industrial robots, assembly lines etc.). The share of engineering products in total Swedish exports accounted for more than 45%.

Figure V.1 illustrates the development in 1980-1986 of the world market share in "traditional" branches - metalworking and non-electrical machinery and equipment. Figure V.2 shows the development in the other, more knowledge-intensive engineering branches.

By international standards, the Swedish engineering industries invest heavily in research and development activities (R and D). Of total manufacturing industries, engineering R and D expenditures represent 70-75% (electrical and electronic engineering R and D accounts for 25-30%, followed by transport equipment with a nearly 25% share). Engineering research concentrates both on the further development of existing products and on new products with an enlarged scale of functional capabilities. Regarding the introduction of new automated technologies such as industrial robots, NC-machine tools and CAD/CAM systems, Sweden, over the period under review, became one of the world's leaders in their relative use (number of installations per 10,000 production workers).

In parallel with its research, production and trade activities, Sweden devotes significant resources to the improvement of energy management (more than 60% of the energy consumed by the engineering industries is in the form of oil), the protection of the environment, including the work environment (replacement of workers in monotonous and heavy jobs), the introduction of new organizational and managerial structures (combining higher productivity with a sound environment and modern job design), as well as to further development of training and retraining schemes needed for the adaptation of workers and engineers to the climate of new technologies. In this respect, the continuing growth of direct and indirect labour costs should also be mentioned (wages and salaries, working hours, social security etc.). In other words, the role of engineering industries as the main driving force in the national economy is becoming even more pronounced.

V.11. UKRAINIAN SOVIET SOCIALIST REPUBLIC

In the Ukrainian SSR, the period 1979-1985 was marked by further development of the national economy and improvements in the standard of living . [V.19]

The Ukrainian SSR is now concentrating on improvements in the efficiency of production, the raising of product quality and the preferential application of intensive production factors. The main role in the attainment of these objectives belongs to the engineering industries, the sector which embodies scientific and technological ideas in new, more efficient machines and equipment and creates the material and technological base of production.

The engineering industries are one of the most advanced and important sectors of the Ukrainian economy. In the period under review, as in earlier periods, they developed significantly faster than other industrial sectors (table V.18).

Within the sector, engineering output accounts for approximately 80% and metalworking output for 20% of the total production of the engineering industries.

In 1970, the engineering industries accounted for 17.1% of the country's total industrial output. The proportion rose to 21.2% in 1975, 25.9% in 1980 and 29.4% in 1985 (table V.19).

The Ukrainian SSR has built up a large production potential. Fixed capital formation in the engineering sector increased in 1980-1985 by 44%, whereas the corresponding increase for industry as a whole was of 32% (table V.20).

The rate of growth of the capital/labour ratio is almost 1.5 times higher for the production workers in the engineering sector than for Ukrainian industry as a whole (table V.21).

As in previous years, an extensive set of measures was taken to mechanize and automate manufacturing processes. Table V.22 provides figures for the growth in Ukrainian industry as a whole and in the engineering sector in particular of fully mechanized and automated production lines.

As may be seen from table V.23, labour productivity has grown without interruption both in industry as a whole and in the engineering industries.

The structure of the engineering sector was improved. The production of instruments, machine tools, and electrical and chemical machinery grew rapidly. There were also improvements in the structure of machines, equipment and tools produced.

The rates of development of new types of machinery and equipment, apparatus, instruments and means of automation remained high. They even accelerated in the fields of electrical engineering and the manufacture of electrical apparatus and instruments, metalworking machines, chemical industry equipment, process-monitoring and -control devices and means of automation in general (tables V.24 and V.25).

The technical standard of the equipment produced rose. There was a reduction of almost 10% in the average age of equipment in service, which was, in 1985, 6.5 years.

Ukrainian industry, including the engineering sector, is systematically innovating its product range (table V.26).

Table V.27 shows the changes in output of the main types of engineering products for the period 1979-1985.

In 1985, total output of the Ukrainian engineering industries and total output of Ukrainian industry as a whole were, respectively, 107 and 103.2% of what they had been in 1984. Some 800 new types of machines, equipment, apparatus, instruments or means of automation were produced (see also tables V.24 and V.25).

New types of metalworking equipment with numerical control entered production. Series manufacture was initiated of shield tunnellers for the non-explosive drivage and the support of horizontal roadways through carbonaceous manganese ores. In addition, production commenced of bucket-wheel excavators with a capacity of 630-1,250 cubic metres per hour. Ukrainian industry also mastered manufacture of machines for intensive crop husbandry and of combined units for secondary tillage, application of fertilizers and sowing. Production of selected types of engineering products in 1985 is shown in table V.28.

V.12. UNION OF SOVIET SOCIALIST REPUBLICS

The USSR economy has a well developed and diversified engineering industry. [V.20] Table V.29 gives figures showing the proportion of total industrial output of various sectors of industry. It can be seen from the table that the engineering industries have played a major role in intensifying production: their share of total output rose by 3.1 percentage points over 1980-1985, and amounted in 1985 to 27.4%, so exceeding appreciably the corresponding figures for the building materials and chemical industry sectors (24.5%) and the fuel and power sector (11.1%).

Because of its special place and role in the national economy, engineering has grown rapidly. The figures given in table V.30 show that its output has increased faster than that of industry as a whole and faster than that of most other branches. The index of industrial output as a whole in 1986 was 126 (1980 = 100), while that of engineering output was 145.

In 1986, the Soviet economy underwent positive changes and its efficiency increased. The process of reform in Soviet society steadily gathered momentum. Such long-term factors of economic growth as the speeding-up of scientific and technical progress, the improvement of the structure of production, the restructuring of the investment process, the improvement of product quality and the economizing of resources began to make themselves more fully felt. [V.21]

The key to the intensification of production and the improvement of its technical level was scientific and technological progress. This was faster than before because of the radical transformation of engineering. Updating of the range of engineering products continued to develop: new products being manufactured in the USSR for the first time accounted for 4% of total engineering output in 1986, as against 3.1% in 1985. [V.22] Production began of: personal computers for the automation of administrative and management work, scientific research and computer-aided design, new types of flexible manufacturing units and cells; cryogenic equipment; machine tools for electro-physical and other methods of working; equipment for the manufacture of precision implements and of composite materials.

The changes in the range of industrial products were not, however, as fast as had been planned. The technical level and quality of many products remain low. The proportion of total output in the top-quality category is about 15%. [V.23] State quality control has a special role to play in the campaign to improve product quality; it has now been introduced at 1,500 enterprises making products of vital importance.

More industrial enterprises were equipped with modern means of mechanization and automation. Over 200 flexible manufacturing centres and systems for various types of work, 13,000 industrial robots and manipulators, and 11,000 mechanized transfer lines and automated and rotor production lines were installed. Seven thousand sections, workshops and processes were converted to fully mechanized or automated working, and 249 computer-aided design systems have been introduced. The use of advanced technology increased. However, the numbers of new machines, equipment and techniques are still too low. On the whole, the pace of scientific and technological development remains too slow.

The increasing use of technology in industrial production has led to annual savings of over 4 billion roubles in production costs and to manpower savings equivalent to over 700,000 jobs. [V.24]

The links between science and industry have been strengthened. For industry as a whole the number of research and production complexes in industry increased by 100 compared with 1985; 95% of this increase came in the engineering industries, where there were substantial rises in most branches. On an overall basis, the engineering industries account for roughly half the combined research and production complexes in industry. [V.25] Intersectoral scientific and technical groupings have been set up to develop new generations of equipment, technologies and materials. The targets set by the CMEA countries' Comprehensive Programme for Scientific and Technological Progress have been met. In 1986, 15 basic scientific discoveries were registered and 3,500 new designs of machinery, equipment, apparatus, instruments and automation devices were developed. However, many of the research and development results will be implemented over a long period. A significant proportion of the new designs will only be used by industry in three or more years' time.

The pace of scientific and technological progress in the economy is largely dependent on the development of the engineering industry. In 1986, engineering growth was 1.3 times higher than the growth of industry as a whole. The production of equipment playing a key role in speeding up technological progress grew significantly. The fastest growth rates (see table V.31) were in the production of numerically controlled machine tools (23% growth) and industrial robots (14%). There were also increases in the production of machining centres, carousel- and carousel-and-conveyor-equipped production lines, flexible manufacturing cells and various other types of high-productivity equipment. Rapid growth in the production of modern machines and equipment accounted for over a third of the increase in the volume of engineering output.

The engineering industries, which form the material basis for the acceleration of scientific and technological progress and the technological revamping of the economy, should account for two thirds of the increase in output of heavy industry over the period 1986-1990. In this connection, priority is to be given to speeding up the development of the machine-tool industry, computer technology, precision instrument manufacture and electronic engineering. The growth in output from these branches should exceed by 33-60% the average for engineering as a whole. [V.26]

The main objective of the economic policy at this stage is to improve the quality of products and intensify and speed up production. For this, the prime requisite is growth in labour productivity. Labour productivity is growing faster in engineering than in any other sector of the economy and than in industry as a whole. In 1986, the index of labour-productivity growth compared with 1985 was 104.6% for industry as a whole, and 106.9% - significantly higher than for the other sectors of industry - for engineering [V.27] This higher level is explained by the use of modern technology and more efficient equipment and by the increase in the capital-to-labour ratio. The growth in labour productivity in 1981-1986 yielded significant increases in industrial output (table V.32).

In 1985, fixed capital in Soviet industry was valued at 766 billion roubles. [V.28] The proportion of the total attributable to engineering, which grew throughout the period under review, amounted in 1985 to 24.7%, or 189.2 billion roubles, with buildings and installations (the passive constituents of fixed capital) accounting for 42.6% of this sum and machinery and equipment (the active constituent) the remaining 57.4% (see table V.33).

The total capital investment in Soviet industry under the eleventh five-year (1981-1985) plan amounted to 300.7 billion roubles. The share for engineering and metalworking was 24.3%, or 73 billion roubles, which exceeded the amounts devoted to light industry, the food industry and chemicals and petrochemicals by factors of 6.6, 4.0 and 3.2, respectively. [V.29] In 1986, capital investment in engineering industries rose by 14% - i.e., by far more than investment in the economy as a whole. [V.30] During the twelfth five-year plan (1986-1990), the development of this branch will be allocated a larger amount of investment than ever before, nearly twice as much as during the eleventh five-year plan. The money will be used primarily to re-equip or rebuild plants.

V.13. UNITED KINGDOM

The United Kingdom is almost unique amongst industrialized ECE countries in that, over the period covered by this study, the percentage contributions of various sectors of industry to GDP have shown considerable changes, as can be seen from table V.34. [V.31]

It will be seen that the discovery and subsequent extraction of oil in the North Sea resulted in an increase of from less than 1% in 1976 to some 6% in 1985 in the contribution of the oil and gas sector to GDP. The major counterpart to this was the reduction from 29 to 24% in the contribution of the manufacturing industries. In 1986, the contribution of oil and natural gas to GDP fell back to 3%. The other major change was the growth of the contribution to GDP of the services sector.

Within manufacturing, the contribution of the engineering industries declined, as can be seen from table V.35.

Between 1978 and 1986, GDP increased by 14%, but the increase in production was only 8%; within production, output of the energy and water sectors increased by some 40%, but engineering output was 7% lower in 1986 than in 1978.

There have been considerable differences in the experience of the various sectors within engineering. This is shown in table V.36, which gives the index of production (1980 = 100) for the years 1978-1986 at the 2-digit (class) level of the United Kingdom Standard Industrial Classification (SIC(80)).

It will be seen that there was a decline in production from 1978 to 1981, followed by a recovery which accelerated in 1984 and 1985, with a slight fall in 1986. However, even at this level of aggregation, there was a diversity of experience. Apart from a fall in 1981, output of the office machinery and computer industry rose steadily and rapidly between 1978 and 1985 and the fall in output in this sector largely accounted for the decline in the output of total engineering in 1986; at the opposite extreme, output of the motor vehicle industry fell between 1978 and 1982 and did not show any significant recovery in the period under consideration.

Even within these groups, there were considerable variations in the experience of the individual industries. [V.32] These differences are illustrated in table V.38, where the index of production is shown at the 4-digit (activity heading) level of SIC(80).

Table V.37 contains a detailed correlation between the 4-digit level of ISIC (Rev.2) and the 4- and 5-digit level of SIC(80) for the engineering industries and a key to the descriptions of the activity headings (AH) in tables V.38-V.47. Where 5-digit levels are given in table V.37, they may be aggregated to form the 4-digit heading, e.g. AH 3245 is made up of 3245.1 from ISIC 3824 and 3245.2 and 3245.3 from ISIC 3829.

Tables V.39-V.47 give information at the AH level for sales of principal products, imports, exports, estimated home market, import penetration, exports as a percentage of manufacturers' sales, producer price indices and crude and normalized trade balances for the years 1978 to 1986.

Table V.48 gives information on employment at the 2-digit level of SIC(80). This is the lowest level at which a run of such information is available for the period under consideration. It will be seen that there has been a general decline in the numbers employed, except in office machinery and data processing equipment, where there was an increase of some 40%. [V.33]

V.13. UNITED STATES

Table V.49 illustrates the development of selected United States engineering industries between 1979-1986 (in terms of shipments). It complements the information provided above (e.g. in chapter III) on the ongoing structural development in industrially developed countries which favours high-technology industries, such as the manufacture of computers and communication equipment, electronic components, precision instruments and aerospace. Among the decling branches one can find agricultural machinery, construction machinery, mining and oilfield machinery and even such traditional fields as electrical and non-electrical machinery, as well as shipbuilding or railroad transport equipment.

The United States contribution to the *Annual Review 1986* includes more detailed analysis of the recent development of the main economic indicators (production, investmens, R and D expenditures, employment, international trade, short- and medium-term prospects) in the following branches:

- Computer equipment and software
- Radio communication and detection equipment
- Telephone equipment
- Microelectronics
- Industrial and scientific instruments
- Medical equipment
- Photographic equipment and supplies
- Motor vehicles
- Aerospace [V.34]

Table V.1

Results achieved by the engineering industries in Belgium, 1983-1986

(Billions of Belgian francs)

	1983	*1984*	*1985*	*1986* [a]
Orders in hand	774.9	858.4	924.5	980.7
Deliveries	789.8	825.0	931.6	986.4
Exports	523.8	780.1	878.5	949.6
			(Thousands of employees)	
Employment	251.5	249.1	246.3	242.3

a Provisional figures.

Table V.2

Capital investment in modernization and rebuilding in engineering subsectors as a percentage of total capital investment in industry in Bulgaria, 1979-1985

Sector	*1979*	*1980*	*1981*	*1982*	*1983*	*1984*	*1985*
Industry as a whole,	100.0	100.0	100.0	100.0	100.0	100.0	100.0
of which:							
Mechanical engineering and metalworking	19.4	21.0	29.4	20.8	23.7	28.7	29.6
Electrical engineering and electronics	6.5	9.4	8.4	11.1	16.8	13.0	11.3

Source: Statisticheski godishnik NRB (Statistical Yearbook of Bulgaria), 1985, 1986.

Table V.3

Development indices of capital spending in Bulgaria on industrial modernization and rebuilding in engineering subsectors, 1979-1985

(1979 = 100)

Sector	*1979*	*1980*	*1981*	*1982*	*1983*	*1984*	*1985*
Industry as a whole	100	124	124	144	126	111	113
of which:							
Mechanical engineering and metalworking	100	124	137	152	141	166	172
Electrical engineering and electronics	100	165	162	243	324	225	197

Source: As for table V.2.

Table V.4

Engineering workers as a percentage of the work-force in the national economy and in industry in Bulgaria, 1979-1985

Sector	*1979*			*1980*			*1985*		
	I	*II*	*III*	*I*	*II*	*III*	*I*	*II*	*III*
Engineering industries, ***total***	7.0	9.0	26.9	7.2	9.0	27.0	8.0	9.8	28.6
of which:									
Mechanical engineering and metalworking	4.6	5.9	17.6	4.7	5.9	17.7	5.0	6.2	18.0
Electrical engineering and electronics	2.4	3.1	9.3	2.5	3.1	9.3	3.0	3.6	10.6

Notes: I Relative to population of working age; II Relative to national workforce; III Relative to industrial workforce.
Source: As for table V.2.

Table V.5

Structure of the work-force by category in industry and engineering in Bulgaria, 1980 and 1985

(Percentage)

Year	*Sector*	*Total work-force*	*Production staff*	*Manual workers*	*Engineers and technicians*	*Others*	*Non-production staff*
1980	Industry as a whole	100	93.91	81.19	7.51	5.21	6.09
	of which:						
	Mechanical engineering and metalworking	100	96.42	80.46	10.21	5.75	3.58
	Electrical engineering and electronics	100	96.33	82.32	9.26	4.75	3.67
1985	Industry as a whole	100	93.87	80.97	7.81	5.09	6.13
	of which:						
	Mechanical engineering and metalworking	100	96.87	80.17	10.62	6.08	3.63
	Electrical engineering and electronics	100	96.91	83.04	9.28	4.59	3.09

Source: As for table V.2.

Table V.6

Structure of Bulgarian exports of specialized machinery and equipment by subsector, 1975-1985

(Percentage)

Subsector	*1975*	*1980*	*1985*
Total	100.0	100.0	100.0
of which:			
Computer equipment	32.0	24.7	36.0
Hoisting and transport machinery	33.2	37.3	23.9
Communication equipment	7.0	5.8	8.3
Tractors and agricultural machinery	7.8	4.8	5.7
Power generation and electrical engineering machinery	3.9	7.2	9.2
Equipment for the light and food industries	2.3	2.4	2.5
Transport equipment	10.2	6.7	6.6
Other	3.6	11.1	7.8

Source: Calculated from Vnshna trgovia na NRB (Foreign Trade of Bulgaria), Central Statistical Directorate, Sofia, 1986.

Table V.7

Index of the gross output in mechanical engineering in the German Democratic Republic, 1980-1985

(1980 = 100)

Year	*1980*	*1981*	*1982*	*1983*	*1984*	*1985*
Index	100	106	111	115	119	126

Table V.8

Development of industrial production in mechanical engineering and its share in total industry in the German Democratic Republic, 1981-1985

Year	*1981*	*1982*	*1983*	*1984*	*1985*
Billion mark	78.2	86.5	88.7	90.2	99.2
Percentage share[a]	22.8	20.8	20.5	19.3	18.9

a See ref. II.12.

Table V.9

Development of the number of employees in mechanical engineering (excluding apprentices) in the German Democratic Republic, 1981-1985

Year	*1981*	*1982*	*1983*	*1984*	*1985*
Total number of employees	936 488	942 142	952 613	957 768	973 283
of which:					
Shop workers	552 395	556 786	562 346	561 205	567 110
Female employees	288 088	289 256	289 790	290 048	292 242

Table V.10

Production of selected engineering devices and products in the German Democratic Republic, 1981-1985

Device/product	*1981*	*1985*	*Index 1985/1981*
Metalcutting machine tools (million mark)	2 176	2 762	126.9
NC-machine tools (units)	935	1 284	137.3
Lorries (units)	39 396	45 305	115.0
Passenger cars (units)	180 233	210 370	116.7
Agricultural machines (million mark)	5 226	5 485	105.0
Hoisting and handling machinery (million mark)	3 149	3 229	102.5
Industrial gears (million mark)	571	727	127.3
Armatures (million mark)	1 191	1 352	113.5
Hydraulic devices (million mark)	875	1 105	126.3

Table V.11

Percentage share of export of machines, equipment and means of transport in total industry exports in the German Democratic Republic, 1981-1985

Year	*1981*	*1982*	*1983*	*1984*	*1985*
Percentage	48.9	48.5	47.8	47.0	46.6

Table V.12

Percentage share of import of machines, equipment and means of transport in total industry imports in the German Democratic Republic, 1981-1985

Year	*1981*	*1982*	*1983*	*1984*	*1985*
Percentage	32.0	32.3	29.9	26.0	26.8

Table V.13

Share of engineering industries in the performance of total industry in Hungary, 1980-1985

(Total industry = 100)

Indicator	*1980*	*1981*	*1982*	*1983*	*1984*	*1985*
Gross output	23.0	23.6	24.0	24.1	23.9	24.7
Employment	32.0	32.0	32.3	31.9	31.7	32.2
Gross value of fixed assets	18.5	18.6	18.7	18.5	18.2	...
Total sales	23.0	22.9	23.2	23.2	23.1	23.9
Total exports	41.5	42.1	43.4	42.4	41.0	43.8

Table V.14

Branch pattern of Hungarian engineering industries, 1980-1984

(Percentage)

	Indicator					
	Gross output		*Gross value of fixed assets*		*Employment*	
Subsector (branch)	*1980*	*1984*	*1980*	*1984*	*1980*	*1984*
Total engineering industries	100.0	100.0	100.0	100.0	100.0	100.0
of which:						
Machinery and equipment	21.7	23.4	23.2	23.6	24.7	26.2
Transport equipment	28.3	27.6	30.3	29.7	20.0	18.9
Electrical machinery and apparatus	14.6	13.6	11.9	12.1	12.2	12.3
Telecommunication equipment and vacuum engineering	15.5	16.4	15.7	16.0	19.4	19.9
Precision instrument industry	8.8	8.8	8.1	8.2	11.3	11.0
Metal products industry	11.1	9.7	9.2	9.3	11.5	11.0
Research and development	0.0	0.5	1.6	1.1	0.9	0.7

Table V.15

Index of employment by individual engineering subsector in Hungary, 1985

(1984 = 100)

Subsector (branch)	*Total employment*	*of which:* *Blue collar*	*White collar*
Total industry	99.4	99.1	100.5
Engineering industries	99.3	99.1	99.7
Machinery and equipment	99.2	99.1	99.5
Transport equipment	99.2	99.6	98.1
Electrical machinery and apparatus	99.8	100.0	99.4
Telecommunication equipment and vacuum engineering	99.1	98.3	101.4
Precision instrument industry	98.0	97.7	98.7
Metal products industry	100.2	99.9	101.0

Table V.16

Indices of industrial production by engineering subsectors in Italy, 1975-1984

(1980 = 100)

Year	*Metal products*	*Agricultural and industrial machinery*	*Office equipment and precision instruments*	*Electrical equipment*	*Means of transport*
1975	84.8	83.7	44.8	76.8	72.6
1976	92.2	85.9	52.4	83.2	78.3
1977	93.7	90.9	54.7	86.9	81.3
1978	88.9	88.0	72.1	88.5	85.7
1979	90.0	89.1	94.2	89.6	88.1
1980	100.0	100.0	100.0	100.0	100.0
1981	99.0	96.0	98.6	95.6	102.6
1982	87.2	84.4	97.2	100.8	99.1
1983	79.6	79.2	102.7	102.3	103.2
1984	77.4	79.0	113.3	108.2	102.2
1984/1975 x 100	91.27	94.38	252.90	140.89	140.77

Source: Bank of Italy - ISTAT (1985) data.

Table V.17

Pattern of growth and sectoral breakdown of the engineering industries' output in Poland, 1982-1985

(1982 prices)

Sector/subsector	*Index (1982 = 100)*		*Share share in %*	
	1984	*1985*	*1982*	*1985*
Total engineering industries	*111*	*123*	*100.0*	*100.0*
Metal products	110	116	20.6	19.5
Machinery and equipment	113	118	28.6	27.5
Precision instruments	116	131	4.4	4.6
Transport equipment	118	122	27.2	27.0
Electrical equipment and electronics	123	137	19.2	21.4
of which:				
Electrical equipment	115	126	10.1	10.3
Electronics	132	149	9.1	11.1

Table V.18

Growth index of total industrial and engineering output in the Ukrainian SSR, 1981-1985

(1980 = 100)

Sector/subsector	*1981*	*1982*	*1983*	*1984*	*1985*
Industry as a whole	103	106	110	115	119
Engineering industries	105	110	117	125	135
of which:					
Machinery and equipment	106	110	118	127	...

Table V.19

Share of individual sectors in total industrial output of the Ukrainian SSR, 1970-1985 [a]

(Percentage)

Sector	1970	1975	1979	1980	1981	1982	1983	1984	1985
Industry as a whole	100	100	100	100	100	100	100	100	100
of which:									
Electric power	3.4	3.4	-	3.4	3.3	3.3	3.2	...	3.4
Fuel	9.4	7.7	-	6.7	6.4	6.3	6.1	...	5.9
Chemicals and petrochemicals	4.5	5.4	-	5.9	6.1	6.1	6.2	...	6.2
Engineering industries	17.1	21.2	28.8	25.9	26.5	26.9	27.4	28.3	29.4
Light industry	13.5	12.4	-	12.8	12.8	12.4	12.1	...	11.7
Food industry	24.1	21.2	-	18.9	18.7	19.1	19.3	...	18.4

a Total output evaluated as at 1 January 1982 in producers' wholesale prices.

Table V.20

Growth index of fixed capital formation in total industry and in the engineering industries in the Ukrainian SSR, 1981-1985

(1980 = 100)

Sector/subsector	1981	1982	1983	1984	1985
Industry as a whole	106	112	118	125	132
Engineering industries	108	117	125	133	144
of which:					
Machinery and equipment	109	118	127	136	146

Table V.21

Growth index of the capital/labour ratio for production workers in the engineering industries in the Ukrainian SSR, 1981-1985

(1980 = 100)

Sector	1981	1982	1983	1984	1985
Industry as a whole	106	111	117	123	129
Engineering and metalworking	107	115	122	129	137

Table V.22

Numbers of mechanized and automated production lines at manufacturing enterprises in the Ukrainian SSR, 1975-1985

	Mechanized production lines					*Automated production lines*				
Sector	*1975*	*1979*	*1981*	*1983*	*1985*	*1975*	*1979*	*1981*	*1983*	*1985*
Industry as a whole	22 098	25 715	27 701	29 069	30 515	2 639	3 703	4 250	4 831	5 396
of which:										
Engineering industries	5 959	7 744	8 644	9 258	9 822	1 091	1 711	2 001	2 448	2 934

Table V.23

Index of labour productivity in total industry and in the engineering industries in the Ukrainian SSR, 1981-1985

(1980 = 100)

Sector/Subsector	*1981*	*1982*	*1983*	*1984*	*1985*
Industry as a whole	102	104	108	111	115
Engineering industries	104	107	113	120	127
of which:					
Machinery and equipment	104	107	114	121	...

Table V.24

Numbers of new types of machinery and equipment developed in the Ukrainian SSR, 1979-1985

Type of machinery and equipment	*1979*	*1980*	*1981*	*1982*	*1983*	*1984*	*1985*
Total	590	568	550	594	641	614	639
of which:							
Power industry equipment	5	13	7	8	15	11	12
Electrotechnical equipment	86	73	70	87	132	106	95
Metalworking machine tools	43	36	39	43	43	49	35
Forging and pressing equipment	27	26	28	23	22	14	14
Foundry equipment	13	9	10	10	11	13	10
Metallurgical and mining equipment	37	41	28	43	51	50	38
Fuel industry equipment	12	9	8	-	18	9	12
Transport and lifting equipment	27	34	23	29	16	10	11
Motor vehicles, tractors and related equipment	3	12	12	3	24	22	17
Agricultural machinery	28	33	21	14	21	33	45
Chemical industry equipment, pumps and compressors	45	34	37	80	59	48	69
Building and roadmaking machinery	14	5	9	6	13	14	9
Woodworking, pulp and paper industry equipment	3	5	7	3	27	4	10
Equipment for light industry	5	6	13	13	16	4	8
Food-processing equipment	40	31	36	42	48	44	53
Others	202	201	220	190	125	183	201

Table V.25

Numbers of new types of instruments and means of automation developed in the Ukrainian SSR, 1979-1985

Type of equipment/instrument	*1979*	*1980*	*1981*	*1982*	*1983*	*1984*	*1985*
Total	246	221	185	199	178	237	151
of which:							
Optical instruments and apparatus	16	11	5	8	6	4	19
Electrical measuring instruments	17	26	16	22	28	34	16
Radio measuring instruments	4	-	1	3	6	-	-
Data-processing and office equipment	37	51	37	23	31	16	9
Process-monitoring and -control instruments	60	42	61	91	62	102	41
Instruments for measuring physical properties	19	22	9	13	14	11	2
Instruments for measuring mechanical properties	82	58	41	36	29	40	31
Other instruments	11	11	15	3	2	30	33

Table V.26

Introduction of new types of industrial products and withdrawal from production of obsolete designs of machines, equipment, apparatus, instruments or parts in the Ukrainian SSR, 1979-1985

	1979	*1980*	*1981*	*1982*	*1983*	*1984*	*1985*
Introduction of new types of industrial products	1 011	1 107	968	1 001	1 047	1 149	1 053
Withdrawal from production of obsolete designs of machines, equipment, apparatus, instruments or parts	371	438	453	471	515	610	697

Table V.27

Production of selected engineering products in the Ukrainian SSR, 1979-1985

	1979	*1980*	*1981*	*1982*	*1983*	*1984*	*1985*
Machine tools (thousands)	34.7	32.9	31.9	31.0	30.3	31.7	31.1
Forging and pressing machines (units)	7 918	8 428	8 861	9 231	9 536	9 597	9 907
Industrial robots and programmable manipulators (units)	-	-	229	692	1 535	1 845	1 843
Generators for turbines (million kW)	3.0	2.3	1.6	2.1	2.0	1.7	2.0
Heavy electrical machinery (thousands)	8.6	9.1	8.5	8.9	8.9	9.6	9.5
AC electric motors over 100 kW (thousands)	1.5	1.4	1.4	1.3	0.9	0.3	0.4
(thousand kW)	220	211	216	193	135	51	58
AC electric motors from 0.25 to 100 kW (thousands)	2 545	2 639	2 733	2 757	2 796	2 745	2 836
(thousand kW)	8 592	9 545	10 013	10 633	10 363	10 433	10 415
Flame-proof electric motors (thousands)	246	248	243	240	236	238	247
(thousand kW)	2 317	2 372	2 456	2 587	2 893	3 011	3 234
Electric cutting motors (units)	1 986	2 300	2 133	2 195	2 536	2 687	2 779
Power transformers (million kVA)	69.8	72.1	73.2	74.2	70.5	73.0	76.9
Blast-furnace and steel-making equipment (thousand tonnes)	103.4	102.9	96.2	90.9	98.3	86.2	96.2
Rolling-mill equipment (thousand tonnes)	57.1	60.1	57.5	52.8	57.6	47.6	57.7
Continuous miners (units)	1 166	1 252	1 232	1 174	1 132	1 047	1 027
Mine lifting machines (units)	341	341	357	361	322	296	290
Mine loading machines (units)	242	264	277	291	308	262	229
Electric ore-haulage locomotives (units)	729	797	824	926	941	990	1 029
Tractors (thousands)	142	135.6	133.1	131.8	134.7	128.8	135.9
Tractor-drawn seed drills (thousands)	86.6	78.0	79.6	70.3	68.7	72.3	75.2
Side-delivery reapers (thousands)	74.4	67.6	66.0	70.8	76.5	73.5	73.3
Beet harvesters (thousands)	10.6	9.5	9.3	9.7	8.9	6.6	5.0
Agricultural sprinkler units (thousands)	...	9.0	9.2	8.3	7.2	7.6	7.8
Bearings (millions)	130.4	130.2	137.8	142.8	145.5	149.8	145.0
Electric welding equipment (thousands)	45.9	44.0	43.4	43.8	44.9	40.0	42.0
Excavators (units)	9 725	9 874	10 122	9 830	9 435	9 328	10 006
Bulldozers (units)	24 515	23 975	23 994	18 100	15 438	15 009	2 986
Tractor-drawn scrapers (units)	8 415	8 016	7 910	6 277	7 762	7 633	6 558
Refrigerators (thousands)	...	702	691	697	695	693	743
Washing machines (thousands)	...	273	288	290	310	330	372
Bicycles and mopeds (thousands)	...	979	994	979	924	952	984
Motorcycles (thousands)	...	84.5	89.9	99.3	96.4	90.0	100.0

Table V.28

Selected types of engineering products produced in the Ukrainian SSR in 1985

Type of product	*Produced in 1985*	*Index (1984 = 100)*
Machine tools (million roubles)	420	104
of which:		
Numerically controlled tools	110	117
Forging and pressing machines (thousands)	10.0	104
Instruments, means of automation and spare parts therefor (billion roubles)	1.2	106
Data processing equipment and spare parts therefor (billion roubles)	1.0	114
Tractors (million horsepower)	12.6	109
Excavators (thousands)	10.0	107

Table V.29

Share of individual industries in total industrial output of the USSR, 1980-1985

(Percentage)

Industry/sector	*1980*	*1984*	*1985*
Industry as a whole	100	100	100
Heavy industry	66.7	68.3	68.6
- Engineering industries	24.3	26.4	27.4
- Building materials and chemical products	25.0	24.6	24.5
- Fuel and power	12.1	11.5	11.1
Light industry	16.2	14.7	14.6
Food industry	15.4	15.4	15.2

Source: Narodnoe khozyaistvo SSSR v 1985g (The USSR Economy in 1985), Moscow, 1986, p. 101.

Table V.30

Index of industrial output by individual industries in the USSR, 1985-1986

	Index (1980 = 100)		*1986/1985 annual growth rate (%)*
Industry/Sector	*1985*	*1986*	
Industry as a whole	120	126	4.9
Heavy industry	123	131	5.7
- Engineering industries	135	145	7.3
- Fuel and power	111	114	3.5
- Metallurgy	111	115	4.0
Light industry	108	110	2.0
Food industry	118	124	5.0

Source: SSSR v tsifrakh v 1986g. (The USSR in figures in 1986), Moscow, 1987, p. 92.

Table V.31

Production of selected engineering products in the USSR in 1986

Type of product	*Produced in 1986*	*Growth 1986/1985 (percentage)*
Turbines (million kW)	20.9	104
Turbine generators (million kW)	14.9	121
AC electric motors (million kW)	55.7	102
Metalcutting machine tools (billion roubles)	2.9	109
of which: numerically controlled	1.3	123
Forging and pressing machines (million roubles)	687	105
Industrial robots and manipulators (thousand)	15.1	114
Automation instruments and equipment and spare parts (billion roubles)	4.8	105
Computer equipment and spare parts (billion roubles)	4.8	113
Oil-production machinery (million roubles)	247	108
Chemical equipment and spare parts (million roubles)	996	103
Tractors (million hp)	54.5	103
Agricultural machines (billion roubles)	4.0	100.9
Excavators (thousand)	42.9	100.9

Source: Pravda, 18 January 1987, p.2.

Table V.32

Increases in industrial production due to growth in labour productivity in the USSR, 1981-1986

(Percentage)

Industry/sector	*1981-1985*	*1986*
Industry as a whole	86	96
Heavy industry	84	93
- Engineering industries	87	95
- Fuel and power sector	25	77
- Metallurgy	82	100
- Building materials	83	90
Light industry	100	100
Food industry	94	100

Source: Vestnik statistiki, No. 5, 1987, p.61.

Table V.33

Breakdown of USSR industrial productive capital by industry and type of capital at 1 January 1986

(Percentage share)

	Type of capital										
					Machinery and equipment of which:						
	Industrial productive fixed capital Total	Buildings	Installations	Transmission lines	Machinery and equipment Total	Power plant and equipment	Working machinery and equipment	Measuring and regulating instruments and devices and laboratory equipment	Computer equipment	Transport equipment	Other fixed capital
Industry as a whole	100	27.6	19.0	10.6	39.9	7.5	29.3	1.6	1.1	2.2	0.7
Heavy industry	100	26.3	20.2	11.5	39.0	8.1	27.7	1.8	1.1	2.2	0.8
- Engineering industries	100	36.3	6.3	3.6	49.8	2.3	41.3	3.0	2.7	2.1	1.9
- Chemicals and petrochemicals	100	30.9	14.3	11.2	41.2	3.0	34.7	2.4	0.8	1.9	0.5
- Forestry, woodworking and paper and pulp	100	32.0	16.7	4.6	40.6	4.6	34.9	0.4	0.5	5.6	0.5
- Building materials	100	39.1	18.0	4.5	34.7	3.1	30.5	0.7	0.3	3.0	0.7
- Power engineering	100	14.9	15.7	31.9	37.0	33.6	1.4	1.2	0.5	0.4	0.1
- Fuel	100	7.4	55.4	12.4	22.7	2.5	18.6	1.2	0.3	1.8	0.3
Light industry	100	41.8	5.0	2.9	47.7	2.0	43.7	0.5	1.2	1.4	1.2
Food industry	100	33.9	10.6	2.9	48.1	3.1	43.5	0.7	0.4	4.0	0.5

Source: *Narodnoe khozyaistvo SSSR v 1985g*, Moscow, 1986g. pp. 118-119.

Table V.34

Value added analysed by industry in the United Kingdom, 1976-1986 [a]

	1976	*1977*	*1978*	*1979*	*1980*	*1981*	*1982*	*1983*	*1984*	*1985*	*1986*
Agriculture, forestry and fishing	2.8	2.6	2.4	2.3	2.2	2.2	2.3	2.1	2.2	1.8	1.8
Extraction of mineral oil and natural gas	0.5	1.6	1.9	3.3	4.4	5.5	5.8	6.2	7.0	6.2	2.8
Other energy and water supply	5.6	5.5	5.4	5.2	5.7	5.6	5.4	5.5	3.9	4.8	4.7
Manufacturing (revised definition)	28.7	29.5	29.5	28.4	26.6	24.8	23.6	23.6	23.6	23.8	24.3
Construction	6.5	8.1	8.2	8.1	6.0	5.8	5.8	6.0	6.1	6.1	6.2
Distribution, hotels and catering; repairs	12.4	12.8	13.1	12.9	12.5	12.4	12.2	12.3	12.8	13.2	14.0
Transport	5.5	5.4	5.5	5.3	4.8	4.7	4.5	4.5	4.5	4.4	[illegible].5
Communication	2.9	2.5	2.5	2.4	2.4	2.6	2.7	2.5	2.6	2.6	2.7
Banking, finance, insurance, business services and leasing [b]	10.8	11.5	11.4	12.2	12.4	12.7	13.6	13.7	14.2	14.8	15.8
Ownership of dwellings	6.1	5.8	5.8	6.0	6.1	6.3	6.3	6.3	6.1	5.9	5.8
Public administration, national defence and compulsory social security	7.7	7.0	6.7	6.6	7.3	7.4	7.3	7.3	7.3	7.2	7.2
Education and health services	9.3	8.7	8.4	8.2	9.0	9.4	8.9	9.0	8.8	8.6	9.1
Other services	5.2	5.3	5.4	5.4	5.3	5.5	5.5	5.7	5.9	6.0	6.4
Adjustment for financial services [b]	-4.1	-4.4	-4.1	-4.3	-4.7	-4.8	-4.9	-4.5	-4.9	-5.1	-5.3
Gross domestic product at factor cost (income-based)	100.0	100.0	100.0	100.0	100.0	100.0	100.0	100.0	100.0	100.0	100.0

a Before providing for depreciation but after deducting stock appreciation.
b The contribution of the banking, finance, insurance etc. industries is measured before deducting industries' net receipts of interest. This is offset in the aggregate gross domestic product by the item "Adjustment for financial services" equal to the net interest receipts.

Table V.35

Index of industrial production in the United Kingdom, 1978-1986

(1980 = 100)

	1978	*1979*	*1980*	*1981*	*1982*	*1983*	*1984*	*1985*	*1986*
GDP	99.7	102.4	100.0	99.1	100.7	104.0	106.5	110.3	113.7
Index of industrial production	103.1	107.0	100.0	96.6	98.4	101.9	103.3	108.0	111.0
of which:									
Engineering	109.6	107.1	100.0	91.8	92.9	94.9	99.5	103.9	102.6
Energy and water	85.0	100.5	100.0	103.8	110.0	115.9	110.2	120.1	125.2

Table V.36

Index of production by individual engineering subsectors in the United Kingdom, 1978-1986

(1980 = 100)

Class		*1978*	*1979*	*1980*	*1981*	*1982*	*1983*	*1984*	*1985*	*1986*
31	Metal goods	123.8	121.6	100.0	92.2	92.8	95.1	101.5	99.4	97.6
32	Mechanical engineering	112.8	108.9	100.0	89.2	90.5	87.0	88.3	91.5	90.4
33	Computers and office machinery	77.8	93.4	100.0	86.2	95.8	144.7	206.1	263.7	241.9
34	Electrical engineering	103.9	103.1	100.0	94.3	98.4	104.7	114.4	115.8	117.3
35	Motor vehicles	118.9	115.7	100.0	82.9	80.1	84.0	80.8	87.4	82.5
36	Other transport equipment	95.8	92.4	100.0	103.5	100.8	95.0	91.5	94.4	96.5
37	Instrument engineering	104.3	102.8	100.0	101.3	94.4	95.4	102.3	111.1	112.2
	Total engineering	109.6	107.1	100.0	91.8	92.9	94.9	99.5	103.9	102.6

Table V.37

Transformation key between ISIC (Rev.2) and SIC(80) classifications of engineering products

SIC(80)		*Notes*
ISIC Group 3811 - Manufacture of cutlery, hand tools and general hardware		
3161	Hand tools and implements	Excludes manufacture of saw blades for machines. Includes plated ware.
3162	Cutlery, spoons, forks and similar tableware; razors	
3169.1	Locks etc.	
3169.2	Needles, pins and other metal small wares	Includes manufacture of paper fasteners etc.
ISIC Group 3812 - Manufacture of furniture and fixtures primarily of metal		
3166.1	Metal furniture	
ISIC Group 3813 - Manufacture of structural metal products		
3142	Metal doors, windows etc.	
3204	Fabricated constructional steelwork	Includes manufacture of structural steelwork and fabrication of oil platforms
3205	Boilers and process plant fabrication	
ISIC Group 3819 - Manufacture of fabricated metal products except machinery and equipment n.e.s.		
2234	Drawing and manufacture of steel wire and steel wire products	Excludes manufactufacture of uninsulated non-ferrous wire
3137	Bolts, nuts, washers, rivets springs and non-precision chains	Includes manufacture of non precision chains
3138	Heat and surface treatment of metals, including sintering	
3163	Metal storage vessels	
3164.1	Metal cans and boxes	
3164.2	Metal kegs, drums and barrels	
3164.3	Metallic closures	
3164.5	Other packaging products of metal	
3165	Domestic heating and cooking appliances (non-electrical)	
3166.2	Safes etc.	
3167	Domestic and similar utensils of metal	
3169.3	Base metal fittings and mountings for furniture, builders joinery, leather and travel goods n.e.s.	Excludes manufacture of needles, pins and paper fasteners, etc.
3169.4	Miscellaneous finished metal products n.e.s.	
ISIC Group 3821 - Manufacture of engines and turbines		
3281.1	Industrial internal combustion engines	
3281.3	Other prime movers	

Table continued.

Table V.37 (continued)

ISIC Group 3822 - Manufacture of agricultural machinery and equipment		
3211	Agricultural machinery	
3212	Wheeled tractors	
ISIC Group 3823 - Manufacture of metal and wood-working machinery		
3221	Metalworking machine tools	Excludes manufacture of gas cutting and welding equipment
3222	Engineers' small tools	Includes manufacture of saw blades for machines
3275.1	Woodworking machinery	
3285.2	Portable power tools	
3286.2	Machinery for foundries and rolling mills	
3286.3	Manufacture of other machinery and mechanical equipment n.e.s.	
ISIC Group 3824 - Manufacture of special industrial machinery and equipment except metal and wood-working machinery		
3230	Textile machinery	
3244	Food and drink and tobacco processing machinery	
3245.1	Chemical industry machinery	Includes manufacture of centrifuges
3246	Process engineering contractors	
3251	Mining machinery	
3254	Construction and earth-moving equipment	
3275.2	Rubber and plastics working machinery	
3275.3	Leather working and footwear making and repairing machinery	
3275.4	Paper making machinery	
3275.5	Glass and brick making and similar machinery	
3276	Printing, bookbinding and paper goods machinery	
ISIC Group 3825 - Manufacture of office, computing and accounting machinery		
3285.1	Scales and weighing machinery	
3301	Office machinery	
3302	Electronic data-processing equipment	

Table continued.

Table V.37 (continued)

ISIC Group 3829 - Manufacture of machinery and equipment except electrical n.e.s.		
3245.2	Furnaces and kilns	
3245.3	Gas, water and waste treatment plants	
3255	Mechanical lifting and handling equipment	
3261	Mechanical power transmission equipment	
3262	Ball, needle and roller bearings	
3275.6	Laundry and dry-cleaning equipment	
3283	Compressors and fluid power equipment	
3284	Refrigerating machinery, space heating, ventilating and air conditioning equipment	
3286	Other industrial and commercial machinery	Includes manufacture of slot and gaming machines, lawn mowers, testing and vending machines
3287	Pumps	
3288	Industrial valves	
3289.2	Precision components for engines and machinery n.e.s.	
3289.3	Mechanical engineering n.e.s.	Includes manufacture of gas cutting and welding equipment
3290	Ordnance, small arms and ammunition; tracked armoured fighting vehicles	Excludes manufacture of guided missiles etc. and wheeled armoured fighting vehicles
ISIC Group 3831 - Manufacture of electrical industrial machinery and apparatus		
3420	Basic electrical equipment	Excludes manufacture of electrical welding apparatus and battery chargers
3434	Electrical equipment for motor vehicles, cycles and aircraft	
3442	Electrical instruments and control systems	
3454.2	Other electronic equipment n.e.s.	Includes manufacture of electron microscopes
ISIC Group 3832 - Manufacture of radio, television and communication equipment and apparatus		
3433	Alarms and signalling equipment	
3441	Telegraph and telephone apparatus and equipment	
3443	Radio and electronic capital goods	Includes manufacture of flight simulators and hearing aids
3444	Components other than active components, mainly for electronic equipment	
3452	Gramophone records and pre-recorded tapes	
3453	Active components and electronic sub-assemblies	
3454.1	Electronic consumer goods	
ISIC Group 3833 - Manufacture of electrical appliances and housewares		
3460	Domestic type electrical appliances	

Table continued.

Table V.37 (continued)

ISIC Group 3839 - Manufacture of electrical apparatus and supplies n.e.s.		
3410	Insulated wires and cables	
3432	Batteries and accumulators	
3435	Electrical equipment for industrial use	Includes manufacture of battery chargers, brazing equipment catering equipment, dielectric heating equipment, electro-chemical apparatus, electrolytic chemical. electroplating equipment and soldering and welding equipment
3470	Electric lamp and other lighting equipment	
ISIC Group 3841 - Shipbuilding and repairing		
3281.2	Marine engines	
3289.1	Marine engineering	
3610	Shipbuilding and repairing	Includes manufacture of floating oil rigs
ISIC Group 3842 - Manufacture of railroad equipment		
3620	Railway and tramway vehicles	
ISIC Group 3843 - Manufacture of motor vehicles		
3510	Motor vehicles and their engines	Includes manufacture of wheeled armoured fighting vehicles
3521	Motor vehicle bodies	
3522	Trailers and semi-trailers	
3523	Caravans	
3530	Motor vehicle parts	
ISIC Group 3844 - Manufacture of motorcycles and bicycles		
3633	Motor cycles and parts	
3634	Pedal cycles and parts	
ISIC Group 3845 - Manufacture of aircraft		
3640	Aerospace equipment manufacturing and repairing	Includes manufacture of guided missiles and weapons and spacecraft etc.
ISIC Group 3846 - Manufacture of transport equipment n.e.s.		
3650	Other vehicles	Includes manufacture of powered invalid carriages

Table continued.

Table V.37 (concluded)

ISIC Group 3851 - Manufacture of professional and scientific, measuring and controlling equipment n.e.s.		
3710	Measuring, checking and precision instruments	Excludes manufacture of cyclotrons, dental consumable goods and meteorological instruments
3720	Medical and surgical equipment and orthopaedic appliances	Includes manufacture of opthmalic instruments
ISIC Group 3852 - Manufacture of photographic and optical goods		
3731	Spectacles and unmounted lenses	Excludes manufacture of opthalmic instruments, meteorological instruments and electron microscopes
3732	Optical precision instruments	
3733	Photographic and cinematographic equipment	
ISIC Group 3853 - Manufacture of clocks and watches		
3740	Clocks, watches and other timing devices	

Table V.38

Index of production of the engineering industries, by product, in the United Kingdom, 1978-1986

(1980 = 100)

Activity heading	*1978*	*1979*	*1980*	*1981*	*1982*	*1983*	*1984*	*1985*	*1986*
2234	117	118	100	93	91	91	82	85	84
3161	137	125	100	91	74	78	78	77	77
3162	105	105	100	89	80	74	75	74	67
3163	90	108	100	80	67	70	67	58	66
3164	120	122	100	93	94	93	96	98	98
3165	112	117	100	83	81	87	95	96	98
3166	105	115	100	90	88	95	105	109	112
3167	133	120	100	93	83	88	76	73	71
3169	124	115	100	98	108	120	127	131	134
3204	102	116	100	91	111	105	98	102	101
3205	98	107	100	105	110	105	98	93	91
3211	178	135	100	86	97	101	90	96	86
3212	124	122	100	92	88	86	99	96	84
3221	108	103	100	69	64	58	65	78	76
3222	117	109	100	76	73	71	80	85	78
3230	134	123	100	84	77	76	88	86	89
3244	115	111	100	86	84	80	80	80	80
3245	127	116	100	82	82	76	73	78	87
3251	98	108	100	76	85	71	58	67	84
3254	118	107	100	91	78	71	73	85	78
3255	105	109	100	83	87	88	91	93	93
3261	114	112	100	85	89	89	94	107	105
3262	113	109	100	80	73	67	68	73	77
3275	119	113	100	82	81	78	78	85	89
3276	101	102	100	84	78	71	73	88	98
3281	127	103	100	93	97	79	90	82	75
3283	121	108	100	82	80	75	86	90	89
3284	100	110	100	85	84	83	86	88	91
3285	85	103	100	85	84	90	97	107	109
3286	102	107	100	81	80	76	76	89	106
3287	114	103	100	94	92	80	78	89	89
3288	93	96	100	95	91	89	81	89	94
3289	120	114	100	95	95	86	96	95	94
3290	107	93	100	97	109	120	104	104	106
3301	119	117	100	89	103	104	106	121	141
3302	74	91	100	86	95	149	218	287	260
3410	123	117	100	97	89	91	83	93	93
3420	120	102	100	85	91	90	92	85	88
3432	111	104	100	93	98	81	84	81	78
3433	101	108	100	86	92	101	106	111	106
3434	128	124	100	71	75	82	92	102	101
3435				Included in AH 3433					
3441	86	89	100	103	103	104	106	115	119
3442	97	104	100	96	106	120	130	139	137
3443	82	91	100	104	110	113	126	131	127
3444	87	101	100	94	99	106	123	106	115
3452	115	113	100	108	121	143	165	186	218
3453	113	115	100	93	98	126	156	164	162
3454				Included in AH 3452					
3460	108	109	100	96	98	115	116	120	129
3470	114	111	100	96	92	98	104	103	106
3510	124	115	100	81	68	79	78	82	82
3521	84	97	100	82	63	72	67	74	77
3522	135	138	100	78	77	96	98	99	86
3523	107	114	100	78	74	80	86	88	97
3530	117	118	100	88	85	88	90	89	85
3610	114	107	100	104	111	100	92	86	76
3620	85	90	100	99	101	86	77	73	72
3633	131	115	100	65	69	67	59	50	42
3634				Included in AH 3633					
3640	88	87	100	107	99	96	96	105	114
3650	99	107	100	101	86	116	133	130	121
3710	104	101	100	114	91	97	100	115	115
3720	94	101	100	98	102	103	116	116	120
3731	99	104	100	96	107	123	138	151	145
3732				Included in AH 3731					
3733	100	97	100	72	64	66	63	62	80
3740	148	1 310	100	82	119	57	61	49	56

Table V.39
Grossed up principal product sales in the United Kingdom, 1978-1986

(Millions of pounds at current prices)

Activity heading	*1978*	*1979*	*1980*	*1981*	*1982*	*1983*	*1984*	*1985*	*1986*
2234	746	835	778	725	780	722	774	850	858
3161	167	178	172	191	190	211	215	236	229
3162	74	74	79	80	81	89	101	109	105
3163	60	66	66	68	71	77	64	63	71
3164	1 085	1 212	1 132	1 108	1 202	1 311	1462	1 539	1 600
3165	200	223	226	242	251	279	291	319	350
3166	293	376	373	356	370	413	457	483	540
3167	207	214	220	171	166	176	183	192	216
3169	1 282	1 474	1 425	1 632	1 859	2 000	2 199	2 363	2 499
3204	773	957	916	1 013	1 286	1 346	1 365	1 512	1 474
3205	914	1 053	1 097	1 251	1 598	1 864	1 621	1 577	1 782
3211	297	308	277	247	277	322	279	302	255
3212	806	973	908	914	878	916	1 016	980	892
3221	601	664	705	565	598	562	683	820	831
3222	545	617	698	580	566	556	649	770	789
3230	282	294	255	256	230	219	306	307	325
3244	391	429	457	430	493	521	521	538	575
3245	351	380	395	357	388	397	395	439	505
3251	501	599	643	599	720	630	507	631	789
3254	869	954	941	1 023	940	839	876	1 011	1 008
3255	044	1 163	1 218	1 038	1 166	1 305	1 451	1 573	1 758
3261	420	449	524	508	474	472	552	648	684
3262	195	214	248	214	210	195	206	229	253
3275	321	349	354	259	328	324	350	458	485
3276	297	346	342	351	363	364	433	580	638
3281	751	663	786	879	889	764	920	877	778
3283	538	549	623	549	593	713	778	801	
3284	771	867	858	801	961	1 052	1 135	1 210	1 326
3285	181	230	248	249	122	142	233	268	271
3286	458	515	555	539	608	630	650	780	863
3287	335	352	387	394	409	384	409	474	472
3288	290	332	360	406	432	414	401	454	474
3289	632	680	724	777	749	818	1 105	1 156	1 508
3290	563	533	575	622	742	847	835	806	788
3301	116	123	121	113	114	120	139	169	216
3302	761	984	1 053	1 047	1 278	1 535	2 519	3 459	2 978
3410	666	795	877	910	966	1 051	1 141	1 256	1 274
3420	1 758	1 799	2 040	1 949	2 186	2 193	2 343	2 367	2 904
3432	302	329	343	309	335	298	324	367	361
3433	139	171	191	188	184	237	246	301	338
3434	462	495	480	417	456	513	613	711	710
3435	205	238	238	225	252	241	299	339	320
3441	591	693	944	1 128	1 275	1 418	1 475	1 668	1 750
3442	522	598	667	737	837	974	1 081	1 301	1 353
3443	988	1 173	1 481	1 906	2 104	2 311	2 633	2 851	2 969
3444	509	617	748	664	754	831	1 064	1 128	1 154
3452	143	154	135	130	129	148	194	236	304
3453	664	758	712	655	676	869	1 092	1 139	1 147
3454	545	551	523	545	586	735	873	885	989
3460	799	890	936	915	920	1 083	1 140	1 208	1 330
3470	409	452	483	522	567	635	700	726	798
3510	4 425	4 761	4 603	4 420	4 842	4 775	5 493	6 746	7 255
3521	199	201	225	210	206	237	243	290	285
3522	271	303	232	185	192	245	286	295	294
3523	160	180	176	147	148	172	194	206	241
3530	2 907	3 152	3 113	2 841	2 894	3 177	3 367	3 428	3 649
3610	760	830	954	648	1 072	1 101	1 037	1 381	575
3620	216	308	341	355	344	286	347	242	274
3633	26	22	21	11	10	5	2	6	8
3634	111	108	113	82	86	107	103	87	74
3640	2 181	2 5528	3 701	3 015	4 076	4 048	4 272	4 905	5 741
3650	40	49	48	45	57	60	73	85	84
3710	636	735	840	817	928	1 088	1 239	1 433	1 501
3720	165	198	215	243	293	307	341	355	370
3731	68	77	92	98	95	117	140	143	133
3732	67	71	84	91	107	140	148	166	178
3733	233	265	261	212	215	199	193	180	244
3740	143	120	105	97	134	111	117	100	92

Table V.40

Imports of engineering products to the United Kingdom, 1978-1986

(Millions of pounds at current prices)

Activity heading	*1978*	*1979*	*1980*	*1981*	*1982*	*1983*	*1984*	*1985*	*1986*
2234	47	56	65	65	78	89	107	122	140
3161	62	79	78	75	85	96	119	131	139
3162	34	39	37	40	40	44	54	54	55
3163	0	1	1	1	1	1	1	1	1
3164	41	49	72	55	67	84	103	111	124
3165	30	40	37	33	45	56	71	80	81
3166	40	46	51	55	70	81	99	125	138
3167	37	47	42	51	45	49	51	52	6[illegible]
3169	204	247	254	258	309	387	468	514	595
3204	33	42	52	53	66	83	106	126	147
3205	60	140	166	366	207	303	232	136	127
3211	121	150	131	137	183	215	233	263	240
3212	166	219	186	186	257	319	325	363	389
3221	261	339	331	267	292	249	323	387	512
3222	98	102	129	115	135	156	230	230	228
3230	115	120	104	93	103	117	150	167	206
3244	157	182	208	209	255	312	358	354	401
3245	57	65	68	70	81	90	115	137	150
3251	11	11	16	10	15	18	16	21	22
3254	274	314	283	288	356	404	434	464	455
3255	217	249	264	249	297	310	381	444	498
3261	51	53	53	60	66	74	94	111	110
3262	93	105	100	91	105	108	143	169	173
3275	168	221	197	192	214	240	334	379	368
3276	152	183	209	188	212	264	339	387	419
3281	139	144	133	163	181	200	230	261	275
3283	96	113	108	122	156	165	197	227	242
3284	124	162	163	163	199	229	285	315	366
3285	59	71	72	78	99	107	130	137	153
3286	214	248	267	250	307	354	480	572	587
3287	98	107	101	92	108	117	141	159	167
3288	51	63	70	78	00	110	118	156	140
3289	34	42	47	47	56	57	76	86	91
3290	50	62	72	62	72	87	79	105	132
3301	154	181	185	172	197	250	267	293	312
3302	830	1 024	1 080	1 317	1 737	2 519	3 561	3 920	3 918
3410	37	52	67	76	101	133	188	229	249
3420	314	348	378	405	519	624	821	888	943
3432	47	47	51	65	84	89	110	125	137
3433	13	17	18	22	25	34	42	42	57
3434	82	86	77	82	88	120	129	145	151
3435	70	84	86	86	109	113	135	151	143
3441	54	61	70	96	125	200	234	318	365
3442	285	338	414	475	592	701	866	996	997
3443	227	287	332	473	545	657	776	819	867
3444	124	146	155	172	225	277	391	425	452
3452	40	45	50	63	70	112	151	169	202
3453	428	489	528	624	763	1 040	1 542	1 722	1 744
3454	365	499	497	819	1 065	1 229	1 056	1 1196	1 284
3460	252	303	324	428	467	601	734	834	930
3470	85	84	85	88	108	132	154	182	217
3510	2 069	3 008	2 509	2 518	3 411	4 375	4 507	5 096	5 978
3521	3	9	11	9	18	18	15	19	18
3522	24	26	20	20	32	47	48	64	58
3523	3	6	6	9	7	5	8	10	9
3530	628	815	695	813	1 039	1 326	1 479	1 773	2 113
3610	113	161	145	233	267	110	299	149	268
3620	27	17	16	20	21	18	18	30	
3633	127	116	143	131	98	100	81	92	94
3634	27	36	46	39	50	72	83	51	57
3640	817	1 033	1 820	1 508	1 755	1 899	2 288	2 787	2 420
3650	99	107	100	101	86	116	133	130	121
3710	105	107	120	126	162	206	245	258	260
3720	73	84	80	103	129	163	198	216	223
3731	35	40	42	40	45	57	71	88	90
3732	37	37	40	48	49	70	81	94	99
3733	276	312	327	391	418	496	562	602	669
3740	161	176	159	163	167	184	225	249	270

Table V.41

Exports of engineering products from the United Kingdom, 1978-1986

(Millions of pounds at current prices)

Activity heading	*1978*	*1979*	*1980*	*1981*	*1982*	*1983*	*1984*	*1985*	*1986*
2234	97	105	92	100	106	105	127	144	135
3161	89	88	95	102	104	95	108	119	111
3162	49	46	52	54	59	57	63	69	60
3163	0	1	1	1	1	1	0	0	1
3164	87	106	102	99	114	112	135	157	172
3165	14	17	18	18	19	22	23	23	21
3166	60	54	56	55	64	60	56	64	61
3167	31	32	36	27	29	28	28	29	27
3169	297	295	347	318	338	344	390	446	424
3204	291	267	293	275	339	306	283	329	250
3205	252	300	283	219	277	273	268	256	250
3211	142	139	131	123	127	114	139	157	148
3212	621	696	727	692	631	602	848	847	702
3221	279	289	365	353	353	285	320	372	380
3222	129	137	162	165	172	165	212	227	204
3230	201	215	242	225	212	175	194	220	248
3244	251	278	298	284	312	329	346	380	411
3245	149	169	188	191	256	222	212	276	261
3251	48	92	100	72	86	71	108	111	105
3254	738	731	844	924	922	732	778	966	858
3255	366	393	460	436	412	366	444	499	582
3261	74	77	97	102	102	92	103	112	105
3262	86	100	116	112	110	105	130	156	170
3275	211	231	275	260	267	229	234	317	320
3276	167	182	214	217	236	255	326	392	428
3281	420	404	486	531	510	429	514	567	536
3283	188	168	209	253	257	265	311	351	375
3284	150	172	175	173	186	192	209	225	223
3285	100	105	117	109	111	107	121	146	147
3286	320	335	379	440	467	421	512	551	610
3287	187	182	228	242	234	200	211	252	270
3288	119	129	162	201	198	188	186	209	209
3289	90	93	126	136	137	124	148	173	168
3290	205	185	220	331	456	408	366	371	434
3301	141	133	160	139	148	174	189	229	250
3302	630	834	936	917	1 184	1 647	2 599	3 314	3 094
3410	155	146	164	177	235	236	212	281	211
3420	806	779	949	1 043	1 201	1 166	1 212	1 337	1 467
3432	90	79	106	101	116	105	122	130	134
3433	39	40	50	60	65	80	81	86	100
3434	76	72	80	80	73	72	67	82	75
3435	71	70	103	87	93	99	119	133	153
3441	98	111	97	131	144	157	183	242	219
3442	320	362	443	485	583	711	926	1 171	1 258
3443	408	407	461	653	769	974	1 001	1 162	1 240
3444	75	99	120	117	129	145	181	217	222
3452	61	52	49	60	72	82	120	158	189
3453	466	423	555	535	677	754	1 076	1 350	1 369
3454	123	112	134	109	143	174	292	356	448
3460	185	178	197	163	152	158	164	190	201
3470	143	118	129	110	114	115	127	148	147
3510	1 562	1 522	1 583	1 609	1 609	1 582	1 655	1 915	1 853
3521	9	10	12	7	8	10	8	8	14
3522	71	62	60	55	53	49	49	54	54
3523	49	48	39	24	18	16	16	18	24
3530	1 369	1 503	1 550	1 570	1 526	1 573	1 835	2 070	2 131
3610	300	296	306	149	135	209	463	206	199
3620	69	114	100	140	129	118	109	70	74
3633	29	30	16	12	10	21	21	19	24
3634	59	46	51	36	32	33	39	37	33
3640	1 164	1 311	1 931	2 339	2 844	2 895	3 181	3 797	4 292
3650	14	14	14	12	13	14	14	15	14
3710	175	192	236	267	288	293	309	340	338
3720	115	135	146	164	191	215	250	305	328
3731	15	18	23	23	29	37	45	50	41
3732	32	33	38	38	49	54	58	91	82
3733	246	303	366	348	367	339	372	326	350
3740	68	71	58	65	69	48	66	63	71

Table V.42

Estimated home market for engineering products in the United Kingdom, 1978-1986

(Millions of pounds at current prices)

Activity heading	*1978*	*1979*	*1980*	*1981*	*1982*	*1983*	*1984*	*1985*	*1986*
2234	696	786	751	690	739	706	754	828	863
3161	139	169	155	164	171	212	226	248	257
3162	59	67	64	67	62	76	92	94	100
3163	60	66	66	68	71	77	65	64	71
3164	1 039	1 155	1 102	1 064	1 155	1 283	1 430	1 493	1 552
3165	216	246	245	257	277	313	339	376	410
3166	273	368	368	356	376	434	500	544	617
3167	213	229	226	195	182	197	206	215	257
3169	1 189	1 426	1 332	1 572	1 830	2 043	2 277	2 431	2 670
3204	515	732	675	791	1 013	1 123	1 188	1 309	1 371
3205	722	893	980	1 398	1 528	1 894	1 585	1 457	1 659
3211	276	319	277	261	333	423	373	408	245
3212	351	496	384	408	504	633	493	496	579
3221	583	714	671	479	537	526	686	835	963
3222	514	582	665	530	529	547	667	773	813
3230	196	199	117	124	121	161	262	254	283
3244	297	333	367	355	436	504	533	512	565
3245	259	276	275	236	213	265	298	300	354
3251	464	518	559	537	649	577	415	541	706
3254	405	537	380	387	374	511	532	509	605
3255	895	1 019	1 022	851	1 051	1 249	1 388	1 518	1 674
3261	397	425	480	466	438	454	543	647	689
3262	202	219	232	193	205	198	219	242	256
3275	278	339	276	191	275	335	450	520	533
3276	282	347	337	322	339	373	446	575	629
3281	470	403	433	511	560	535	636	571	517
3283	446	494	522	418	492	489	599	654	668
3284	745	857	846	791	974	1 089	1 211	1 300	1 469
3285	140	196	203	218	110	142	242	259	277
3286	353	428	443	349	448	563	618	801	840
3287	246	277	260	244	283	301	339	381	369
3288	222	266	268	283	324	336	333	401	405
3289	576	629	645	688	668	751	1 033	1 069	1 431
3290	408	410	427	353	358	526	548	540	476
3301	129	171	146	146	163	196	217	232	278
3302	961	1 174	1 197	1 447	1 831	2 407	3 481	4 065	3 802
3410	548	701	780	809	832	948	1 117	1 204	1 312
3420	1 266	1 368	1 469	1 311	1 504	1 651	1 951	1 918	2 380
3432	259	297	288	273	303	282	312	362	364
3433	113	148	59	150	144	191	207	257	295
3434	468	509	477	418	471	560	675	774	786
3435	204	252	221	224	268	255	315	357	310
3441	547	643	917	1 093	1 256	1 461	1 526	1 744	1 896
3442	487	574	638	725	846	964	1 021	1 126	1 092
3443	807	1 053	1 352	1 726	1 880	1 994	2 348	2 508	2 596
3444	558	664	783	719	850	963	1 273	1 336	1 384
3452	122	147	136	133	127	178	225	247	317
3453	626	824	685	744	762	1 128	1 558	1 511	11 522
3454	787	938	886	1 255	1 508	1 790	1 637	1 725	1 825
3460	866	1 015	1 063	1 180	1 235	1 526	1 710	1 852	2 059
3470	351	418	439	520	573	652	727	760	868
3510	4 932	6 247	5 529	5 329	6 644	7 568	8 345	9 927	11 380
3521	193	200	224	212	216	245	250	301	289
3522	224	267	192	150	171	243	285	305	298
3523	117	138	143	132	137	161	186	198	226
3530	2 166	2 464	2 258	2 084	2 407	2 930	3 011	3 131	3 631
3610	573	695	793	732	1 204	1 002	873	1 330	644
3620	174	211	257	235	236	189	156	190	230
3633	124	108	148	130	98	84	62	79	78
3634	79	98	108	85	104	146	147	101	98
3640	1 835	2 250	3 590	2 184	2 987	3 052	3 379	3 895	3 869
3650	28	38	37	38	50	54	68	81	84
3710	566	650	724	676	802	1 001	1 175	1 351	1 423
3720	123	157	149	182	231	255	289	266	265
3731	88	99	111	115	111	137	166	181	182
3732	72	75	86	101	107	156	171	169	195
3733	263	274	222	255	266	356	383	456	563
3740	236	225	206	195	232	247	276	286	291

Table V.43

Imports of engineering products as a percentage of estimated home market in the United Kingdom, 1978-1986

Activity heading	*1978*	*1979*	*1980*	*1981*	*1982*	*1983*	*1984*	*1985*	*1986*
2234	7	7	9	9	11	13	14	15	16
3161	44	47	50	46	50	45	53	53	54
3162	57	59	58	60	65	58	59	58	55
3163	0	2	2	1	1	1	1	1	1
3164	4	4	7	5	6	7	7	7	8
3165	14	16	15	13	16	18	21	21	20
3166	15	13	14	15	19	19	20	23	22
3167	17	21	19	26	25	25	25	24	24
3169	17	17	19	16	17	19	21	21	22
3204	7	6	8	7	7	8	9	10	11
3205	8	16	17	26	14	16	15	9	8
3211	44	47	47	52	55	51	62	64	98
3212	47	44	48	46	51	50	66	73	67
3221	45	47	49	56	54	47	47	46	53
3222	19	18	19	22	26	29	35	30	28
3230	59	60	89	75	85	73	57	66	73
3244	53	55	57	59	58	62	67	69	71
3245	22	24	25	30	38	34	39	46	42
3251	2	2	3	2	2	3	4	4	3
3254	68	58	74	74	95	79	82	91	75
3255	24	24	26	29	28	25	27	29	30
3261	13	12	11	13	15	16	17	17	16
3262	46	48	43	47	51	55	43	69	68
3275	60	65	71	101	78	72	74	78	69
3276	54	53	62	58	63	71	76	67	67
3281	30	36	31	32	32	37	36	46	53
3283	22	23	21	29	32	34	33	35	36
3284	17	19	19	21	20	21	24	24	25
3285	42	36	35	36	90	75	54	53	55
3286	61	58	60	72	69	63	78	71	70
3287	40	39	39	38	38	39	42	42	45
3288	23	24	23	28	28	33	35	39	35
3289	6	7	7	7	8	8	7	8	6
3290	12	15	17	18	20	17	14	19	28
3301	119	106	127	118	121	128	123	126	112
3302	87	87	90	91	95	105	102	96	103
3410	7	7	9	9	12	14	17	19	19
3420	25	25	26	31	34	38	42	46	40
3432	18	16	18	24	28	32	35	35	38
3433	12	12	11	15	17	18	20	16	19
3434	18	17	16	20	19	21	19	19	19
3435	34	33	39	38	41	44	43	42	46
3441	10	10	8	9	10	14	15	18	19
3442	59	59	65	66	70	73	85	88	91
3443	28	27	25	27	29	33	33	33	33
3444	22	22	20	24	26	29	31	32	33
3452	33	31	37	47	55	63	67	68	64
3453	68	77	77	84	100	92	99	114	115
3454	47	53	66	65	71	69	65	69	70
3460	29	30	30	36	38	39	43	45	45
3470	24	20	19	17	19	20	21	24	25
3510	42	48	45	47	51	58	54	51	52
3521	2	4	5	4	8	7	6	6	6
3522	11	10	11	13	19	19	17	21	19
3523	3	4	4	7	5	3	4	5	4
3530	29	33	31	39	43	45	49	57	58
3610	20	23	18	32	22	11	34	11	42
3620	16	8	6	9	8	11	12	9	13
3633	102	107	97	101	100	119	131	116	121
3634	34	37	43	46	48	49	57	50	58
3640	45	46	51	69	59	62	68	72	63
3650	7	8	8	11	12	15	13	14	17
3710	19	16	17	19	20	21	21	19	18
3720	59	57	54	57	56	64	69	81	84
3731	40	40	38	35	41	42	43	49	49
3732	51	49	47	48	46	45	47	56	51
3733	105	114	147	153	139	147	132	119	
3740	68	78	77	84	72	74	82	87	93

Table V.44

Exports of engineering products as a percentage of grossed up principal product sales in the United Kingdom, 1978-1986

Activity heading	*1978*	*1979*	*1980*	*1981*	*1982*	*1983*	*1984*	*1985*	*1986*
2234	13	13	12	14	14	15	16	17	16
3161	54	49	55	53	55	45	50	50	48
3162	66	62	66	67	73	64	62	63	57
3163	-	2	2	1	1	1	0	0	1
3164	8	9	9	9	9	9	9	10	11
3165	7	8	8	7	8	8	8	7	6
3166	20	14	15	15	17	15	12	13	11
3167	15	15	16	16	17	16	15	15	13
3169	23	20	24	19	18	17	18	19	17
3204	38	28	32	26	26	23	21	22	17
3205	28	28	26	18	17	15	17	16	14
3211	48	45	47	50	46	35	50	45	58
3212	77	72	80	76	72	66	83	86	79
3221	46	44	52	62	59	51	47	45	42
3222	24	22	23	28	30	30	33	29	26
3230	71	73	95	88	92	80	63	72	76
3244	64	65	65	66	63	63	66	71	71
3245	42	44	48	54	66	56	54	51	52
3251	10	15	16	12	12	11	21	25	13
3254	85	77	90	90	98	87	89	96	85
3255	35	34	38	42	35	28	31	32	33
3261	18	17	19	20	22	19	19	17	15
3262	44	47	47	52	52	54	63	68	67
3275	66	66	78	100	81	71	67	69	66
3276	56	53	63	62	65	70	75	68	67
3281	56	61	62	60	57	56	56	65	69
3283	35	31	34	46	43	44	44	45	47
3284	19	20	20	22	19	18	18	19	17
3285	55	46	47	44	50	76	52	54	54
3286	70	65	68	82	77	67	79	71	71
3287	56	52	59	61	57	52	52	53	57
3288	41	39	45	50	46	45	46	46	44
3289	14	14	17	18	18	15	13	15	11
3290	36	35	38	53	61	48	44	46	55
3301	122	108	132	123	130	145	136	136	116
3302	83	85	89	88	93	107	103	96	104
3410	24	18	19	20	24	22	19	22	17
3420	46	43	47	54	55	53	52	56	51
3432	30	24	31	33	35	35	38	35	37
3433	28	23	26	32	35	34	33	29	30
3434	16	15	17	19	16	14	11	12	11
3435	35	29	43	39	37	41	40	39	48
3441	17	16	10	12	11	11	12	15	13
3442	61	61	66	66	70	73	86	90	93
3443	41	35	31	34	37	42	40	41	42
3444	15	16	16	18	17	17	17	19	19
3452	44	34	36	46	56	55	62	67	62
3453	70	56	78	82	100	87	99	119	119
3454	23	20	26	20	24	24	33	40	45
3460	23	20	21	18	17	15	14	16	15
3470	35	26	27	21	20	18	18	20	18
3510	35	32	36	33	29	33	30	28	26
3521	5	5	5	3	4	4	3	3	5
3522	26	20	26	30	28	20	17	18	18
3523	31	27	22	16	12	9	8	9	10
3530	47	48	50	55	53	50	54	60	58
3610	40	36	32	23	13	19	45	15	35
3620	22	37	29	39	38	36	44	29	27
3633[a]	112	136	76	109	100	420	1 050	317	300
3634	53	43	45	44	37	31	38	43	45
3640	53	52	52	78	70	72	74	77	75
3650	35	29	29	27	23	23	19	18	17
3710	28	26	28	33	31	27	25	24	23
3720	70	68	68	67	65	70	73	86	89
3731	22	23	25	23	31	32	32	35	31
3732	48	46	45	42	46	39	39	55	46
3733	95	114	140	164	171	170	193	181	143
3740	48	59	55	67	51	43	56	63	77

a The figures for this activity heading are included for completeness, but because the exports figure includes re-exports and United Kingdom production is comparatively small, this ratio has no real significance.

Table V.45

Producer price index for selected engineering products in the United Kingdom, 1978-1986 [a]

(1980 = 100)

Activity heading	*1978*	*1979*	*1980*	*1981*	*1982*	*1983*	*1984*	*1985*	*1986*
3161	73.5	83.2	100.0	113.4	121.7	128.2	134.5	141.1	148.5
3163	77.8	86.0	100.0	108.6	120.3	131.6	146.1	158.4	168.2
3164	79.4	87.5	100.0	103.2	109.5	116.3	124.1	130.2	133.8
3165	78.8	86.1	100.0	113.7	122.8	129.9	133.7	139.3	145.9
3169	76.4	87.1	100.0	105.7	110.2	115.3	123.0	130.2	136.3
3211	77.1	88.3	100.0	104.9	110.4	114.2	119.8	126.7	131.5
3221	74.8	85.0	100.0	107.4	115.9	123.1	133.2	146.9	153.7
3230	76.3	86.0	100.0	106.9	111.3	113.3	118.9	122.0	125.5
3244	75.2	86.4	100.0	111.0	125.7	136.2	142.2	152.2	161.7
3251	76.5	85.9	100.0	109.8	116.8	133.2	139.9	148.3	152.7
3254	82.8	90.4	100.0	108.9	116.4	119.7	124.0	131.8	136.1
3255	79.0	88.2	100.0	108.2	115.9	120.9	125.1	129.7	135.4
3261	74.7	84.6	100.0	107.0	108.0	113.5	118.9	129.3	136.0
3262	73.0	83.3	100.0	110.2	119.2	123.7	127.9	132.8	138.2
3283	75.7	85.4	100.0	109.2	119.4	127.8	134.9	142.8	151.0
3285	76.9	87.1	100.0	108.8	116.0	120.2	125.4	128.8	131.3
3287	74.5	85.1	100.0	106.4	113.5	121.6	127.7	134.0	141.9
3288	76.8	86.0	100.0	111.0	127.4	136.1	145.6	157.2	167.6
3301	79.9	86.9	100.0	114.8	110.1	112.1	118.4	123.8	128.0
3410	66.9	80.8	100.0	102.8	111.6	124.8	138.2	144.2	143.6
3432	76.8	90.5	100.0	99.5	103.3	109.1	112.0	127.3	131.7
3434	75.9	85.3	100.0	111.7	117.1	122.7	128.4	133.4	142.8
3442	79.4	86.8	100.0	106.2	112.9	120.6	127.3	136.4	142.9
3460	79.9	88.1	100.0	105.2	103.2	106.3	109.6	114.8	119.9
3510	79.1	88.8	100.0	108.8	115.4	119.9	127.4	136.2	145.7
3710	79.2	88.1	100.0	111.2	122.1	132.1	140.8	147.7	156.1

a Producer price index not available for activity headings not listed.

Table V.46

Crude engineering trade balance in the United Kingdom, 1978-1986

(Millions of pounds at current prices)

Activity heading	*1978*	*1979*	*1980*	*1981*	*1982*	*1983*	*1984*	*1985*	*1986*
2234	50	49	27	35	28	16	20	22	-5
3161	28	9	17	27	19	-1	-11	-12	-28
3162	15	7	15	14	19	13	9	15	5
3163	0	0	0	0	0	0	-1	-1	0
3164	46	57	30	44	47	28	32	46	48
3165	-16	-23	-19	-15	-26	-34	-48	-57	-60
3166	20	8	5	0	-6	-21	-43	-61	-77
3167	-6	-15	-6	-24	-16	-21	-23	-23	-34
3169	93	48	93	60	29	-43	-78	-68	-171
3204	258	225	241	222	273	223	177	203	103
3205	192	160	117	-147	70	-30	36	120	123
3211	21	-11	0	-14	-56	-101	-94	-106	-92
3212	455	477	541	506	374	283	523	484	313
3221	18	-50	34	86	61	36	-3	-15	-132
3222	31	35	33	50	37	9	-18	-3	-24
3230	86	95	138	132	109	58	44	53	42
3244	94	96	90	75	57	17	-12	26	10
3245	92	104	120	121	175	132	97	139	111
3251	37	81	84	62	71	53	92	90	83
3254	464	417	561	636	566	328	344	502	403
3255	149	144	196	187	115	56	63	55	84
3261	23	24	44	42	36	18	9	1	-5
3262	-7	-5	16	21	5	-3	-13	-13	-3
3275	43	10	78	68	53	-11	-100	-62	-48
3276	15	-1	5	29	24	-9	-13	5	9
3281	281	260	353	368	329	229	284	306	276
3283	92	55	101	131	101	100	114	124	133
3284	26	10	12	10	-13	-37	-76	-90	-143
3285	41	34	45	31	12	0	-9	9	-6
3286	106	87	112	190	160	67	32	-21	23
3287	89	75	127	150	126	83	70	93	103
3288	68	66	92	123	108	78	68	53	69
3289	56	51	79	89	81	67	72	87	77
3290	155	123	148	269	384	321	287	266	302
3301	-13	-48	-25	-33	-49	-76	-78	-64	-62
3302	-200	-190	-144	-400	-553	-872	-962	-606	-824
3410	118	94	97	101	134	103	24	52	-38
3420	492	431	571	638	682	624	391	449	524
3432	43	32	55	36	32	16	12	5	-3
3433	26	23	32	38	40	46	39	44	43
3434	-6	-14	3	-2	-15	-48	-62	-63	-76
3435	1	-14	17	1	-16	-14	-16	-18	10
3441	44	50	27	35	19	-43	-51	-76	-146
3442	35	24	29	10	-9	10	60	175	261
3443	181	120	129	180	224	317	285	343	373
3444	-49	-47	-35	-55	-96	-132	-210	-208	-230
3452	21	7	-1	-3	2	-30	-31	-11	-13
3453	38	-66	27	-89	-86	-286	-466	-372	-375
3454	-242	-387	-363	-710	-922	-1 035	-764	-840	-836
3460	-67	-125	-127	-265	-315	-443	-570	-644	-729
3470	58	34	44	22	6	-17	-27	-34	-70
3510	-507	-1 486	-926	-909	-1 802	-2 793	-2 852	-3 181	-4 125
3521	6	1	1	-2	-10	-8	-7	-11	-4
3522	47	36	40	35	21	2	1	-10	-4
3523	46	42	33	15	1	11	8	8	15
3530	741	688	855	757	487	247	356	297	18
3610	187	135	161	-84	-132	99	164	57	-69
3620	42	97	84	120	108	97	91	52	44
3633	-98	-86	-127	-119	-80	-79	-60	-73	-70
3634	32	10	5	-3	-18	-33	-44	-14	-24
3640	347	278	111	831	1 087	996	893	1 010	1 872
3650	12	11	11	7	7	6	5	4	0
3710	70	85	116	141	126	87	64	82	78
3720	42	51	66	61	62	52	52	89	105
3731	-20	-22	-19	-17	-16	-20	-38	-49	
3732	-5	-4	-2	-10	0	-16	-23	-3	-17
3733	-30	-9	39	-43	-51	-157	-190	-276	-319
3740	-93	-105	-101	-98	-98	-136	-159	-186	-199

Table V.47

Normalized engineering trade balance in the United Kingdom, 1978-1986 [a]

Activity heading	*1978*	*1979*	*1980*	*1981*	*1982*	*1983*	*1984*	*1985*	*1986*
2234	.35	.30	.18	.21	.15	.08	.09	.08	-.02
3161	.19	.05	.10	.15	.10	-.01	-.05	-.05	-.11
3162	.18	.08	.17	.15	.19	.13	.08	.12	.04
3164	.36	.37	.17	.29	.26	.14	.13	.17	.16
3165	-.36	-.40	-.35	-.29	-.41	-.44	-.51	-.55	-.59
3166	.20	.08	.05	0	-.04	-.15	-.28	-.32	-.39
3167	-.09	-.19	-.08	-.31	-.22	-.27	-.29	-.28	-.39
3169	19	.09	.15	.10	.04	.-06	-.09	-.07	-.17
3204	.80	.73	.70	.68	.67	.57	.46	.45	.26
3205	.62	.36	.26	-.25	.14	-.05	.07	.31	.33
3211	.08	-.04	0	-.05	-.18	-.31	-.25	-.25	-.24
3212	.58	.52	.59	.58	.42	.31	.45	.40	.29
3221	.03	-.08	.05	.14	.09	.07	0	-.02	-.15
3222	.14	.15	.11	.18	.12	.03	-.04	-.01	-.06
3230	.27	.28	.40	.42	.35	.20	.13	.14	.09
3244	.23	.21	.18	.15	.10	.03	-.02	.04	.01
3245	.45	.44	.47	.46	.52	.42	.30	.34	.27
3251	.63	.79	.72	.76	.70	.60	.74	.68	.65
3254	.46	.40	.50	.52	.44	.29	.28	.35	.31
3255	.26	.22	.27	.27	.16	.08	.08	.06	.08
3261	.18	.18	.29	.26	.21	.11	.05	0	-.02
3262	.04	-.02	.07	.10	.02	-.01	-.05	-.04	-.01
3275	.11	.02	.17	.15	.11	-.02	-.18	-.09	-.07
3276	.05	0	.01	.07	.05	-.02	-.02	.01	.01
3281	.50	.47	.57	.53	.48	.36	.38	.37	.32
3283	.32	.20	.32	.35	.24	.23	.22	.21	.22
3284	.09	.03	.04	.03	-.03	-.09	-.15	-.17	-.24
3285	.26	.19	.24	.17	.06	0	-.04	.03	-.02
3286	.20	.15	.17	.28	.21	.09	.03	-.02	.02
3287	.31	.26	.39	.45	.37	.26	.20	.23	.24
3288	.40	.34	.40	.44	.38	.26	.22	.15	.20
3289	.45	.38	.46	.49	.42	.37	.32	.34	.30
3290	.61	.50	.51	.68	.73	.65	.64	.56	.53
3301	-.04	-.15	-.07	-.11	-.14	-.18	-.17	-.12	-.11
3302	-.14	-.10	-.07	-.18	-.19	-.21	-.16	-.08	-.12
3410	.61	.47	.42	.40	.40	.28	.06	.10	-.08
3420	.44	.38	.43	.44	.40	.30	.19	.20	.22
3432	.31	.25	.35	.22	.16	.08	.05	.02	-.01
3433	.50	.40	.47	.46	.44	.40	.32	.34	.27
3434	-.04	-.09	.02	-.01	-.09	-.25	-.32	-.28	-.34
3435	.01	-.09	.09	.01	-.08	-.07	-.06	-.06	.03
3441	.29	.29	.16	.15	.07	-.12	-.12	-.14	-.25
3442	.06	.03	.03	.01	-.01	.01	.03	.08	.12
3443	.29	.17	.16	.16	.17	.19	.16	.17	.18
3444	-.25	-.19	-.13	-.19	-.27	-.31	-.37	-.32	-.34
3452	.21	.07	-.01	-.02	.01	-.15	-.11	-.03	-.03
3453	.04	-.07	.02	-.08	-.06	-.16	-.18	-.12	-.12
3454	-.50	-.63	-.58	-.77	-.76	-.73	-.57	-.54	-.48
3460	-.15	-.26	-.24	-.45	-.51	-.58	-.63	-.63	-.64
3470	.24	.17	.21	.11	.03	-.07	-.10	-.10	-.19
3510	-.14	-.33	-.23	-.22	-.36	-.47	-.46	-.45	-.53
3521	.50	.05	.04	-.13	-.38	-.29	-.30	-.41	-.13
3522	.49	.41	.50	.47	.25	.02	.01	-.08	-.04
3523	.88	.78	.73	.45	.03	.52	.33	.29	.45
3530	.37	.30	.38	.32	.19	.09	.11	.08	0
3610	.45	.30	.36	-.22	-.33	.31	.22	.16	-.15
3620	.44	.74	.72	.75	.72	.70	.59	.42	
3633	-.63	-.59	-.80	-.83	-.69	-.65	-.59	-.66	-.59
3634	.37	.12	.05	-.04	-.22	-.30	-.36	-.16	-.27
3640	.18	.12	.03	.22	.24	.21	.16	.15	.28
3650	.75	.65	.65	.41	.37	.27	.22	.15	0
3710	.25	.28	.33	.36	.28	.17	.12	.14	.13
3720	.22	.23	.29	.23	.19	.14	.12	.17	.19
3731	-.40	-.38	-.29	-.27	-.22	-.21	-.22	-.28	-.37
3732	-.07	-.06	-.03	-.12	0	-.13	-.17	-.02	-.09
3733	-.06	-.01	.06	-.06	-.06	-.19	-.20	-.30	-.31
3740	-.41	-.43	-.47	-.43	-.52	-.59	-.55	-.60	-.58

a Defined as (Exports - Imports) / (Exports + Imports).

Table V.48

Structure of employment in the engineering industries in the United Kingdom, 1978-1986

(Thousands)

Group		1978	1979	1980	1981	1982	1983	1984	1985	1986
31	Metal goods	518	508	472	410	379	346	333	321	308
32	Mechanical engineering	1 043	1 045	1 000	913	844	775	757	75 7	730
33	Office machinery and data data processing equipment	67	75	77	78	79	80	85	91	92
34	Electrical and electronic electronic engineering	759	764	735	675	635	615	607	594	570
35	Motor vehicles and parts	474	465	426	357	318	295	279	268	252
36	Other transport equipment	392	391	381	367	349	327	302	290	280
37	Instrument engineering	126	130	124	115	111	104	105	105	105

Table V.49

Value of shipments by selected industries in the United States, 1979-1986

(Shipments in millions of 1982 US dollars - industry data)

SIC	Industry/type of product	1979	1982	1984	1986[a]	Index 1986-1979
3541, 3542	Metalworking machine tools	8 193	5 841	4 337	4 496	54.9
3531	Construction machinery	21 259	11 658	12 108	12 160	57.2
3585	Air conditioning, refrigeration and heating equipment	15 142	12 390	14 980	15 611	103.1
3523	Farm machinery and equipment	17 617	10 743	9 337	7 209	40.9
3532, 3533	Mining and oilfield machinery	11 884	13 304	7 565	6 279	52.8
361, 3621, 3622	Electrical equipment	21 596	18 497	19 457	18 891	87.5
3573	Electronic computing equipment [b]	21 466	36 767	53 524	53 244	248.0
3662	Radio and television communication equipment	24 562	33 032	37 311	43 709	178.0
3661	Telephone and telegraph apparatus	12 319	13 394	15 183	16 689 [c]	135.5 [d]
367	Electronic components and accessories	23 104	34 517	48 290	54 282	234.9
234.9						
381, 3822, 3823, 3824, 3829, 383	Precision instruments [e]	13 568	15 341	16 594	17 947	132.3
3711	Motor vehicles and car bodies	107 658	70 740	112 404	123 025	114.3
3465, 3592, 3647, 3691, 3694, 3714	Motor vehicle parts and stampings	85 469	54 204	78 300	87 374	102.2
372, 376	Aerospace	66 742	66 425	71 276	86 675	129.9

Source: 1988 U.S. Industrial Outlook, U.S. Department of Commerce, Washington D.C.

a Estimated except for exports and imports.
b Value at current prices.
c 1985 shipments.
d Index 1985-1979.
e Includes scientific instruments, optical devices, measuring and controlling instruments with the exception of instruments to measure electricity (SIC 3825).

REFERENCES

V.1. The original contributions as provided by Governments in 1986 were considered in their entirety by the second *ad hoc* Meeting for the present study, held in February 1987. Individual national contributions were issued in the three working languages of ECE (English, French and Russian) under the following symbols: ENG.AUT/AC.10/R.2/Add.18 (Ukrainian SSR), Add.19 (Austria), and Add.20 (Czechoslovakia). Document ENG.AUT/AC.10/R.2/Add.21 (USSR) is available in English only.

V.2. These national contributions, edited by the secretariat, were presented to and discussed at the third *ad hoc* Meeting for the study, held in December 1987. The contribution of Sweden was issued in English only (ENG.AUT/AC.10/R.3/Add.17). Other texts are available in the three ECE working languages with document symbols allocated as follows: ENG.AUT/AC.10/R.3/Add.9 (Belgium), Add.10 (Poland), Add.11 (German Democratic Republic), Add.12 (Italy), Add.13 (Bulgaria) and Add.16 (updated information for USSR).

V.3. The contribution of the United Kingdom was considered by the third *ad hoc* Meeting for the study in the form of a conference room paper (ENG.AUT/AC.10/CRP.4/Add.1).

V.4. The information provided by individual ECE member countries in answer to regular questionnaires for *Annual Reviews of Engineering Industries and Automation* includes, in addition to statistical data, a verbal part on recent developments. These national statements on the preceding year's achievements in national engineering industries form a special chapter in the *Annual Review*.

V.5. Transmitted by the Government of Austria in November 1986.

V.6. Joseph A. Schumpeter (1983-1950), Austrian economist. According to Schumpeter, economic development proceeds by fits and starts. Dynamic entrepreneurs take up new inventions and use them to introduce innovations into their companies, thus gaining an advantage of time over their competitors. With a time lag, other entrepreneurs imitate these innovations, which process of adaptation will eventually lead to an equilibrium on a higher level. However, this equilibrium will be disturbed by further innovations.

V.7. Transmitted by the Government of Belgium in September 1987.

V.8. Based on a text provided by the Government of Bulgaria (see ECE document ENG.AUT/AC.10/R.3/Add.13). Further modifications and editorial changes were introduced by A. Alexandrov, General Director of the Industry Development Institute in Sofia, and by by S.A. Elekoev of the IMEMO Institute, USSR Academy of Sciences in Moscow, consultant to the ECE secretariat.

V.9. Transmitted by the Government of Czechoslovakia in November 1986.

V.10. For detailed information, see also technical papers presented by Czechoslovak experts to the ECE Seminar on Industrial Robotics '86 - International Experience, Developments and Applications, held at Brno (Czechoslovakia) in February 1986. The list of papers is annexed to the Seminar report (document ENG.AUT/SEM.5/4).

V.11. See also Theses on Engineering and Automation in Czechoslovakia, presented by I. Angelis, Deputy Director of the Research Institute for Foreign Economic Relations in Prague, to the ECE Symposium on East-West Business Opportunities and Trade Prospects held at Thessaloniki (Greece) in September 1986.

V.12. Based on a national contribution prepared by Professor H. Weber of the Technical University at Karl-Marx-Stadt. Note that, by definition, mechanical engineering in the German Democratic Republic does not include electrical machinery or electronics (see also data presented in chapter III of the present study).

V.13. Transmitted by the Government of Hungary in December 1987. This text was not included in the draft version of the present study.

V.14. Transmitted by the Government of Italy in August 1987.

V.15. Based on a paper on the role of engineering industries in the Polish economy transmitted by the Government of Poland in June 1987 (ECE document ENG.AUT/AC.10/R.3/Add.10). Further editorial changes were introduced by R. Bandorowicz, Director of the PROMASZ Institute in Warsaw.

V.16. Problems related to the Polish and the world-wide shipbuilding industries were considered at the ECE Seminar "Shipbuilding 2000 Maritime Conference - BALTEXPO '88", held at Gdansk (Poland) in September 1988 (report ENG.AUT/SEM.7/3).

V.17. For more information, see "Polish Metallurgy and Engineering Industry", special edition of *Polish Engineering*, available from the editors, Marszalkowska Street 124, P.O. Box 726, 00-950 Warszawa (Poland).

V.18. Compiled by the ECE secretariat on the basis of publications provided by the Government of Sweden: *Engineering Industries - Structure and Development* (in Swedish), Statens Industriverk SINDDATA 1987, *State of Competition 1987* (in Swedish), Statens Industriverk SINDCOM 1987, and *The Swedish Engineering Industry*, reports published jointly by The Swedish Engineering Employers' Association and The Swedish Association of the Mechanical and Electrical Industry.

V.19. Transmitted by the Government of the Ukrainian SSR in October 1986 and in February 1987 (1985 data).

V.20. Compiled by S.A. Elekoev, consultant to the ECE secretariat (see also ref. V.8). The complete text of the USSR contributions was issued in documents ENG.AUT/AC.10/R.2/Add.21 and ENG.AUT/AC.10/R.3/Add.16.

V.21. See also the USSR national contribution as published in the *Annual Review of Engineering Industries and Automation, 1986*, vol.I, chapter I V. UN/ECE publication, Sales No. E.88.II.E.16.

V.22. *Pravda*, Moscow 18 January 1987. (In Russian)

V.23. *USSR in Figures 1986* (SSSR v tsifrakh v 1986 godu), p.67, Moscow 1987. (In Russian)

V.24. Same source as V.22

V.25. Same source as V.23, p.69.

V.26. *Statistical Bulletin* (Vestnik Statistiki), No.5, p.62, Moscow 1987. (In Russian).

V.27. Same source as V.23, p.98.

V.28. *USSR National Economy* (Narodnoe khozyaistvo SSSR v 1985 godu), p.48 Moscow 1986. (In Russian)

V.29. Same source as V.28, p.368.

V.30. *Kommunist*, No.4, p.57, Moscow 1987. (In Russian)

V.31. Based on the information transmitted by the Government of the United Kingdom of Great Britain and Northern Ireland in December 1987. The contribution was updated by D.H. Hewer of the Department of Trade and Industry in London.

V.32. Tables V.37-V.48 are published as received. Their eventual shortening would mispresent the global, consistent picture of the structural development. The situation in the United Kingdom's engineering industries to some extent reflects general trends registered in other major engineering-oriented countries.

V.33. For more information see, for instance,

- National Contribution of the United Kingdom as published in the *Annual Review of Engineering Industries and Automation, 1986*, vol.I, chapter IV. UN/ECE publication, Sales No. E.88.II.E.16;

- *Review of Changes in Overall Policies and of Agreements on Scientific and Technological Co-operation: Review of Changes in the United Kingdom 1983-1987.* ECE document submitted to the sixteenth session of the Senior Advisers to ECE Governments on Science and Technology (SAST) in September 1988 (document SC.TECH/R.221/Add.9);

- *Sector Profiles,* individual issues as published by the Engineering Industry Training Board (EITB). Available from EITB Publications, P.O. Box 75, Stockport, Cheshire, United Kingdom.

V.34. Compiled by the secretariat on the basis of the United States contribution to the *Annual Review of Engineering Industries and Automation, 1986*, vol.I, chapter IV.. UN/ECE publication, Sales No. E.88.II.E.16; and the *1988 U.S. Industrial Outlook.* Department of Commerce, Washington, D.C.

CHAPTER VI CURRENT TRENDS AND FORESEEABLE DEVELOPMENTS IN ENGINEERING INDUSTRIES AND CONCLUSIONS OF THE STUDY

VI.1. GENERAL TRENDS AND OUTLOOK

Throughout the post-war period, the engineering industries of ECE countries have developed more rapidly than the economies as a whole. As a result, the share of the engineering industries in the total volume of industrial production increased in the great majority of countries and is now on average 30-40%. Some convergent trends have occurred in the structure of the engineering industries in all the major ECE countries.[VI.1]

The structure of the engineering industries in western Europe and North America has been affected by the intensification of international competition, especially from Japan and the newly industrialized countries, the conversion of national markets virtually into international markets, and the expansion of direct foreign investment participation in the fixed capital of subsectors of the engineering industries.

Formation of the new production and organizational structure of all spheres of the economy in ECE countries, which began in the mid-1970s, will continue for the next few years. Apparently, however, transformation of the economic sectors will no longer require the increased volume of production in the engineering industries to outstrip the increase for industry as a whole in the leading countries. A state of relative saturation of markets with investment and consumer goods has been reached, and the development of the engineering industries will be not so much in terms of quantity as of quality. Competition, the individualization of demand, and greater insistence on innovation and quality in engineering output will stimulate a shift of emphasis from economies of scale in production to economies in the diversity of commodities produced in smaller batches.

An analysis of investment flows indicates that the rate of increase of investment was in general reduced in the 1980s, but that the modernization of existing capacity proceeded more rapidly than the growth of production capacity. On average, 60-70% of gross capital investment went to replace withdrawals and for modernization: in the motor vehicle industry of the United States, for example, between 87 and 98% of capital investment in the years 1980-1985 was expended on modernization.[VI.2]

Expansion of production capacity predominated in centrally planned economies and only some 40% of total capital investment went on re-equipment and reconstruction.

What is extremely important in the current reconstruction of production facilities in the engineering industries is that its technical basis is fundamentally new. High-technology production accounts for up to 40% of total capital investment in machinery and equipment. Moreover, since the productivity of the new technology is greater than that of the technology which it replaces, and since the cost of a unit of effective capacity is, as a rule, lower, less machinery and equipment will be needed for the same volume of output. This may be illustrated by the fact that the growing use of numerically controlled machine tools and flexible manufacturing units, and reduction in the material consumption of production, enabled a number of countries with large numbers of machine tools in use (e.g. United States, United Kingdom) to reduce the numbers of metalcutting machine tools by some 15-20% during the years 1973-1983, while increasing the volume of production and maintaining the former replacement coefficients.

As regards the actual structure of the engineering industries, the prospect is apparently one of an increase in the share of the engineering industries in industry as a whole, except for electrical and electronic engineering and the manufacture of precision instruments.

One factor that is holding back a further increase in the share of the engineering industries in industry as a whole is the process of hiving off as service industries such functions of the production infrastructure as the programming and servicing of computers and automated manufacturing equipment, and of automated design and control systems; the design of integrated manufacturing systems and local communication networks; the rendering of services in engineering, leasing, the training of key workers, and consultancy services in other branches, etc. Thus, whereas 15 years ago engineering companies themselves produced 85-90% of the application software for their computer-controlled equipment, some 50% of the software is now bought from outside. In 1985, sales of computer programs by the specialized sector in the United States reached 21.5 billion dollars. There are now some 4,300 suppliers providing services in this area in the United States, 3,000 in the United Kingdom and more than 2,500 in Japan. The scale of leasing is approaching 20% of the volume of capital investment by industry in machinery and equipment in a number of the leading market economies.

The high-technology segments of the engineering industries have had undoubted priority in development over the past 10 years and they have been decisive for the rate and scale of scientific and technical development in every country. During the 1960s and 1970s, engineering companies concentrated on studying demand and refining their market strategy. In the 1980s, technological competition occupied the foreground, with stress on innovation and the rapidity with which production geared to individual demand could be mastered, and on the organization of pre-sale and after-sale servicing for customers. The crucial factors for success then became the scale and rate of the technological re-equipping of production based on the most recent scientific and technological developments, accelerated development of new commodities ("from the idea to the marketing of the commodity") and a shortening of the production cycle ("door to door"), increased R and D expenditure and expanded scientific and technical co-operation, the establishment of venture companies, and purposeful stimulation of creative and innovative work.

The economic effectiveness and competitive status of countries are increasingly dependent on advances in microelectronics, computer technology and telecommunications; on the development of automated equipment and integrated manufacturing systems; on the mastering of technologies that economize on the use of materials and of laser, plasma and other non-traditional technologies; on progress in the development of machine intelligence, and of new monitoring devices, materials etc. Although all these technologies are basically of importance, an idea of the scale of the effort being made in some countries may be gleaned from the information in tables VI.1-VI.3. [VI.3]

The following should be included among the main trends in the technological and organizational reconstruction of the engineering industries: [VI.4]

- Computerization of manufacturing and control processes, and of the related research and development sphere;
- Transition from the introduction of separate units of automated equipment to integrated automation, further development of computer-integrated manufacturing systems, automated production shops and factories, and the imparting of features of continuous production to the engineering industries;
- The mastering of new technologies, especially those that are sparing in their use of the factors of production (labour, capital, raw materials, semi-products and energy) and that help to increase product quality;
- Accelerated renewal of production facilities and equipment, and reduction of the time taken to design and introduce new products;
- An increased role for small firms, and the perfecting of subcontracting relations; and
- Strengthening of research and development activities, especially for long-term projects with an element of risk. Encouragement for the setting-up of venture companies.

Almost all the trends of scientific and technical development noted in the engineering industries became possible through advances in the electronics industry. This subsector is developing at an extremely rapid rate (10-20% annually) and its share of the gross domestic product of market economies, which was 4.7% in 1985 (2.6% in 1970), is expected to reach 8% by the year 2000. Growth in the manufacture of electronic parts and components is accelerating (average annual growth rates in the period 1976-1984 were 27% in Japan, 20% in the United States and 12-15% in the countries of western Europe). In 1984 alone, the manufacture of integrated circuits increased by 66% in Japan and by 40% in the United States, and the growth rate for the manufacture of equipment to produce integrated circuits and microprocessors outstripped the growth rate for the manufacture of these electronic components themselves. In the United States in 1985, more than 600,000 microprocessors were installed in equipment for the direct control of manufacturing processes.

Nevertheless, computers were the most important output of the sector. The volume of computer manufacture has increased at average annual rates of the order of 30% over the past 10 years, despite the vast scale on which computers and related equipment are produced (table VI.4). Whereas the output of computers in ECE countries and Japan was estimated at practically $100 billion in 1985, it seems likely that by 1990 that level of output will be reached in the United States alone.

It would be impossible to use computers as a tool to increase the efficiency of production, design and control without adequate development of communication systems and equipment. Communications are becoming one of the key elements of the developing unified national industrial information system and, for the future, of a world-wide system. "The new integrated digital telecommunication systems and networks, which are currently under implementation in several ECE countries, represent an investment in infrastructure perhaps as important for industrial and economic development as were earlier investments in railways, roads and electricity transmission". [VI.5] More than 70% of the computers currently in use in the United States are intended for application in conjunction with telecommunication lines.

Average annual rates of increase in the output of computers, telecommunication equipment and related terminals during the years 1976-1985 exceeded 18% in the United States and 10% in Japan. According to [VI.6], total world sales of the products of the information industry reached an estimated 450 billion dollars in 1984 and are expected to rise to $830 billion by 1990.

Computer-aided design (CAD) systems are becoming increasingly prevalent. The development of new products is being appreciably speeded up and the quality of projects is being improved by the provision to engineers and scientists at their place of work of CAD terminals and personal computers connected to CAD systems. In the space of only four years, between 1980 and 1984, the world CAD market increased from $0.6 billion to 2.4 billion, an annual growth rate of some

40%. Some 250,000 flexible manufacturing cells and units have already been set up in the United States.

Growth rates will undoubtedly remain high in the immediate future in the electronic engineering industries of the ECE countries and Japan, especially for the manufacture of computers, telecommunication equipment, electronic components, numerically controlled systems and CAD.

As has already been pointed out in the *Annual Reviews of Engineering Industries and Automation* and in specific ECE studies, [VI.7] the manufacture of automated industrial equipment in all the countries considered is increasing at very high rates, although some slowing of the growth rate has been noted in certain years. The machine-tool industry, which is a key sector for the manufacture of industrial equipment, is becoming increasingly a high-technology sector. The production capacity of the machine-tool industry is being actively modernized, with the introduction of CAD and flexible manufacturing cells, units and systems. In order to speed the development of new types of products, to cheapen R and D and to reduce the level of risk, machine-tool manufacturing companies are expanding co-operation on R and D, strengthening international scientific, technical and licensing connections and creating venture companies. The promising trends in the development of new products in the machine-tool industry include the creation of equipment for the working of non-metallic materials; equipment suited to the needs of small companies; equipment for the integration of production processes; high-precision equipment; and machinery and apparatus for non-traditional technologies.

Numerically controlled machine tools are the most developed and widely employed type of automated equipment. The output of numerically controlled machine tools increased practically tenfold (in cost terms) during the years 1975-1985 in the Federal Republic of Germany, by a factor of 17 in Japan and 3.2 times in the USSR, and by 1985 accounted for 40-60% of the cost of all machine tools produced in all the major countries manufacturing metalworking lathes.

A great deal of literature and much research have been devoted to programmable industrial robots. In the years 1981-1985, the output of this rapidly developing type of equipment increased fourfold in amount and threefold in cost terms in the largest ECE countries, including a tenfold increase for assembly robots.

On average, some 15% of working time in the engineering industries of developed market economies, and up to 50% of working time in sectors manufacturing highly intricate high-technology products, is spent on monitoring and measuring processes. Automated monitoring and measurement techniques are being increasingly used in the engineering industries of ECE countries to save time, shorten the production cycle and increase the quality of engineering output. In the United States, for example, there was in 1973 one installation for the automated checking of the parameters of products for 71 units of metalworking equipment, whereas in 1983 there was already one installation for every 42 metalworking units. [VI.8] Sales of equipment fitted with optical sensor systems amounted to roughly $190 million in the United States in 1985. It is estimated that such sales will reach $1.5-2 billion by 1990. [VI.9]

Laser, plasma, ionic, electron-beam, X-ray, subsonic and ultrasonic technologies are finding increasing application in the engineering industries of the industrialized countries. Molecular epitaxy and ion implantation are methods used in electronics. As examples, somewhat over 1,000 industrial laser devices were in use in the United States in 1984, roughly 500 in Japan, and 200 in the Federal Republic of Germany, but it is expected that, on a world-wide basis, the number of laser devices will increase fivefold every five years between 1985 and 2000. Commercial development of the phenomenon of superconductivity has very great prospects.

Equipment for use in the above technologies is a relatively modest proportion of technological equipment as a whole, no more than 0.3-0.4%, but many types of modern technology could not be developed in its absence.

Flexible manufacturing systems (FMS) are the most intricate form of the integrated automation of technological processes in the engineering industries. It has been estimated that there were 350-400 such systems in the world in 1984-1985, 25% of them in Japan. Sales of FMS in market economies in 1984 amounted to approximately $120 million, half of them in west European countries. The total volume of FMS sales in 1990 may be as high as one billion dollars. The trends in the manufacture and use of FMS are not unequivocal. On the one hand, there is no doubt that the only way to develop computerized integrated manufacturing technologies - fully automated "unmanned" production shops and factories - is through the establishment of large FMS. Such systems would embody a central control computer, machine tools with programmed control, automated lifting and carrying equipment, and automated general stores and tool stores, controlling and measuring devices, test facilities, and swarf collection systems. Consequently, there are a few systems in all countries that are serving as "proving grounds". As a rule, these systems are less economically effective than had been hoped, owing to the extremely high initial capital expenditure, insufficiently reliable equipment, the vast complexity of integrating large systems and the lack of trained personnel.

On the other hand, the desire to achieve a rapid return on capital investment, especially having regard to the accelerated technical obsolescence, is obliging manufacturers to give preference to comparatively simple and reliable manufacturing cells. The conscious decision is often taken not to use such a fundamental property of an FMS as its flexibility, and systems are employed to manufacture no more than three or four parts of standard sizes, i.e. they operate in practice under production-line conditions. When United States engineering companies were questioned on the criteria that led them to introduce automation equipment, flexibility came sixth in order of importance. Increased profits were placed first by 91% of the companies surveyed, and high-quality output second. Those countries that fail to overcome this tendency, which has an adverse effect on the rate of technological development, are in danger of finding themselves outsiders in the competitive struggle in the near future.

Computerization and automation are appreciably affecting the strength and qualifications of the workforce. It has been estimated that roughly one half of all workers in the United States will operate computer or computerized technologies by the end of the 1990s. In the Federal Republic of Germany, 30-40% of office workers and managerial workers will work in automated units connected to data banks by the end of the century. Expenditure on intellectual input is being proportionately increased by automation. Over the past decade there has been a threefold to fourfold increase in the number of computer programmers and operators and of specialists on integrated systems, and this trend continues. At the same time, the use of highly productive automated techniques enables workers directly occupied on technological processes to be released. As yet, the overall effect of automation in terms of loss of jobs is slight, and in no way comparable to the consequences of, for example, structural changes in the economy. It has been calculated that the rates of reduction in the numbers of employees in the engineering industries of market economies will be 0.1-0.3% annually to the year 2000. Specialists in the Federal Republic of Germany consider that there will be a 7% reduction of employment in the engineering industries, excluding electrical engineering, in the period 1986-2010, a 16% reduction in electrical engineering, 4 per cent in the manufacture of computers and office machines, and more than 30% in the automotive industry, mainly on account of the robotization of assembly processes. Nevertheless, it should be borne in mind that there is a considerable proportion of new jobs arising in sectors manufacturing automated equipment and in the sphere of the production infrastructure.

Within the engineering industries, the dividing lines between the established sectors are being eroded. More and more companies, especially the large and medium-sized, are diversifying their activities. Although the desire of companies for financial stability is one of the main reasons for diversification, there are other factors operating in the same direction. Such factors include the individualization of demand; the increasing prevalence of mechatronic technology, i.e. mechanical devices with electronic control systems; and the switch in fixed capital investment from purchases of equipment from a diversity of sources to the practice of acquiring complete manufacturing systems and projects from a single source, a single contractor. Many companies that set about mastering new production methods do so in the not unfounded hope of achieving technological breakthrough at intersectoral boundaries.

Machine-tool manufacturers are producing electronic control systems for NC-machines and industrial robots, while electrical and electronic engineering companies are encroaching upon machine-tool manufacturers in the market for robots, and are organizing production of their own computer-aided design and flexible manufacturing systems. Automotive companies have organized the manufacture of integrated circuits and other types of electronic components and parts for their own needs. Engineering companies in Japan are venturing into the sphere of the creation of new materials - ceramics for example - and even into biotechnology.

The average cost of electronic packages for light motor vehicles had reached $500 dollars in the United States in 1985, which was 10 times greater than in 1970. By the year 2000, this figure may be as high as $1,500. The output of electronic components for motor vehicles in market-economy countries reached $5 billion in 1985 and may possibly exceed $14 billion by 1990.

The manufacture of many new types of household mechatronic appliances - video tape recorders and electronically controlled cine cameras and cameras - is developing even more rapidly than electronic engineering in the automotive industry. Video tape recorder sales in 1985 reached $4.7 billion in the United States, $2.8 billion in Japan and at least $1.5 billion in the Federal Republic of Germany.

Electronic control is one of the main approaches to improving consumer durables and making industrial equipment more effective. Electric motors and fluorescent lamps fitted with control microprocessors are 30-40% more economic than conventional motors and lamps. Sewing machines, washing machines, typewriters, television sets etc. are being fitted with electronic control systems.

Organizational changes in the engineering industries are proceeding along the following main lines: first, the role of small and medium-sized enterprises is being increased by the individualization of demand, more rapid updating of products and the developing use of automated process-control facilities; these trends are enabling the optimum size of enterprises to be reduced. Secondly, co-operative relations between companies are being expanded, with the creation of joint CAD systems, data banks and quality control systems. Longer-term and more durable relationships are being established between contractors and sub-contractors. Large and flexible integrated intersectoral groupings are being set up on this basis. Thirdly, in the face of the expansion of R and D and the increased level of risk in developing new products, competition has given rise to the form of organization known as venture, or risk, capital. The number of small trend-setting companies and the volume of venture capital are growing rapidly (total venture capital in the United States is put at $40 billion). The setting up of venture companies within large corporations is a new phenomenon. This trend is likely to continue and to intensify.

VI.2. EXPECTED DEVELOPMENT IN CENTRALLY PLANNED ECONOMIES

Common features of industrial strategies for 1986-1990 in centrally planned economies can be summarized as follows:[VI.10]

- Acceleration of industrial growth, particularly in terms of net industrial output and a transition to a predominantly intensive path of industrial growth;
- Increased industrial efficiency stemming from constrained production inputs, mainly on the side of labour resources, energy and material inputs;
- Accelerated structural changes which are to reflect more closely the changes in external and domestic demand and are designed to contribute more to efficiency growth;
- The speeding-up of scientific and technological progress and improvements in the quality of industrial goods; and
- The deepening international division of labour and export promotion.

Within the industrial development plans for 1986-1990, the engineering industries have been accorded high priority. The trend of a faster rate of growth of engineering than total industrial output is to continue (table VI.5).

Two main factors are expected to contribute to the rapid expansion of engineering and in particular of high-technology branches:

- The revival of aggregate investment levels between 1981-1985 and 1986-1990 in seven countries combined at an annualized rate of over 4% annually (compared with 2% between 1976-1980 and 1981-1985) should act as an important factor on the demand side; and
- Export promotion measures to increase the share of engineering products in exports, together with a policy to improve supplies of consumer durables, will add to a growing demand for engineering products.

Against this, the expansion in volume of engineering output is likely to be hampered by the slow-down in investment in the engineering industries in 1981-1985. Full attainment of the quality improvements called for could also be jeopardized by the deterioration in the age structure of the capital stock. Moreover, official statements generally stress provisions to improve the quality of existing fixed assets, to cut down the share of new construction in total capital expenditures, and to re-equip or expand existing capacities. All this implies that the share of machinery and equipment in total investment will rise. Hence, the tasks laid before the engineering branch with regard to supplies of investment goods for domestic use are very demanding. It is likely that, if investment plans are to be realized, imports of engineering products will also have to increase. Indeed, they already rose strongly in eastern Europe, but not in the Soviet Union, in 1986. A sustained revival will, however, depend on a recovery of convertible currency export earnings. At all events, the falling share of machinery and equipment in total imports during 1981-1985 recorded in most east European countries, noted earlier, is likely to be reversed.

Industrial strategies embodied in five-year plans for 1986-1990 give particular prominence to high-technology branches such as microelectronics, computers and industrial robots. This is also in line with the "Comprehensive Programme of Scientific and Technological Progress up to the year 2000" adopted by the CMEA countries in 1985.[VI.11]

Complete and comparable information for all countries under review is lacking but a number of sub-branch objectives have been published. In *Bulgaria,* such high-technology branches as information technology, telecommunications, electronics, biotechnology and industrial robots are to expand rapidly (by 15-25% of average annual rate of growth). In *Czechoslovakia,* electrical and electronic engineering are to be the most dynamic engineering sub-branches with an expected average annual rate of growth of about 10%. In the *German Democratic Republic,* emphasis is placed on microelectronics, information technology and computers, and industrial automation, including robotics. Rapid expansion of production in these branches reflects efforts to accelerate scientific and technological progress and to boost the share of high-technology products suitable for the domestic market and export. Thus, electronics and electrical engineering output is to grow at an average annual rate of 8.6% in 1986-1990. In the *Hungarian* five-year plan for 1986-1990, quantified industrial production targets are scarce. Within engineering, and subject to market conditions, the output of the following sub-branches should develop faster than average: road vehicles, vacuum-technology products, communication equipment, health and education equipment, agricultural and food-processing machinery and consumer durables. In *Poland,* electrical engineering and electronics outputs are expected to increase by 5.7% annually. The upsurge foreseen in engineering outputs in *Romania* is linked with the diffusion of technological change. Priority is to be given to electronics, automation, precision machinery, hydraulic and pneumatic equipment and highly productive machine tools. The main elements of the intensification strategy in the *Soviet Union* (speeding up of scientific and technological progress, the retooling and technical re-equipment of industry, accelerated structural change and higher efficiency in the utilization of production inputs) are closely linked with the performance of the engineering industry. Within it, the following sub-branches are to grow faster than total engineering output: machine tools, instrument-making, electronics and electrical engineering, information technology and computers. Production of such high-technology items as industrial robots, flexible manufacturing systems and computers is to grow from 2.2 to 3 times. This implies average annual rates of growth within a range of 17-25%. Moreover, the plan envisages raising considerably the quality of machinery and equipment.

All countries provide for some upturn in the level of total investment in 1986-1990 (table VI.6). While few data on investment in engineering industries are available, the increase in investment allocations, together with the expected key role of engineering industries in ensuring the "intensification" of economic development, suggests that investment in the branch will rise faster than the 4% growth rate retained for total capital expenditure. This would, for most countries, contrast with performance in 1981-1985. Electrical engineering (including electronics) will receive special attention in Czechoslovakia - investment volumes are expected to double compared with 1981-1985. Engineering investment as a whole is planned to rise by 80% in the Soviet Union - more than three times faster than the planned growth of total investment. However, it should be borne in mind that, in all countries, investments in the engineering industries in 1986-1990 will, for the most part, hardly affect output capacity until after 1990.

Table VI.1

A forecast of production of selected industries in Japan, 1984-2000

(Trillions of yen)

Industry	*1984*	*1990*	*2000*
Semiconductors	2.5	7.3	26
Computers	3.3	7.4	25
Household electronic equipment	4.5	6.6	14
Automated programming means	0.5	1.2	11
Telecommunication equipment	1.6	4.6	11
Office automation	1.0	3.0	6.0
Numerically controlled systems	0.5	0.9	1.6
Industrial robots	0.2	0.6	1.4
CAD	0.1	0.4	1.0
Very-high-purity ceramic materials	0.3	1.5	5.0
New alloys	0.2	0.3	1.5
Bio-sensors and equipment for biotechnological processes	-	0.5	1.8
Total	10.4	34.5	100.8

Source: Toe keidzai tokei geppo, No. 4, 1986. (In Japanese)

Table VI.2

Estimated structure of sales of selected automation equipment and systems in the United States, 1986

Type of equipment	*Sales (billion dollars)*	*As a percentage of the total*
Control computers, various	21.0	56
Computer-aided engineering	6.6	18
Automated lifting and moving equipment	6.5	17
CAD systems	2.7	7
Programmable industrial robots	0.7	2
Total	37.5	100

Source: Electronic Business, 15 February 1986.

Table VI.3

Market trends for automation-oriented systems and equipment in the Federal Republic of Germany, 1970-2000

	1970-1980	*1980-1990*	*1990-2000*
Automation equipment Manufacturing trends	Separate NC-machine tools - Increased productivity of machine tools	FMS and factories - Increased equipment utilization - Automation - Refinement of information	Automated factory shops - Systems optimization - Organization of production
Factors making for success in the market	- Precision of equipment - Product quality	- Quality of electronic systems - Commodity price	- Integrated solution of production problems

Source: VDG Zeitschrift, 1985, Vol. 127, No.15, pp. 583-9.

Table VI.4

Production of computers in selected market economies, 1976-1986

(Billions of US dollars at current prices)

Country/region	1976	1980	1984	1985	1986
United States	10.4	26.6	52.5	56.7	65.8
Japan	2.4	5.7	11.8	15.8	...
Western Europe	...	9.0	18.8	22.9	26.5
of which:					
Germany, Federal Republic of	...	2.6	4.7	5.6	6.7
United Kingdom	...	2.0	3.3	4.8	5.5
France	...	2.0	3.3	3.7	4.1

Sources: *US Industrial Outlook*, 1981; *Census of Manufacturers*, 1982; *Japan Electronic Almanac*, 1982; *Mackintosh Yearbook of West European Electronic Data*, 1980, 1986.

Table VI.5

Planned industrial and engineering output growth in centrally planned economies, 1986-1990

(Average annual percentage change)

Country/region	Industry	Engineering industries
Bulgaria	4.9	8.5
Czechoslovakia	3.0	5.4
German Democratic Republic	3.7-4.1	5.4-5.7
Hungary	2.7-3.0	5.4-5.7 [a]
Poland	3.0	4.7
Romania	7.5-8.3	10.3
Eastern Europe	4.4	...
Soviet Union	4.6	7.4
Eastern Europe and the Soviet Union	4.5	...

Source: Five-year plan documents.
a Manufacturing industries.

Table VI.6

Gross fixed investment volume in centrally planned economies, 1986-1990 [a]

(Annualized percentage volume changes) [b]

Country/region	1976-1980	1981-1985	Plan 1986-1990
Bulgaria	6.9	4.5	6.5 *
Czechoslovakia	5.8	-0.4	2.1
German Democratic Republic	5.0	0.1	1.7 *
Hungary	5.5	-2.0	0.8 *
Poland	6.0	-7.5	4.4
Romania	10.9	1.3	3.6
Eastern Europe	6.9	-1.3	3.4
Soviet Union	5.0	3.3	4.3
Eastern Europe and the Soviet Union	5.6	1.9	4.1

Source: ECE secretariat Common Data Base, derived from national statistics, plan documents and plan fulfilment reports.
a Based on data at constant prices of 1980 (Bulgaria), 1977 (Czechoslovakia), 1980 (German Democratic Republic), 1981 (Hungary), 1982 (Poland), 1981 (Romania) and 1976 (Soviet Union).
b Data are annualized percentage volume changes compared with totals in the preceding five-year period.

VI.3. SUMMARY OF FORESEEABLE DEVELOPMENTS BY INVIDIVUAL SUBSECTORS

Trends towards acceleration of the development of all subsectors of the engineering industries are continuing to a greater or lesser extent in all countries of the ECE region and Japan. The metal products industry is probably the least sophisticated sector of the engineering industries, being traditionally labour intensive and incorporating only low value added. This sector is increasingly being challenged by the development of new synthetic and other materials which are cheaper to produce and whose properties are equal to or even better than those of the original product. Traditional steel coil springs, for example, could soon be replaced by plastic units. A key test facing the metal products industry will be its ability to introduce flexible labour-saving and highly productive technologies in order to assure competitiveness on the world market. R and D must also be devoted to improving the quality of products and to developing new materials and products which meet best the specific requirements of the users. The ECE region's share in total world production of metal products, however, remains very high. [VI.12]

In the manufacture of non-electrical machinery, NC-machines comprise an increasingly important share of total machine-tool production, with particular emphasis on expanding the use and production of flexible manufacturing systems (FMS), automated guided vehicles (AGV), industrial robots etc. Among the conclusions of the ECE Seminar on Industrial Robotics '86, held in Czechoslovakia in February 1986, it was projected that the significant technical advances which are being made in the design and capabilities of robots would accelerate their diffusion in both engineering and non-engineering industries. Software is expected to play an increasingly important role in the further development of robots, and in extending their capabilities and their integration into large computer-based manufacturing systems. [VI.13]

The vast range of products that are included under the subsector of electrical machinery encompasses an array of new developments. For example the need for economical, very efficient electric motors has already led to the development in the United States of a new electronically controlled hybrid AC/DC electric motor, and further improvements are likely. Similarly, improvements can be expected in the production of generators and transformers, including the introduction of new magnetic materials such as the so-called amorphous alloys now beginning to be used in transformer cores, as well as the development of more energy-efficient lighting control systems for both incandescent and high-intensity discharge lamps.

In the highly competitive electronic components field, new advanced technologies are being developed in order to increase production efficiency, such as the computerized mounting of electronic components. In the telecommunications area, in addition to the introduction of optical-fibre cable systems (both single-mode and multi-mode), considerable emphasis is being placed on the development and improvement of transmitting and receiving equipment, including emitters, photodetectors, connectors, repeaters etc. The field of electromedical equipment, and that of digital image processing (DIP) in particular, is one of the sectors of the engineering industries with large market potential, high R and D input and direct socio-economic impact, all of which should ensure a fast pace of development. The vast array of home appliances, and in particular audio-visual equipment, will continue to proliferate, with high-definition television and probably digital audio-tape systems likely to explode on the market in the short term.

In the transport equipment subsector, the weight of the automotive industry will continue to be significant and competition keen. Improvements under way include the installation of CAD/CAM systems, the application of FMS, industrial robotics, automated handling and storage devices, etc. In the rail-related part of this sector, high-speed inter-urban transport facilities will continue to see new development, possibly with the introduction of a coal-electric locomotive. The shipbuilding picture on the whole remains relatively bleak, with some relief possible in the area of specialized vessels.

Further developments in aircraft design and manufacture comprise the extended use of advanced materials, such as carbon-fibre composites, and manufacturing techniques, particularly CAD/CAM; further significant advances in flight control can be derived from the introduction of advanced electronic flight decks. One unexpected by-product of the United States space shuttle failure may be to stimulate a further revival of interest in a second-generation supersonic transport to succeed Concorde in the late 1990s, and possibly a horizontal take-off and landing (hotol) vehicle that could travel into orbit from a runway.

The expanding market for industrial and scientific instruments reflects the current level of capital investment in plant and equipment, in combination with the large amounts spent on R and D. Development in electronic testing and measuring instruments are determined by trends in the electronic component and manufacturing equipment industries. Greater density in semiconductor chips and circuit boards is leading to the development of more sophisticated and more costly test systems that are used for both product design and quality control. In the photographic sector of the industry, the trend is towards more electronic components, with filmless cameras and computer-linked SLRs being the latest developments, as is further miniaturization of camcorders. In the watch industry, the trend is towards wrist watches equipped with a quartz analogue movement for high precision. Work will continue on providing longer-life batteries.

VI.4. CONCLUSIONS OF THE STUDY

The basic conclusions that can be drawn from the present study are as follows: [VI.14]

1. It is obvious that the engineering industries, as dominant producers, suppliers and users of machinery and equipment, constitute the core of the economy in all industrially developed countries. The level reached in their development determines the economic and technological potential of any country, as well as its role in the world-wide division of labour. In other words, the level of engineering production and exports has proved to be a sensitive and reliable indicator of the economic and social climate in individual ECE member countries.

2. Engineering industries, in their capacity as the suppliers of capital (investment) goods and intermediate products (semi-finished products, parts and components) to all economic sectors, significantly influence the growth of productivity, technological innovation, structural changes in industry etc. In addition, as the major producers of consumer durables for individual and home use, as well as of equipment for all kinds of services (banking, health care, education, transport etc.), they are important contributors to the improvement of living standards and the satisfaction of social expectations.

3. The ECE region continues to occupy a dominant position in the output of world engineering industries despite the fact that, during the past 10 years, its share has decreased from 80 to 77%. The most solid position of the ECE region is in the metal products industry (due to high-quality goods in this traditional subsector with low value added) and in the precision instruments subsector (high value added, considerable sophistication and intellectual input). At the same time, the share of the ECE region has been considerably reduced in the world production of electrical and electronic goods - the subsector characterized by rapid technological advancement. It must be recognized that ECE countries are in a vulnerable position in competition with Japan and also with some of the newly industrialized countries active in this field. To some extent, the same applies to traditional industries producing non-electrical machinery and transport equipment (e.g. metalworking machine tools, automotive industry, shipbuilding etc.).

4. The period 1979-1986 was marked by substantial growth of automation-oriented and information technologies and innovations based on these technologies. Several countries which have recently mastered the production of equipment and software for information technologies showed remarkably high growth rates of output, averaging from 20 to 40% a year, in individual areas (computing facilities including personal computers, telecommunication equipment, robotics etc.). However, the implementation of some of the most advanced approaches, such as flexible manufacturing systems, automated assembly and computer-integrated manufacturing, expanded much more slowly than was forecast, owing to the need for heavy initial investments, insufficient reliability of new equipment necessitating costly maintenance and the lack of highly skilled and experienced manpower. It would appear that, in order to maintain the general competitiveness of ECE member countries on world markets, further advances in the application of automation-oriented and information technologies, as well as of new materials, will be a crucial factor. [VI.15]

5. It is generally accepted that wide-ranging, well-established and flexible R and D is the prerequisite of technological progress in the engineering industries as well as in other sectors of the economy. The share of engineering industries in total industrial R and D expenditure of the ECE region currently exceeds 70%. The leading position of the region is also reflected in the fact that, in 1985, almost 90% of all patents issued in this field originated in the region. However, the really successful and competitive position of some countries in basic engineering R and D has seriously deteriorated owing to the relatively slow and limited commercial implementation of new developments, often resulting from the need for high initial investments or from the insufficient adaptability of industry in accepting new technological approaches.

6. The late 1970s and the 1980s saw the widespread introduction of new labour-, material- and energy-saving technologies in the engineering industries of all industrialized countries. The development and use of new materials such as thermoplasts, composites, monocrystals, amorphous metal alloys, ceramics and optical fibres has also helped to conserve resources (inputs) and to increase the efficiency of engineering technologies and products. Optimistic expectations are connected with the possible development of commercial superconductors in the near future. The utilization of new materials is expanding, but not as fast as expected: steel consumption by the engineering industries has remained high.

7. Widespread application as from the mid-1970s of microelectronic controls, along with the curtailment of energy-intensive industries and the shift to non-waste technologies, gave these countries the opportunity to decrease considerably their raw-material and energy consumption per unit of GDP. In this respect, the engineering industries have also played a dominant role in the field of water- and air-pollution abatement.

8. In many countries of the region, while the engineering industries have managed to meet the demand for goods and consumer durables, relatively insufficient investment (in comparison with other industrial and non-industrial sectors) in this vitally important industrial sector has led to negative tendencies such as a low equipment discard rate (resulting in the aging of the total stock of ma-

chinery and equipment) and the deterioration of R and D potential.

9. The share of the engineering industries in the total industrial gross output of many countries accounted for some 30-35%. In terms of value added, their share was even higher - by some 5-7 percentage points. Consequently, the engineering industries' share in the total industrial employment of these countries also varied from 35 to 50%. The quality and qualification requirements of manpower have increased as a result of technological progress and structural shifts to high-technology industries. Production workers now account for some 65-70% of all persons employed in engineering industries (40-50% in some new high-tech industries), the remainder being engineers and managerial staff.

10. It is expected that, within the next 5 to 10 years, about half of all jobs in the engineering industries will be directly or indirectly connected with the operation of computerized equipment. In the absence of proper management from now on, the social consequences of this trend could be perilous. At the present time, the shortage of highly skilled specialists is already a limiting factor in the implementation of information and other new technologies.

11. In the large majority of ECE countries, the growth of labour productivity during the period under review was significantly higher than in total industry. However, in many of these countries the increased application of automated technologies led to important reductions in the engineering work-force and contributed to an increase in unemployment. In other countries, for several reasons (including social security), employment in the engineering industries remained high and the expected growth of labour productivity did not come about.

12. A number of ECE member countries have benefited from the availability of highly skilled, experienced and reliable domestic manpower. With this traditional advantage, it would seem preferable that they should concentrate on the manufacture and assembly of new and sophisticated machinery, equipment and instruments, characterized by high value added and intellectual input. The proliferation of new engineering technologies gives rise to such social problems as a sudden need for manpower training and retraining, mobility, labour safety and health protection, attractiveness of workplaces, man-machine interaction etc. An international exchange of studies and experience, including the development of manpower structures, would enable ECE countries better to justify their strategies and national priorities in this field.

13. The engineering industries of the majority of ECE countries are heavily export-oriented. During the period under consideration, their share of total industry exports amounted to 30-50% (some 70% in Japan). The years 1981-1985 were, however, marked by a slight drop in the growth rate of exports of engineering goods. This was to some extent a reaction to the financial difficulties experienced by several ECE countries, to the severity of international competition resulting in part from the increased output of the newly industrialized countries and to the relative saturation of many segments of the market. The situation gave impetus to the expansion of some new and non-traditional forms of trade and other international economic relations, such as trade in patents, licences and know-how, multilateral co-operation in R and D, manufacturing and marketing, including joint ventures (both east-west and north-south).

14. As a result of the continuing innovation in the design of new equipment and of engineering products in general (in particular in the field of information technologies), the role of international standardization has significantly increased. In some areas, a new generation (basic functional innovation) of hardware and software appears nearly every 4-7 years, necessitating special efforts in developing compatibility solutions, new interfaces, new operating approaches etc. The stock of traditional means of production is aggravated and the possibility of exporting outdated products is reduced.

15. In the ECE region, a really powerful R and D potential and accumulated unique know-how have been created, especially in basic research. On the other hand, the region still lags behind (primarily Japan) in the application and commercialization of new developments and innovations. Improvement is needed in technology transfer and spin-off from universities and laboratories to industry as well as in comprehensive information exchange. In this respect, new forms of international co-operation, such as the EUREKA programme or CMEA's complex programme of technological progress of countries up to the year 2000, are expected to bring practical and valuable results.

16. In order to make possible the continuation of useful international comparative studies, properly reflecting recent developments, there is an urgent need to modernize and update existing statistical nomenclatures and data banks on production and trade in engineering products. In this context, the ECE Working Party on Engineering Industries and Automation, in close co-operation with other international and regional organizations in the field, might play an important role - not only in the improvement of compatible east-west engineering statistics, but also in the preparation of long-term assessments, specialized studies and topical seminars.[VI.16]

REFERENCES

VI.1 For a detailed analysis of the structural development within the engineering industries, see chapters II and III of the present study.

VI.2. *Business Plans for New Plants and Equipment.* McGraw-Hill, New York 1986.

VI.3. During the period under consideration, the ECE Working Party on Engineering Industries and Automation devoted special attention to automation-oriented technologies (several studies and seminars). Since 1983, the Working Party has also studied innovation trends and recent developments in selected traditional engineering fields. In this respect, the field of biomedical engineering, for instance, recorded significant progress towards the integrated use of microelectronics, computer technologies, new materials etc., based on the rapid application of new medical physics approaches.

VI.4. See also subchapter VI.4 summing up the main conclusions of the present study.

VI.5. *Annual Review of Engineering Industries and Automation, 1985* vol.I, chapter II. UN/ECE publication, Sales No. E.86.II.E.30.

VI.6. *The Economist,* 23 November 1985.

VI.7. *Production and Use of Industrial Robots.* UN/ECE publication, Sales No. E.84.II.E.33. *Recent Trends in Flexible Manufacturing.* UN/ECE publication, Sales No. E.85.II.E.35.

VI.8. 11th and 13th *American Machinist* Inventory of Metalworking Equipment, 1973 and 1983.

VI.9. *American Machinist,* June 1986, p.31.

VI.10. Subchapter VI.2 is based on analysis undertaken by the ECE General Economic Analysis Division (GEAD) in 1986.

VI.11. More information on the scope and contents of the CMEA "Comprehensive Programme of Scientific and Technological Progress up to the Year 2000" is provided in chapters II and IV.

VI.12. See table II.44 and subchapter III.2.

VI.13. Report of the ECE Seminar on Industrial Robotics '86 - International Experience, Developments and Applications, Brno (Czechoslovakia) February 1986 (document ENG.AUT/SEM.5/4).

VI.14. At the request of the ECE Working Party on Engineering Industries and Automation (see report of the eighth session, ECE/ENG.AUT/34, para. 10), the secretariat prepared draft conclusions of the study and circulated these for comments and approval to heads of delegations prior to their inclusion.

VI.15. The ECE Working Party has included in its programme of work for 1988-1992 further relevant studies and seminars in order to monitor developments in both traditional and emerging engineering technologies (report of the eighth session, ECE/ENG.AUT/34, annex I).

VI.16. The ECE Working Party on Engineering Industries and Automation constitutes its programme of work on the basis of recommendations of Governments of ECE member countries. New topics on whose inclusion concensus is reached are then inserted in the programme. While carrying out work on individual programme elements (projects), close co-operation is maintained with international governmental and non-governmental organizations active in the field, e.g. in the field of engineering statistics, important co-ordination measures have been agreed upon with OECD.

ANNEX I

SEMINARS SPONSORED BY THE ECE WORKING PARTY ON ENGINEERING INDUSTRIES AND AUTOMATION AND ITS PREDECESSORS (1971-1988)

	Title, place and date of seminar	Symbol of report
1.	Application of Computers as an Aid to Management, Geneva (Switzerland), 11-15 October 1971	AUTOMATION/Working Paper No.3
2.	Application of Metal and Non-Metal Materials in Engineering Industries, Varna (Bulgaria), 28 May-1 June 1973	ENGIN/SEM.1/2
3.	Application of Numerically Controlled Machine Tools, Prague (Czechoslovakia), 12-17 November 1973	AUTOMAT/SEM.1/3
4.	Automated Industrial Production, its Social and Economic Consequences, Lyons (France), 16-21 September 1974	AUTOMAT/SEM.2/2
5.	Techno-Economic Aspects and Results of Anti-Corrosion Measures in Engineering Industries, Geneva (Switzerland), 27-31 January 1975	ENGIN/SEM.2/2
6.	Use of Automated Process Control Systems in Industry, Moscow (USSR), 26-31 May 1975	AUTOMAT/SEM.3/2
7.	Automated Integrated Production Systems in Mechanical Engineering, Prague (Czechoslovakia), 1-6 November 1976	AUTOMAT/SEM.4/2
8.	Industrial Robots and Programmable Logical Controllers, Copenhagen (Denmark), 5-7 September 1977	AUTOMAT/SEM.5/2
9.	Engineering Equipment for Foundries and Advanced Methods for Production of such Equipment, Geneva (Switzerland), 28 November-3 December 1977	ENGIN/SEM.3/2
10.	Techno-Economic Trends in Airborne Equipment for Agriculture and Other Selected Areas of the National Economy (AERO-AGRO '78), Warsaw (Poland), 18-22 September 1978	ENGIN/SEM.4/3
11.	Computer-Aided Design Systems as an Integrated Part of Industrial Production, Geneva (Switzerland), 14-18 May 1979	AUTOMAT/SEM.6/2
12.	Development and Use of Industrial Handling Equipment, Sofia (Bulgaria), 3-8 September 1979	ENGIN/SEM.5/2
13.	Innovation in Engineering Industries: Techno- Economic Aspects of Fabrication Processes and Quality Control, Turin (Italy), 9-13 June 1980	ENGIN/SEM.6/3

	Title, place and date of seminar	Symbol of report
14.	Automation of Welding, Kiev (Ukrainian SSR), 13-17 October 1980	AUTOMAT/SEM.7/3
15.	Automation of Assembly in Engineering Industries, Geneva (Switzerland), 22-25 September 1981	AUTOMAT/SEM.8/3
16.	Present Use and Prospects for Precision Measuring Instruments in Engineering Industries, Dresden (German Democratic Republic), 20-24 September 1982	ENG.AUT/SEM.1/3
17.	Innovation in Biomedical Equipment, Budapest (Hungary), 2-6 May 1983	ENG.AUT/SEM.2/3
18.	Flexible Manufacturing Systems - Design and Applications, Sofia (Bulgaria), 24-28 September 1984	ENG.AUT/SEM.3/4
19.	Development and Use of Powder Metallurgy in Engineering Industries, Minsk (Byelorussian SSR), 25-29 March 1985	ENG.AUT/SEM.4/3
20.	Industrial Robotics '86 - International Experience, Developments and Applications, Brno (Czechoslovakia), 24-28 February 1986	ENG.AUT/SEM.5/4
21.	Automation Means in Preventive Medicine '87, Piestany (Czechoslovakia), 28 September-2 October 1987	ENG.AUT/SEM.6/3
22.	Shipbuilding 2000 Maritime Conference BALTEXPO '88, Gdansk (Poland), 5-9 September 1988	ENG.AUT/SEM.7/3

ANNEX II

ECE SALES PUBLICATIONS ON ENGINEERING INDUSTRIES AND AUTOMATION 1979-1987*

The Telecommunication Industry - Growth and Structural Change (ECE/ENG.AUT/30), Sales No. E.87.II.E.35, New York, 1987. $50.00.

Software for Industrial Automation (ECE/ENG.AUT/29), Sales No. E.87.II.E.19, New York, 1987. $35.00.

Digital Imaging in Health Care (ECE/ENG.AUT/25), Sales No. E.86.II.E.29, New York, 1987. $48.00.

Recent Trends in Flexible Manufacturing (ECE/ENG.AUT/22), Sales No. E.85.II.E.35, New York, 1986. $33.00.

Production and Use of Industrial Robots (ECE/ENG.AUT/15), Sales No. E.84.II.E.33, New York, 1985. $25.00.

Measures for Improving Engineering Equipment with a view to More Effective Energy Use (ECE/ENG.AUT/16), Sales No. E.84.II.E.25, New York, 1984. $16.

Engineering Equipment and Automation Means for Waste-Water Management in ECE Countries, vols. I and II (ECE/ENG.AUT/18), Sales Nos. E.84.II.E.13 and E.84.II.E.23, New York, 1984. $12.50 (vol.I) and $8.50 (vol.II).

Techno-Economic Aspects of the International Division of Labour in the Automotive Industry (ECE/ENG.AUT/11), Sales No. E.83.II.E.14, New York 1983. $23.00.

Development of Airborne Equipment to Intensify World Food Production (ECE/ENG.AUT/4), Sales No. E.81.II.E.24, New York 1981. $16.00.

Annual Reviews of Engineering Industries and Automation

1986 (ECE/ENG.AUT/32), Sales No. E.88.II.E.16. $38.00 (vols. I and II)
1985 (ECE/ENG.AUT/26), Sales No. E.86.II.E.30. $38.00 (vols. I and II)
1983-1984 (ECE/ENG.AUT/19), Sales No. E.85.II.E.43. $27.00 (vols. I and II)
1982 (ECE/ENG.AUT/17), Sales No. E.84.II.E.12. $11.00
1981 (ECE/ENG.AUT/10), Sales No. E.83.II.E.20. $11.00
1980 (ECE/ENG.AUT/7), Sales No. E.82.II.E.18. $13.50
1979 (ECE/ENG.AUT/3), Sales No. E.81.II.E.16. $8.00

Bulletins of Statistics on World Trade in Engineering Products

1986, Sales No. E/F/R.88.II.E.14. $46.00
1985, Sales No. E/F/R.87.II.E.10. $45.00
1984, Sales No. E/F/R.86.II.E.10. $38.00
1983, Sales No. E/F/R.85.II.E.11. $35.00
1982, Sales No. E/F/R.84.II.E.5. $38.00
1981, Sales No. E/F/R.83.II.E.8. $38.00
1980, Sales No. E/F/R.82.II.E.5. $26.00
1979, Sales No. E/F/R.81.II.E.13. $26.00

* Sales publications and documents out of print may be requested in the form of microfiches. Price per microfiche $1.65; printed on paper $0.15 per page.